Juan Jairo Inga Charaja

Inverse Dynamic Game Methods for Identification of Cooperative System Behavior

Karlsruher Beiträge zur
Regelungs- und Steuerungstechnik
Karlsruher Institut für Technologie

Band 10

Inverse Dynamic Game Methods for Identification of Cooperative System Behavior

by
Juan Jairo Inga Charaja

KIT Scientific Publishing

Karlsruher Institut für Technologie
Institut für Regelungs- und Steuerungssysteme

Inverse Dynamic Game Methods for Identification
of Cooperative System Behavior

Zur Erlangung des akademischen Grades eines Doktor-Ingenieurs
von der KIT-Fakultät für Elektrotechnik und Informationstechnik des
Karlsruher Instituts für Technologie (KIT) genehmigte Dissertation

von Juan Jairo Inga Charaja, M.Sc.

Tag der mündlichen Prüfung: 16. Oktober 2020
Hauptreferent: Prof. Dr.-Ing. Sören Hohmann
Korreferent: apl. Prof. Dr.-Ing. Daniel Görges

Impressum

KIT Scientific Publishing

Karlsruher Institut für Technologie (KIT)
KIT Scientific Publishing
Straße am Forum 2
D-76131 Karlsruhe

KIT Scientific Publishing is a registered trademark
of Karlsruhe Institute of Technology.
Reprint using the book cover is not allowed.

www.ksp.kit.edu

Print on Demand 2021 – Gedruckt auf FSC-zertifiziertem Papier

ISSN 2511-6312
ISBN 978-3-7315-1080-2
DOI 10.5445/KSP/1000128612

Preface

This thesis is the result of my work as a research assistant at the Institute of Control Systems (IRS) of the Karlsruhe Institute of Technology (KIT). This thesis would not exist if it were not for the support of many people. First and foremost, I would like to express my deepest gratitude towards Prof. Dr.-Ing. Sören Hohmann, who gave me the opportunity to work on this project under his supervision. I especially treasure your trust, support, as well as all inspiring discussions we had during these years. I would also like to kindly thank apl. Prof. Dr.-Ing. Daniel Görges for the assessment of this thesis and for his participation in the evaluation commitee of my thesis defense. I very much enjoyed all of our conversations and appreciate your genuine interest in my work.

Many thanks go to all of the IRS staff for the great working atmosphere. In particular, I would like to thank Martin, with whom I shared an office for many years and always with good vibes. I also acknowledge Tim Molloy with whom an outstanding research collaboration unfolded, the results of which are partially appreciable in this thesis. Furthermore, I appreciate the support of the members of my research group at IRS, especially Esther, Florian, Philipp and Simon, all of which proof-read various parts of this thesis, giving me valuable feedback. I am also grateful to all bachelor and master students I supervised and who supported my work with their own thesis projects.

Ohne die Unterstützung weiterer Menschen würde ich diese Zeilen nicht schreiben können. Ein besonderer Dank gilt daher meinem Lehrer Dr. Karl Moesgen, der mich in Lima inspiriert und gefördert hat, außerdem meinen Paten Anneliese und Toni sowie meiner ehemaligen Gastfamilie Seidel, die mir alle dabei geholfen haben, nach Deutschland zu kommen und mich hier gut einzuleben. Schließlich auch meinen ehemaligen Abschlussarbeitsbetreuern Gunter und Michael, die mir gezeigt haben, dass ich an der "Tür der Promotion" klopfen kann.

A todos mis amigos en Karlsruhe les agradezco por colaborar indirectamente a este proyecto por medio de todos los bailes, conversaciones, risas, fiestas, pichangas, chelas, etc. y por ser siempre una valiosa conexión hacia mis orígenes y cultura. Los más grandes agradecimientos van hacia mi familia, especialmente a mis papás Juan y Nadja así como a mi hermana Leda por su incondicional apoyo y sacrificio para que pueda seguir mi propio camino y por el amor que me han dado durante toda mi vida. Aber auch ein sehr besonderer Dank an dich, liebe Lena, für deinen Rückhalt und dafür, dass du in diesen Jahren stets für schöne Momente abseits von mathematischen Formeln gesorgt hast, die mir ein Gleichgewicht verschafft haben.

Heidelberg, October 2020

Flow in the living moment. — We are always in a process of becoming and nothing is fixed. Have no rigid system in you, and you'll be flexible to change with the ever changing. Open yourself and flow, my friend. Flow in the total openness of the living moment. If nothing within you stays rigid, outward things will disclose themselves.

Bruce Lee

Contents

List of Figures

List of Tables

Abbreviations and Symbols

Abbreviations

Abbreviation	Description
ARE	Algebraic Riccati Equation
BFGS	Broyden–Fletcher–Goldfarb–Shanno
DB	Direct Bilevel (method)
FB	Feedback
FNE	Feedback Nash Equilibrium
GT	Ground Truth
HJB	Hamilton-Jacobi-Bellman
IDG	Inverse Dynamic (Differential) Game
ILQDG	Inverse Linear-Quadratic Dynamic (Differential) Game
IOC	Inverse Optimal Control
IRL	Inverse Reinforcement Learning
KKT	Karush-Kuhn-Tucker (conditions)
LQ	Linear-Quadratic
MaxEnt	Maximum Entropy
MDP	Markov Decision Process
MPS	Memoryless Perfect State
NSAE	Normalized Sum of Absolute Errors
OL	Open-Loop
OLNE	Open-Loop Nash Equilibrium
PD	Proportional-Derivative
PE	Persistence of Excitation
PES	Pareto Efficient Solution
QP	Quadratic Program
RDE	Riccati Differential Equation
SD	Standard Deviation
SNR	Signal-to-Noise Ratio
SVD	Singular Value Decomposition
TPBVP	Two-Point Boundary Value Problem
w.r.t.	with respect to

Symbols

Latin Letters

Symbol	Description
$\boldsymbol{A}$	System matrix
$\boldsymbol{b}$	Degrees of freedom in the solution of the unconstrained quadratic program of the residual-based method
$\boldsymbol{B}$	Input matrix
C	Real constant
$\boldsymbol{D}$	Jacobi matrix in MaxEnt IRL
$\boldsymbol{f}$	System dynamics
$\boldsymbol{F}$	Closed-loop system matrix
g	Running costs in a cost function
g_e	Gravitational constant
$\tilde{\boldsymbol{g}}$	First derivative of J w.r.t. controls in MaxEnt IRL
$\boldsymbol{G}$	Control matrix in a control-affine nonlinear system
$\tilde{\boldsymbol{G}}$	Second derivative of J w.r.t. controls in MaxEnt IRL
h	Terminal costs in a cost function
H	Hamiltonian
$\boldsymbol{H}$	Hessian matrix of the QP in an ILQDG with feedback structure
$\boldsymbol{I}$	Identity matrix
i	Player index (exclusive use)
j	General index for enumeration
J	Cost function
k	Discrete time step
k_E	Final time step in a discrete-time dynamic game
K	Total number of data points available in a dynamic game
$\boldsymbol{K}$	Feedback matrix (linear feedback strategy)
l	General index for enumeration
L	Dimension of the parameter vector corresponding to a quadratic cost function where matrix properties are disregarded
$\boldsymbol{L}$	Rectangular matrix for extracting the parameter vector in the residual-based method
m	Dimension of control vector
M	Dimension of the cost function parameter vector / Torque applied by players in ball-on-beam example
$\boldsymbol{M}$	Matrix for the solution of an ILQDG with a feedback information structure
n	Dimension of state vector

n_t	Number of trajectory sets available in MaxEnt IRL
N	Number of players in a game
$\boldsymbol{N}$	Matrix of the LQ optimal control problem of the residual-based method
p	Probability density function
$\boldsymbol{P}$	Riccati matrix
$\boldsymbol{Q}$	Cost function matrix weighting the states
$\boldsymbol{Q}_T$	Cost function matrix weighting the final state
r_C	Residual function corresponding to the control equation in residual-based method
r_L	Residual function corresponding to the costate differential equation in residual-based method
$\boldsymbol{r}$	Residual function in an ILQDG
$\boldsymbol{R}$	Cost function matrix weighting the controls
$\boldsymbol{S}_{ij}$	$\boldsymbol{B}_j\boldsymbol{R}_{jj}^{-1}\boldsymbol{R}_{ij}\boldsymbol{R}_{jj}^{-1}\boldsymbol{B}_j^\top, \quad i,j \in \mathcal{P},\ i \neq j$
$\boldsymbol{S}_j$	$\boldsymbol{B}_j\boldsymbol{R}_{jj}^{-1}\boldsymbol{B}_j^\top$
$\boldsymbol{T}$	Transformation matrix
t	Time variable
T	Duration of a dynamic game
T_{CPU}	Computation time
$\boldsymbol{u}$	Control variable
$\boldsymbol{v}$	"Control variable" of the LQ optimal control problem in the residual-based method
$\boldsymbol{U}$	Left matrix of a Singular Value Decomposition
$\boldsymbol{x}$	State variable
$\boldsymbol{z}$	"State variable" of the LQ optimal control problem in the residual-based method

Greek Letters

Symbol	**Description**
$\boldsymbol{\alpha}$	Variable including cost function parameters and Lagrange multipliers in the residual-based method
β	Discount factor in a discounted cost function
$\boldsymbol{\gamma}$	Strategy in a dynamic or differential game
Γ	Set of possible strategies in a dynamic or differential game
$\delta^{\boldsymbol{x}}$	NSAE error of the states
$\delta^{\boldsymbol{u}_i}$	NSAE error of the controls of player i
$\delta^{\boldsymbol{u}}$	NSAE error of all controls (sum of all $\delta^{\boldsymbol{u}_i}$)
$\delta^{\boldsymbol{\theta}}$	Relative parameter error of identification results

Δ^{θ}	Absolute parameter error of identification results
ϵ	Gaussian white noise
ζ	Set of state trajectories and control trajectories of all players
η	Set-valued function denoting the information available to all players
$\boldsymbol{\theta}$	Vector of cost function parameters
Θ	Set of cost function parameter vectors / Rotational inertia
$\boldsymbol{\Theta}$	Canonical parameter set of a LQ differential game
κ	Algorithm iteration
λ	Eigenvalue
$\boldsymbol{\mu}$	Feature count
ξ	Example trajectory for the introduction of probabilistic approaches in Chapter 6
ρ	Weighting factor in residual-based method
σ	Spectrum (set of eigenvalues)
Σ	Singular value matrix in the residual-based method
τ	Weighting factor in cost functionals for Pareto efficient solutions
$\boldsymbol{\phi}$	Feature or basis function vector
φ	Angle
χ	Variable for information structure definition
$\boldsymbol{\psi}$	Costate variable vector

Calligraphic and other symbols

Symbol	Description
$\mathcal{C}$	Constrained set for Direct Bilevel method
$\mathcal{D}$	Data set of observed trajectories in IRL-based methods
$\mathbb{E}$	Expectation
$\mathcal{F}$	Set of all feedback control matrices which lead to a stable closed-loop system
$\mathcal{J}$	Set of cost functions
$\mathcal{K}$	Set of stages of a dynamic game
$\mathcal{L}$	Likelihood function
$\mathbb{N}$	Set of integer numbers
$\mathcal{P}$	Set of players
$\mathbb{R}$	Set of real numbers
$\mathcal{S}$	Set of solutions associated to a cost function in a dynamic optimization problem
$\mathcal{T}$	Set of points in time where data is available
$\mathcal{U}$	Action space in dynamic games

Indices, exponents and operator names

Symbol	Description
$\dot{\square}$	Time derivative
$\square^{+}$	Moore–Penrose inverse
$\square^{*}$	Variable corresponding to a Nash Equilibrium
$\square_{b}$	Variable corresponding to the ball of a ball-on-beam system
$\square_{D}$	Variable corresponding to a discrete-time (infinite) dynamic game
$\square^{\mathrm{FB}}$	Variable corresponding to a feedback information pattern or strategy
$\square_{i}$	Variable corresponding to player i
$\square_{\neg i}$	Variables corresponding to all players except player i
$\square_{(j)}$	j-th entry of a vector
$\square_{(r,c)}$	Entry of a matrix in the r-th row and the c-th column
$\square^{(k)}$	Variable corresponding to the stage k in a discrete-time dynamic game
$\square^{(s)}$	Variable corresponding to the subject pair s in the experimental results
$\square_{\mathrm{max}}$	Maximum value
$\square_{\mathrm{mean}}$	Mean value
$\square_{\mathrm{median}}$	Median value
$\square^{\mathrm{OL}}$	Variable corresponding to an open-loop information pattern or strategy
$\square^{p}$	Variable corresponding to a Pareto efficient solution
$\square_{p}$	Variable corresponding to a cost function for Pareto efficient solutions
$\square^{S}$	Variable corresponding to a Stackelberg Equilibrium
$\square_{\mathrm{SD}}$	Standard deviation
$\square_{w}$	Variable corresponding to the beam of a ball-on-beam system
$\square_{\boldsymbol{\theta}}$	Variable associated to parameters $\boldsymbol{\theta}$
$\tilde{\square}$	Measured or observed variable
$\hat{\square}$	Estimated variable
$\underline{\square}$	Sequence of variables

1 Introduction

Automatic and intelligent machines have become ever-present in today's society. Previously developed for industrial environments to perform repetitive tasks on their own and out of human reach, the robots and automation systems of today interact closely with humans and several other robotic systems. Current trends of technological development entail an even closer interaction, for instance, at a haptic level. This means that machines physically interact with a cooperation partner, e.g. a human, in order to assist him in the more efficient and safe completion of various tasks. Such a close interaction is given in the fields of cooperative industrial robots, robot-assisted surgery and assistance systems for vehicle control and various other human-machine cooperation settings. Therefore, automated robotic systems increasingly need the ability to predict the behavior of the humans or previously unknown machines that may interact with them. This ability is a crucial part for the design of such *cooperative systems* and for the exploitation of the full potential of cooperation synergies. Hence, adequate modeling and identification methods are essential; such mathematical models and suitable identification approaches can lead to a better general understanding of interacting agents and also to the possibility of implementing model-based control algorithms in a technical device for an adequate behavior during interaction with e.g. a human partner.

The aforementioned situation demands a modeling framework which, on the one hand, serves as a mathematical approach for the design of the automatic controller, but on the other hand, allows the description of human behavior. Descriptive and biologically interpretable models for human behavior have been explored in the biologic and neuroscientific communities. In particular, motor control of humans has been conjectured to arise from minimum principles [NC61]. Several optimality principles have been proposed to explain the generation of a specific trajectory which serves as a command to lower-level biomechanical models (see [Eng01] for an extensive review). Given these optimality criteria, optimal control theory arises naturally as a model for movement planning and generation [Tod04] and has become a widely accepted approach in the neuroscience community. This led to further work which used this approach to model not only different kinds of human movement [MTL10, EHAAM16], but also the behavior of a human controlling a dynamic system [PCC⁺15]. The theory of optimal control itself is one of the most applied concepts in automatic control with numerous applications in engineering. Using this concept, an automatic controller can be described by a particular cost function as this leads to a control law which determines its behavior.

In the general case with humans and machines interacting and cooperating with each other, either in terms of self-positioning (e.g. avoiding collision) or through the control of a dy-

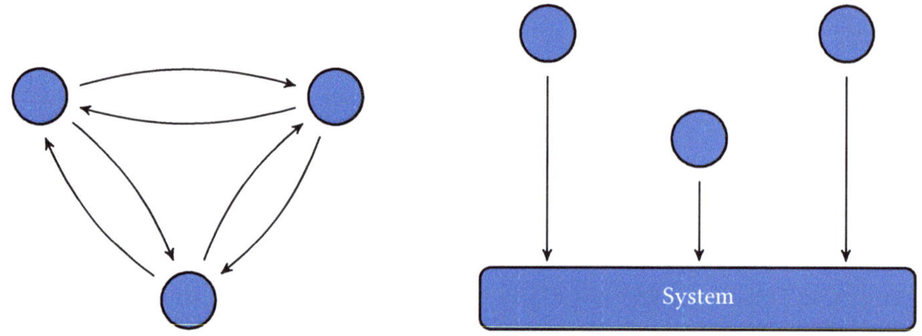

(a) Agents interacting with each other **(b)** Agents interacting through a dynamic system

Figure 1.1: Different scenarios of interaction between several agents

namic system (e.g. haptic shared control of a vehicle), as depicted in Figure 1.1, a possibly conflicting situation emerges. This is due to the fact that human and machine strive each for the optimization of their own individual criterion, thus potentially affecting each other negatively. Conflicts in dynamic situations, the latter of which arise in engineering problems, can be described by *dynamic game theory*, a framework which has been increasingly employed for applications in automatic control [Isa99, RBS16] as well as economics [Doc00] and biology [MGP+18]. In other words, the mathematical framework of dynamic game theory not only includes modeling the behavior of each partner by means of a criterion to be optimized, but also allows for the analysis of the result of their interaction. This result is typically described by an *equilibrium* solution, the computation of which has been the object of considerable efforts (cf. [BO99]). In addition, first studies exist which demonstrate the potential of the so-called *Nash equilibrium* as a descriptive concept for biological systems, for instance, bird collision avoidance behavior [MGP+18] as well as interacting humans in avoidance behavior [TW19] and in haptically coupled scenarios [BOW09, CS17, IFH19].

However, calculating equilibrium solutions in dynamic games demands the knowledge of the criteria each of the players optimize, which in real scenarios are typically unknown. Indeed, intelligent automated systems will usually have incomplete information about other players. Moreover, in human-machine interaction, the objective function of the human partner is usually unknown. For instance, in highly automated driving scenarios, an autonomous driving car would not have knowledge of the objectives of other non-autonomous (human-controlled) vehicles. In these cases, if only measurement data is available, the objectives of the players have to be identified out of a given outcome of the interaction, i.e. players' actions and system states corresponding to a game-theoretic equilibrium. In order to permit a major breakthrough of the application of dynamic game theory, efficient data-based identification of the criteria each of the players optimized becomes essential. This identification problem is denoted as *inverse dynamic game* and its solution is the main research objective of this thesis.

1.1 Research Objective and Contributions

The main objective of this thesis is **the development of methods for solving inverse dynamic game problems**, allowing for an estimation of player objectives from observed interaction behavior. Contrary to the problem of determining equilibrium solutions from known objectives which has been extensively studied, the inverse problem has scarcely been considered in previous work. Most treatments consider special cases, propose computationally heavy methods and do not give further insight on the properties of the problem. Motivated by the aforementioned studies on human-human-interaction, the focus of this thesis are dynamic games where a Nash equilibrium arises and defines the observed behavior. In addition, the efficiency of the methods is endeavoured in view of their utilization in real applications.

In a broad sense, the following contributions are made and presented in this thesis:

1. The development of efficient control-theoretical methods for inverse dynamic games as a means to identify cost functions of interacting players based on given observations. Furthermore, mathematical conditions for successful identification are developed.
2. The development of an inverse dynamic game method based on an approach which stems from computer science and information theory, for which a proof of the unbiasedness of the objective estimation is given.
3. The application of the novel methods using both simulated data from different scenarios and real data from a cooperative steering experiment with 52 participants. The performance of the developed methods and a state-of-the-art approach is compared and thoroughly analyzed.

1.2 Outline

The remainder of this thesis is structured as follows.

In **Chapter 2**, related work and existing literature on the estimation of player objectives in optimal control and dynamic games are reviewed. The research gap is formalized in terms of concrete research questions which shall be answered in this thesis. **Chapter 3** introduces the reader to the necessary mathematical fundamentals of dynamic game theory. In particular, existing results on the determination of equilibrium solutions are reviewed which lay the foundation of the developed inverse dynamic methods of this thesis.

The main theoretical contributions are given in Chapters 4 to 6. **Chapter 4** presents a formal definition of inverse dynamic game problems and presents a control-theoretical approach for open-loop inverse dynamic games. Furthermore, sufficient conditions for succesful identification of unique parameters will be presented. Inverse methods and analysis tools for the class of linear-quadratic (LQ) differential games are presented in **Chapter 5**; necessary and sufficient conditions for identification of unique parameters are also given. In **Chapter 6**, a method based on inverse reinforcement learning is presented and shown to be adequate for solving inverse dynamic games with both open-loop and feedback information structures. The chapter also presents unbiasedness results for the estimation of player objectives with this approach.

The next chapters involve the evaluation of the novel methods in simulations and a real application. First, **Chapter 7** give simulation results to evaluate all presented methods. The properties of each class of method are highlighted and a systematic comparison with a state-of-the-art method is conducted where the quality of the identification, robustness to measurement noise and modeling errors as well as the computational complexity are evaluated. **Chapter 8** shows an application of inverse dynamic games including the identification of human behavior in a haptic shared control task. Similar to Chapter 7, the experimental data is used to compare the methods with respect to the capability of describing observed human cooperative steering behavior.

Finally, **Chapter 9** sums up all insights and results obtained in this thesis.

The structure of the thesis is summarized in Figure 1.2, where the main body is divided into two paths to stress the different principles which underlie the proposed inverse dynamic game methods.

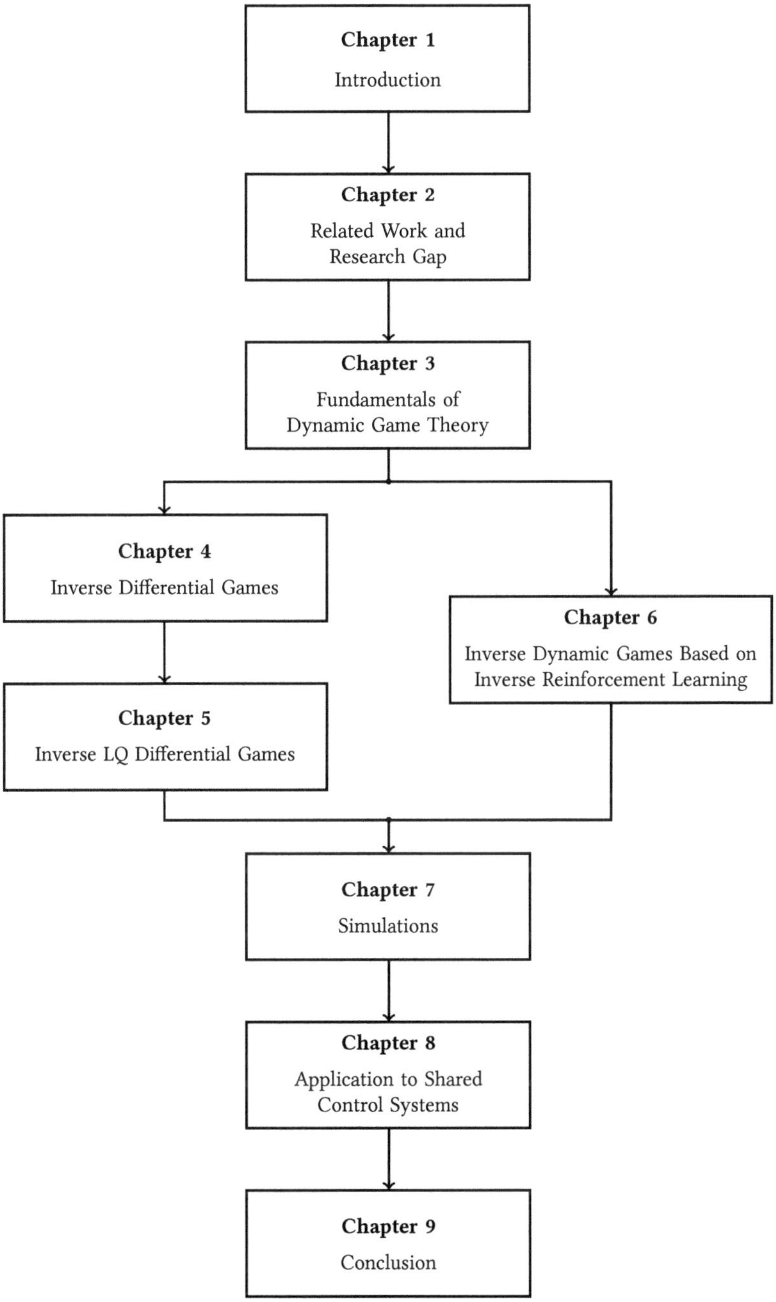

Figure 1.2: Outline of the thesis

2 Related Work and Research Gap

In this chapter, related work concerning methods for the estimation of cost functions is reviewed and the concrete research gap is identified. The majority of related work is concerned with cost function identification in a single-player case, also known as inverse optimal control, both from a control-theoretic and from a computer science point of view. Therefore, this case and its origins are surveyed first to provide an adequate context before covering state-of-the-art methods in a game-theoretical setting. The chapter ends with a discussion on all explored literature, the statement of the research gap and corresponding research questions to be answered in this thesis.

2.1 The Inverse Problem of Optimal Control

The problem of characterizing and describing cost functions corresponding to known optimal solutions was first considered in an optimal control setting, a problem which is known as *inverse optimal control*. The study of inverse problems in optimal control started with Kalman's paper: "When is a linear control system optimal?". The paper introduced conditions for a given linear control law to be optimal with respect to a quadratic performance index in the case of a single-input linear system and also showed that the inverse problem is ill-posed [Kal64]. Further progress was made by [Tha67] and [MA73] which stated similar conditions for a control-affine system and more general performance indices. These conditions serve the characterization of control laws which are optimal, but are not computationally convenient in order to calculate a particular cost function. The computational aspect was addressed in [JK73], where formulas were given for calculating a particular set of cost function matrices based on the known system dynamics and control law. Generalized results were given by [Cas80], where the Hamilton-Jacobi-Bellman equation was proposed as a means to calculate all possible cost function parameters corresponding to a known control feedback law in a linear-quadratic optimal control problem. Similarly, [FN84] extended Kalman's results to the multivariable case and dropping the assumption of a stabilizing control law.

After these initial efforts, inverse optimal control as a means to determine cost functions receded into the background in favor of the development of control synthesis methods. The newly introduced objective of inverse optimal control consisted in the calculation of a control law which is optimal with respect to any cost function, a property which is desirable due to the resulting robust stability of the closed-loop system. An approach was developed in [Fuj87] for the linear-quadratic case. Later, [FK96, KT99] developed an approach for input-affine non-linear systems. Herefor, a link between optimal value functions and Control Lyapunov Functions was established using Sontag's control law [Son89].

Nori and Frezza were the first in the automatic-control community to state a problem which consisted of finding a cost function which explains measured trajectories [NF04], representing a contrast to the first theoretical work and the subsequent approaches focusing on control synthesis. Hence, the "inverse optimal control problem" underwent a shift towards a more application-oriented problem. Most following approaches which can be found under the name of "inverse optimal control" build upon this idea and define the problem as follows:

Definition 2.1 (Inverse Optimal Control Problem)
Let observed state trajectories $\boldsymbol{x}^*(t)$ of a known dynamic system and control trajectories $\boldsymbol{u}^*(t)$ of a controller be given. Determine the cost function J under which the observed trajectories are optimal.

Definition 2.1 assumes the optimality of the observed trajectories, thus intuitively representing the inverse problem to the classical optimal control problem[1] (illustrated in Figure 2.1). Nevertheless, this assumption is sometimes dropped (as e.g. in [NF04]) and therefore, the problem consists of estimating a cost function which best approximates a given set of trajectories.

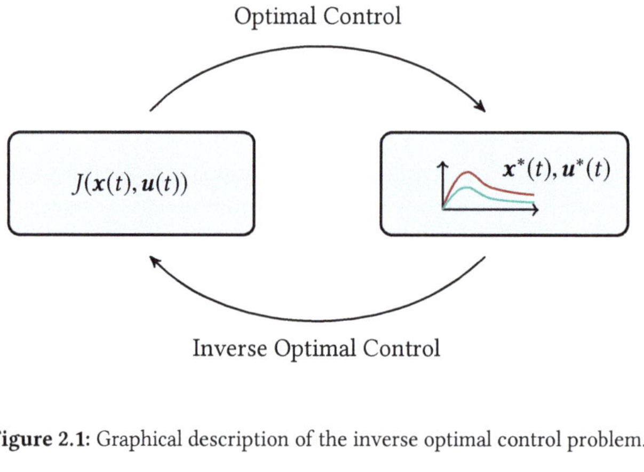

Figure 2.1: Graphical description of the inverse optimal control problem.

Inverse optimal control has been an object of research in the last decades, both from a theoretical and a practical point of view. The variety of methods for solving inverse optimal control problems can be classified into three main groups:

1 In the course of this thesis, the latter problem shall be also referred to as *forward problem* to stress on the contrast to the introduced inverse problem.

1. Direct approaches
2. Inverse Optimal Control (IOC) methods which apply control-theoretical principles
3. Inverse Reinforcement Learning (IRL) techniques which stem from computer science

It must be noted that the classification varies in literature. Indeed, a variety of articles use the term "inverse optimal control" as a term to denote the problem of estimating cost functions from measured data, similar to [NF04] and independently of the applied method. Nevertheless, in this thesis, this classification is proposed and shall be delineated in the following. Almost all articles found in literature present approaches which are based on the assumption of a particular structure of the cost function, e.g. a quadratic cost function. Therefore, the problem of identifying a cost function is reduced to determining parameters $\boldsymbol{\theta}$ such that the observed state and control trajectories are optimal with respect to the resulting cost function $J(\boldsymbol{\theta})$.

The presented method classes are further described in the following.

2.1.1 Direct Approaches

One of the most common ways to solve the inverse optimal control problem is a direct approach, where the cost function is determined iteratively. In each iteration, an optimal control problem is solved in order generate the trajectories which are optimal with respect to the current cost function candidate. These trajectories are then compared to the observed ones. Based on this comparison, which usually includes the calculation of an error measure between trajectories, the cost function can be updated such that the error is reduced. The overall aim of the method is to determine cost function parameters such that the error between both sets of trajectories is minimized. Due to the fact that the solution of the optimal control problem in each iteration can be represented as a "lower" level of the main optimization problem, these kinds of methods are also known as *bilevel methods* [MTL10]. Figure 2.2 shows a schematic diagram of both levels of the direct approach: the upper level, where the cost function of the current iteration κ is updated such that a performance measure, e.g. the error between trajectories is minimized, and the lower level, where an optimal control problem is solved to determine trajectories which are optimal with respect to the current cost function candidate.

The first algorithm of this kind was presented in [MTL10] and applied for human locomotion modeling. Further applications of this approach include driver steering behavior modeling [MFH17], reach-to-grasp human motion [EHAAM16] and human leg movements [BPC+06]. The implementation of the methods usually differ in the techniques for solving the upper level problem. For example, in [EHAAM16], the upper level problem is solved by means of particle swarm optimization. In [BPC+06], a static optimization version of the problem is posed and solved by nonlinear programming techniques. All methods require the repeated solution of optimal control (or static optimization) problems in the lower level and therefore potentially yield large computation times. Therefore, the importance of efficient numerical techniques for the solution of the problems in both levels is often stressed in literature (see e.g. [MTL10]). As a way of mitigating the computational effort, [ARARU⁺11], [HSB12] and, very recently, [ZLH19] replace the lower level problem by its corresponding optimality conditions. As a consequence of the high computation times, the methods are mostly suitable for offline applications only.

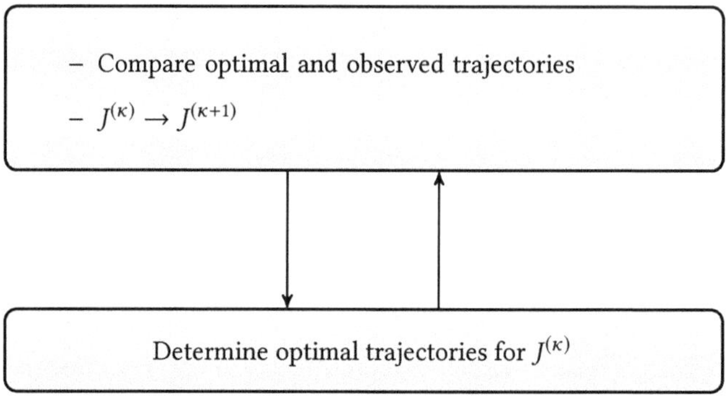

Figure 2.2: Direct bilevel approach for inverse optimal control: The upper level updates the cost function candidate such that an error measure is minimized. The lower level solves an optimal control problem.

2.1.2 Inverse Optimal Control

This class of methods exploits results from optimal control theory and do not rely on the repeated solution of an optimal control problem. The methods are based on the assumption that the observed trajectories are optimal with respect to an (unknown) cost function. With this assumption, optimality conditions are exploited in order to develop computational methods to find the parameters of the cost function which explains observed data. The optimal parameters are determined by minimizing an objective function (usually called *residual function*) which describes the extent to which optimality conditions are violated.

The variety of methods arises from the different kinds of optimality conditions which have been applied. In the continuous-time case, these include the minimum principle of Pontryagin[2] and the resulting Hamilton differential equations [JAB13], the Euler-Lagrange equations [AB14] and the Hamilton-Jacobi-Bellman equation [PHL14]. If time is discretized, then Karush-Kuhn-Tucker (KKT) conditions [KWB11, PJJB12, PR15, PR17] or the discrete-time minimum principle [MTFP16] can be applied.

2 This principle was originally posed in 1955 as a maximum principle given the aim of maximizing an objective function (cf. [Gam99]).

Some work focused on the case where the dynamic system is linear and the cost function structure is quadratic, i.e. an inverse linear-quadratic optimal control problem. This formulation allows for exploiting the arising constant linear feedback matrix if the time horizon tends towards infinity. If this matrix is known, then the cost function parameters can be estimated by solving a linear matrix inequality [Boy94, Section 10.6] or by stating an alternative objective function to be minimized with the algebraic Riccati equations as constraints [PCC+15, FMM+18].

2.1.3 Inverse Reinforcement Learning

Finally, related problems have been tackled in the field of computer science, for which so called inverse reinforcement learning (IRL) techniques have been developed. The IRL problem itself was first introduced by Russell and Ng [Rus98, NR00]. IRL mostly regards a discrete-time Markov Decision Process (MDP), which implies a finite and discrete set of possible control[3] values and states and search for a reward function instead of a cost function.[4] An example scenario (depicted in Figure 2.3) which can be modeled with an MDP is a grid world.[5] The inverse problem consists in finding the cost function if the agent's trajectory from the initial state to the final state, or the optimal strategy, is known. Furthermore, in IRL problems, the strategies and the dynamics of the system are potentially stochastic.

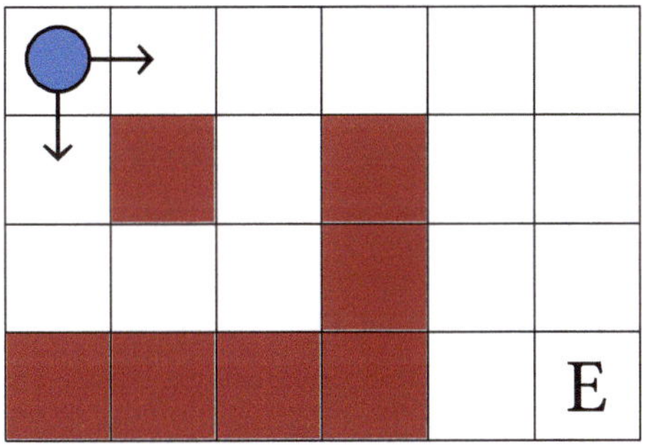

Figure 2.3: Grid world scenario in reinforcement learning, where the aim is to find an optimal policy which leads to the desired final state (E).

There is a vast number of methods which tackle the IRL problem using different principles. Interestingly, many of the methods under the name of IRL which are available in literature

[3] In the IRL literature, the controls are known as actions. In this thesis, both names are used as synonyms.

[4] Minimization of a cost function corresponds to a maximization of the reward function. The maximization problem can be easily cast as a minimization problem by multiplying the reward function with -1. Therefore, in the following, the term "cost function" will be used without loss of generality.

[5] A grid world is the most common test scenario for (inverse) reinforcement learning methods. It describes an agent searching for an optimal strategy which allows him to reach the final state with least cost. In Figure 2.3, this implies avoiding the red blocks which denote a high cost.

are based on a repeated calculation of the control and state sequences based on the current reward function candidate, i.e. the solution of the forward problem. Therefore, the principle is very similar to the aforementioned bilevel method. The methods presented in [AN04, RBZ06, NS07] are exemplarily mentioned. The Bayesian IRL method of [RA07] uses maximum a-posteriori estimation of the cost function which depends on sampling methods and thus demands the repeated estimation of optimal controls. A widespread IRL approach was proposed by Ziebart et al. [ZMBD08]. The idea consists in applying the principle of maximum entropy introduced by Jaynes [Jay57] in order to find a least-biased probability function which explains the observed trajectories.

All of the aforementioned IRL methods consider an MDP as a basis and are therefore limited to discrete-valued and finite states and actions. For large (or even infinite) states and action spaces, these methods suffer from the curse of dimensionality and become highly complex and computationally heavy, especially if they are applied to approximate continuous-valued state and action spaces. Therefore, some effort has been made to develop IRL techniques for continuous-valued spaces, tackling in this way a very similar problem as the literature on cost function identification in a control-theoretical setting. It is conspicuous that these approaches show a strong similarity to the maximum entropy IRL method of [ZMBD08]. For example, [AB11] and [HFKB15] apply a maximum entropy distribution, yet solve the IRL problem using a bilevel structure. On the other hand, [KPRS13] and [LK12] propose a maximum entropy distribution which considers continuous-valued state and action spaces and does not rely on the repeated solution of optimal control problems.

2.2 Inverse Problems in Game Theory

After reviewing literature on cost function identification in a single-player case, this section investigates the extent to which similar problems have been tackled in a game-theoretical scenario, i.e. the identification of cost functions from the observed interaction between several players.

Inverse problems in game theory have received growing attention in the last years, especially for static games. The term *inverse game theory* was introduced in [SC12] to denote the estimation of the actions and cost functions of the adversary, i.e. the other players in the game, in order to obtain better results. Similar work is reviewed in the following.

2.2.1 Inverse Static Games

Even though the concept of inverse game theory initially consisted in estimating adversary cost functions from the point of view of a particular player, its meaning quickly became more general and hence, it gained a strong similarity to the previously introduced inverse optimal control problems. Kuleshov and Schrijvers [KS15] introduce their paper with the words: *"given the observed behavior of players in a game, how can we infer the utilities*[6] *that led to this behavior?"*. They consider parametrizable Bayesian games where players have incomplete information of the opponent's cost function. These are estimated by using data of several realizations of static games. Similar conditions are needed in the approach of Konstantakopoulos et al. [KRJ^{+}18] which leverages necessary and sufficient conditions of each players' cost function to estimate their parameters. In [BGP15], a method based on the solution of variational inequalities is presented to identify cost functions. An application of this work for the optimization of transportation networks is presented in [ZPCP17].

2.2.2 Inverse Dynamic Games

Transferring the problem of Definition 2.1 to a multiplayer (N-player) case leads to the concept of *inverse dynamic games.* A general inverse dynamic game may be defined as follows:

Definition 2.2 (General Inverse Dynamic Game)
Let state trajectories $\mathbf{x}^*(t)$ of a known dynamic system and control trajectories $\mathbf{u}_i^*(t)$ of each player i, $i \in \{1, ..., N\}$ which correspond to a solution of a dynamic game be given. Find the cost functions J_i, for each player i, which generated the trajectories.

In Definition 2.2, the trajectories are generated by several players in a dynamic game acting based on individual cost functions. In addition, the problem is also ill-posed; an evident fact given the ill-posedness of the single-player case. The problem of Definition 2.2 is described as "general" in the sense that the solution type is still unspecified and, contrary to the single-player case, different solution concepts exist which generally lead to different trajectories. If the game is *non-cooperative*, the solution may be a **Nash** or a **Stackelberg** equilibrium depending on the order in which the players act. If the game is *cooperative*, then usually a **Pareto** efficient solution is assumed [ER11]. Literature on dynamic game theory is mostly focused in the concept of Nash equilibria which naturally arises when all players minimize their corresponding cost functions simultaneously. However, there exists a broad class of dynamic games for which the Stackelberg and the Nash solutions coincide.[7]

A literature search reveals that the problem of Definition 2.2 is greatly unexplored as mostly special cases can be found. In the automatic control community, an early work by Fujii and Khargonekar gives an approach to calculate solutions of an inverse linear-quadratic differential game [FK88] with a frequency-domain formulation. The results are similar to the one-player results developed by Kalman in [Kal64]. An inverse two-player zero-sum game has

6 Utility is a term used especially in static game theory to denote a reward as in IRL methods.
7 These concepts will be further explained later in Section 3.5.

been considered in [TMP16] where an approach which exploits necessary conditions for saddle point solutions was presented.[8] In [Wan07], necessary and sufficient conditions for identification in linear-quadratic dynamic games are given. However, these are restricted to the case of a second-order dynamic system and a two-player case. For N-player inverse dynamic game with open-loop strategies, recent results were presented in [MFP17a, MFP17b] where Pontryagin's minimum principle is leveraged. In [MFP17b], a bilevel method analogous to the ones described in Section 2.1.1 was formulated. This is portrayed in Figure 2.4: the upper level, where the N cost functions (denoted by $J_{1:N}$) are updated and the lower level, where a dynamic game is solved to determine trajectories corresponding to the N current cost function candidates.

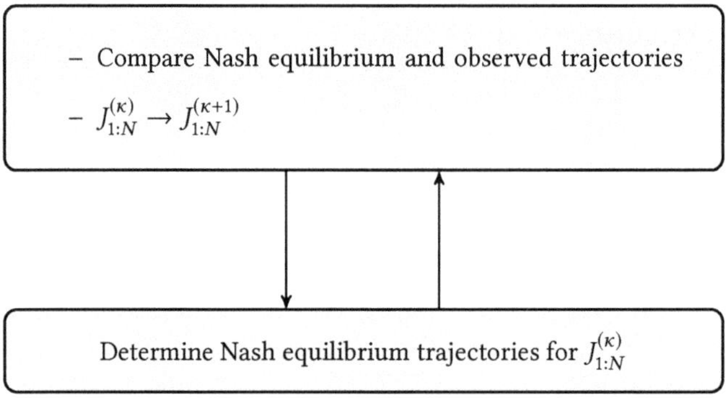

Figure 2.4: Direct bilevel approach for inverse dynamic games: The upper level updates the cost function candidates such that an error measure is minimized. The lower level solves a dynamic game to determine Nash equilibrium trajectories.

Dynamic game theory has been of considerable interest in economics, leading to some proposed methods for the solution of the inverse problem in this field. For example, [BBL07] presented an approach which is based on the estimation of the value of the cost function by means of a Monte Carlo method. The work of Arcidiacono et al. [ABBE16] offers a more efficient method based on least-square estimation and likelihood functions. Both aforementioned methods have the main drawback that the game is limited to discrete-valued strategies and a finite number of possible states. A dynamic game with a linear-quadratic setting was considered in [CFG89], yet restricting the players' cost function matrices to only penalize their own controls and to only have diagonal entries.

As for IRL methods, some methods which aim at extending these techniques to the multi-agent setting were proposed for cases in which all players behave cooperatively [HMRAD16, NKJ+10, ŠKZK17]. On the other hand, IRL-based methods in a noncooperative setting have been proposed in [LBC18, RGZH12]. However, similar to single-agent IRL, all of these methods are based on an MDP and hence are limited to discrete-valued and finite control and state

8 Zero-sum games represent the case where one player strives to minimize a cost function while the second player seeks to maximize the same cost function.

spaces. Literature shows few available work which considers continuous-valued action and state spaces. Two exceptions are [PSS+16], where a cooperative scenario was considered, and [MHLK17], where each agent has an individual cost function, yet not explicitly relating their approach to game-theoretical concepts.

2.3 Discussion

As motivated in Chapter 1, the Nash equilibrium is a promising descriptive concept for the interaction between biological systems and hence potentially adequate for state-of-the-art applications in human-machine interaction. Therefore, this thesis focuses on the solution of inverse dynamic games where the trajectories correspond to a Nash equilibrium. In the following, the term *inverse dynamic game* will refer to this problem.

In order to solve inverse dynamic games, it may appear conceivable to apply a direct bilevel approach analogously to the single-player case (cf. Section 2.1.1). Nevertheless, the lower-level problem would consist in this case in determining the state trajectories and all players' control trajectories corresponding to the dynamic game of the current iteration. Consequently, the method implies the repeated solution of N coupled dynamic optimization problems for each set of cost function candidates. The first evaluation conducted in [MFP17b] presented a simple example where the inverse dynamic game involved the solution of 388 forward dynamic games. Especially for non-linear dynamic games, solving for Nash equilibria is in general computionally heavy and efficient numerical techniques are not available [HdlCIR19][9]. Therefore, applying this approach yields a great risk of huge computation times.

This motivates the need for more efficient methods for inverse dynamic games which do not rely on the repeated solution of a dynamic game. A fast identification of player cost functions allows for an immediate adaptation of automatic controllers based on potential new information, e.g. if the cooperating human changes its behavior. Nevertheless, until now, little effort has been spent in the development of alternative methods for the efficient solution of general N-player inverse dynamic games. Methods which stem from IRL are restricted to discrete-valued and finite states and controls. In addition, IRL methods in a multiplayer setting which consider continuous-valued states and controls are also almost unexplored and their theoretical foundation has not been developed. The situation is similar in the field of automatic control, where only special cases have been treated. Apart from very early work of [CFG89] in an economics-specific scenario, successful attempts to solve general N-player inverse dynamic games have occured only recently ([MFP17a, MFP17b]). This work encourages further effort in exploring alternative techniques for inverse dynamic games which avoid a direct bilevel approach.

[9] A recent study in [HdlCIR19] showed that a nonscalar two-player dynamic game with non-quadratic cost functions can take from 479.11 to 12854 seconds to solve, depending on the applied method.

Finally, almost all of the mentioned approaches, especially in dynamic games, concentrate on delivering a method which is able to estimate a cost function, but do not give further insight on when an estimation is possible. This not less important aspect of the properties of inverse dynamic game problems is almost unaddressed; there is little work on inverse problems in optimal control and dynamic games following the ideas of Kalman and the first theoretical studies (cf. Section 2.1). In addition, the ill-posedness of inverse dynamic games demands further attention. To date, much uncertainty exists concerning the properties of inverse dynamic games as these are still considerably unexplored.

2.4 Conclusion and Research Questions

As discussed in the previous section, the inverse problem of optimal control, i.e. a single-player inverse dynamic game has been investigated from both a theoretical and a computational point of view. However, the problem of modeling and identifying the behavior of several players interacting with each other remains a greatly unexplored field, especially in the case of continuous-valued control and state spaces which is important for many applications. The application of a direct bilevel approach to this problem is inappropriate given the potential complexity of solving for Nash equilibrium trajectories repeatedly. Therefore, the following questions need to be answered:

- How to solve inverse dynamic games efficiently, in particular avoiding the solution of the forward problem?
- Under which conditions can a solution be found and when is this solution unique?

For this purpose, necessary fundamentals concerning dynamic game theory and the forward problem of determining Nash equilibria are introduced in Chapter 3 as a basis for the subsequent result. Afterwards, the posed questions are addressed in Chapters 4 and 5, where methods based on IOC—according to the classification in Section 2.1—are developed, and in Chapter 6 which presents an IRL-based method is introduced as a means to solve inverse dynamic games.

Furthermore, two questions which naturally arise after the development of techniques for solving inverse dynamic games are:

- How do the results of these alternative approaches compare to the results of a direct bilevel approach?
- Which main class of methods, IOC-based, IRL-based or direct bilevel, yields a greater potential for a real application, e.g. in the identification of cooperative systems with humans?

Probably due to the fact that IOC and IRL methods have been studied by different research communities, until now, almost no systematic comparison has been conducted on the performance of these different concepts.[10] Therefore, in Chapter 7, all methods (IOC-based, IRL-based and bilevel methods) are compared to each other using two different major classes of inverse dynamic game problems, where robustness to measurement noise and cost function modeling errors are also examined. Lastly, a first application example is presented in Chapter 8 to evaluate the performance of all methods with real experimental data.

[10] Two notable exceptions are given by [TZ11] and [JAB13]. The first compred bilevel and IOC-similar methods in (single-player) inverse static optimization. The study demonstrated that the alternative method, which was based on optimality conditions, yielded comparable results to the bilevel method with considerably less computational effort. In [JAB13], a single-player inverse optimal control method based on Hamilton differential equations was compared in simulations with the bilevel method [MTL10] and the continuous-time counterparts of the methods presented in [AN04] and [RBZ06]. Their proposed method was shown to perform faster and with less trajectory and parameter error. Nevertheless, all simulated observed trajectories were noise-free.

3 Fundamentals of Dynamic Game Theory

This chapter gives an overview of fundamentals of dynamic game theory. After a short introduction to the general theory of games, non-cooperative dynamic and differential games are introduced. Furthermore, existing solution concepts for the forward problem are introduced and the available means for their calculation are shown. These principles provide a basis for the development of the inverse dynamic game methods proposed in subsequent chapters. The contents of this chapter are based on the books [BO99, Eng05, HKZ12, Tad13].

Game theory can be defined as the theory of mathematical models of decision making to describe situations with conflicts and cooperation between rational players. The conflicts arise from different interests or goals, leading to a strong dependency of each one's individual decisions. The theory emerged from the work of von Neumann [VNM47] and blossomed with the introduction of game equilibria by Nash [Nas51]. Since then, it has been extensively studied such that analytical tools are available for understanding phenomena arising from the interaction between decision makers.

3.1 Introduction to Games

One of the most frequent ways of defining a game is as a normal-form game, described in the following definition.

Definition 3.1 (Game in Normal Form)
A normal-form game is defined by

- *A set of players* $\mathcal{P} = \{1, 2, ..., N\}$.
- *A strategy set* $\mathcal{U}_i$ *for each player* $i \in \mathcal{P}$.
- *A set of cost functions* $\mathcal{J} = \{J_1, J_2, ..., J_N\}$.

A game involves N decision makers called *players* which select particular actions from a possible strategy set. These are chosen such that a specific goal, represented by their individual cost function, is accomplished. Definition 3.1 is very general and allows numerous kinds of games which arise from different properties of the possible actions, strategy sets and cost functions of the players.

If the players act in a self-interested way, i.e. they strive for minimizing their own cost function, regardless of possible negative effects for other players, then the game is called *non-cooperative.* If the players are able to generate binding agreements and act jointly in order to obtain a fair result, then the game is regarded as *cooperative.* If the choice of actions is deterministic, the strategies are called *pure strategies.* The converse is denoted as stochastic or *mixed strategies.* Moreover, games may be *finite* or *infinite*, depending on the strategy set $\mathcal{U}_i$ of each player. If the set of possible strategies $\mathcal{U}_i$ has a finite number of elements for all players, the game is said to be finite. Otherwise, if $\mathcal{U}_i$ is infinite for at least one player, i.e. an infinite number of possible strategies is available for at least one player, the game is infinite.

An important classification of games is based on the number of times a player can choose an action. If the players act only once and independently of each other, the game is *static.* As soon as one player is allowed to act in several *time stages* based on new information resulting from other players' previous actions, then the game is *dynamic.* Therefore, in dynamic games, time plays an important role. The evolution of an infinite dynamic game is naturally described with a difference equation in a discrete-time formulation based on the stages or discrete time steps in which players take action. However, a continuous-time formulation is possible as well, which is also known in literature as a *differential game.*

The results of this thesis are based on **non-cooperative infinite dynamic games** in both discrete and continuous time. Since many results are analogous and comparable, the main aspects of infinite dynamic games will be shown and formalized in this chapter with a formulation in continuous time. Analogous definitions for the discrete-time case can be found in Appendix A.

3.2 Differential Games

The evolution of a differential game depends on the strategies of all players. It can be described by means of the time-dependent state trajectories of a dynamic system defined by differential equations.

Definition 3.2 (Dynamic System in State Space Representation)
A dynamic system is defined by ordinary differential equations and an initial condition given by

$$\dot{\boldsymbol{x}}(t) = \boldsymbol{f}\left(\boldsymbol{x}(t), \boldsymbol{u}_1(t), \ldots, \boldsymbol{u}_N(t), t\right) \tag{3.1a}$$

$$\boldsymbol{x}(0) = \boldsymbol{x}_0, \tag{3.1b}$$

where $\boldsymbol{x}(t) \in \mathbb{R}^n$ *and* $\boldsymbol{u}_i(t) \in \mathbb{R}^{m_i}$, $i \in \mathcal{P}$, *denote the system state vector and the control vector of player i at time step t, respectively. Furthermore,* $\boldsymbol{f} : \mathbb{R}^n \times \mathbb{R}^{m_1} \times \ldots \times \mathbb{R}^{m_N} \times \mathbb{R}_0^+ \mapsto \mathbb{R}^n$ *is a vector function which is continuous in* $t \in [0, T]$ *and globally Lipschitz in* $\boldsymbol{x}$, $\boldsymbol{u}_1, \ldots, \boldsymbol{u}_N$.

The evolution of the differential game is regarded for a time interval $[0, T]$ which represents the **duration of the game**. The vector $\boldsymbol{x}_0$ represents the initial state of the system. The final time T could be $T \to \infty$ or a fixed value depending on the given problem. Lipschitz continuity of $\boldsymbol{f}$ is required to ensure that the initial value problem (3.1) admits a unique solution for every N-tuple $(\boldsymbol{u}_1(t), \ldots, \boldsymbol{u}_N(t))$ of continuous controls $\boldsymbol{u}_i(t)$, $i \in \mathcal{P}$. Each player $i \in \mathcal{P}$ acts upon the system in Definition 3.2 by applying a corresponding input or control trajectory $\boldsymbol{u}_i(t), \forall t \in [0, T]$ which belongs to an action space $\mathcal{U}_i$. Each player's control decision or **strategy**, denoted by $\boldsymbol{\gamma}_i$, is based on the **state information** available to them which is represented by a set-valued function $\eta_i(t)$.[11] The strategy is chosen from a set of available strategies Γ_i and defines a particular control trajectory $\boldsymbol{u}_i(t)$[12], i.e.

$$\boldsymbol{u}_i(t) = \boldsymbol{\gamma}_i(\eta_i(t), t), \quad \boldsymbol{\gamma}_i \in \Gamma_i. \tag{3.2}$$

The strategy and consequently, the control trajectories are determined according to an **individual cost function**

$$J_i = h_i\left(\boldsymbol{x}(T), T\right) + \int_0^T g_i\left(\boldsymbol{x}(t), \boldsymbol{u}_1(t), \ldots, \boldsymbol{u}_N(t), t\right) \mathrm{d}t, \tag{3.3}$$

where h_i denotes costs which arise from the final state or final time and g_i represents running costs which arise for $t \in [0, T]$. The aim of each player i is to minimize the cost function (3.3) by applying appropriate controls $\boldsymbol{u}_i(t)$. This objective is described by the dynamic optimization problem

[11] Different possibilities of player state information and corresponding strategies will be examined later in Sections 3.3 and 3.4.

[12] In the context of dynamic games, actions and strategies are different and have this relationship. On the contrary, in static games these are identical and the terms are therefore not distinguished.

$$\begin{aligned} &\min_{\boldsymbol{u}_i(t)} J_i\left(\boldsymbol{x}(t), \boldsymbol{u}_i(t), \boldsymbol{u}_{\neg i}(t), t\right) \\ &\qquad \text{w. r. t.} \\ &\dot{\boldsymbol{x}}(t) = \boldsymbol{f}\left(\boldsymbol{x}(t), \boldsymbol{u}_i(t), \boldsymbol{u}_{\neg i}(t), t\right) \\ &\boldsymbol{x}(0) = \boldsymbol{x}_0 \end{aligned} \tag{3.4}$$

where $\neg i$ is used as a shorthand notation for "all except i". Therefore, $\boldsymbol{u}_{\neg i}(t)$ denotes the input trajectories of all players except player i.[13] As a result, differential games can be described as N coupled dynamic optimization problems.

To summarize, a definition of differential games which will be used throughout this thesis is given.

Definition 3.3 (Differential Game)
A differential game is defined by

- *A set of players* $\mathcal{P} = \{1, 2, ..., N\}$,
- *A specified time interval* $[0, T]$ *denoting the duration of the game,*
- *An infinite action set* $\mathcal{U}_i$, $\forall i \in \mathcal{P}$,
- *A set-valued function* $\eta_i(t)$, $\forall i \in \mathcal{P}$, *which determines the state information of player* i *at time* t,
- *A dynamic system given by Definition 3.2,*
- *A set of cost functions* $\mathcal{J} = \{J_1, J_2, ..., J_N\}$.

3.3 Information Structures

A relevant characteristic of a differential game is the available information for all players at each time step t. The information set is described by

$$\eta_i(t) \in P_{\neg\emptyset}\left(\{\boldsymbol{x}_0, \boldsymbol{x}(s), \boldsymbol{x}(t)\}\right), \qquad s \in [0, \chi_{i,t}],\ \chi_{i,t} \in [0, t], \tag{3.5}$$

[13] The importance of the uniqueness of the solution of (3.1) for every N-tuple $(\boldsymbol{u}_1, ..., \boldsymbol{u}_N)$ becomes clear at this point. Non-uniqueness is clearly not allowed in a differential game since it would potentially lead to non-uniqueness in the value of the cost functions for a single N-tuple of control trajectories.

where $P_{\neg\emptyset}(\cdot)$ denotes a power set which excludes the empty set and $\chi_{i,t}$ is non-decreasing in t. In a particular time $t \in [0, T]$, player i has knowledge of current or past values of the state $\boldsymbol{x}$. By (3.5), it is possible to describe a variety of information structures which are very common in dynamic game theory. Sometimes partial state information is assumed instead of a complete state information as implied by (3.5) and as considered in this thesis. The next definition lists concrete information structures which shall be focused on in the following.

Definition 3.4 (Information Structure of the Players)
The information structure of player i is said to be

(i) open-loop (OL) pattern if $\eta_i(t) = \{\boldsymbol{x}_0\}$, $t \in [0, T]$.

(ii) memoryless perfect state (MPS) pattern if $\eta_i(t) = \{\boldsymbol{x}_0, \boldsymbol{x}(t)\}$, $t \in [0, T]$.

(iii) feedback (FB) pattern if $\eta_i(t) = \{\boldsymbol{x}(t)\}$, $t \in [0, T]$.

The **open-loop** information pattern describes the situation where all players decide at $t = 0$ the control trajectories $\boldsymbol{u}_i(t)$ to be applied for $t \in [0, T]$ based solely on the initial system state value $\boldsymbol{x}_0$. The control decision remains unchanged for the whole duration of the game, regardless of any possible disturbance on the states. Figure 3.1 shows a graphical representation of a differential game with an open-loop information structure for each player.

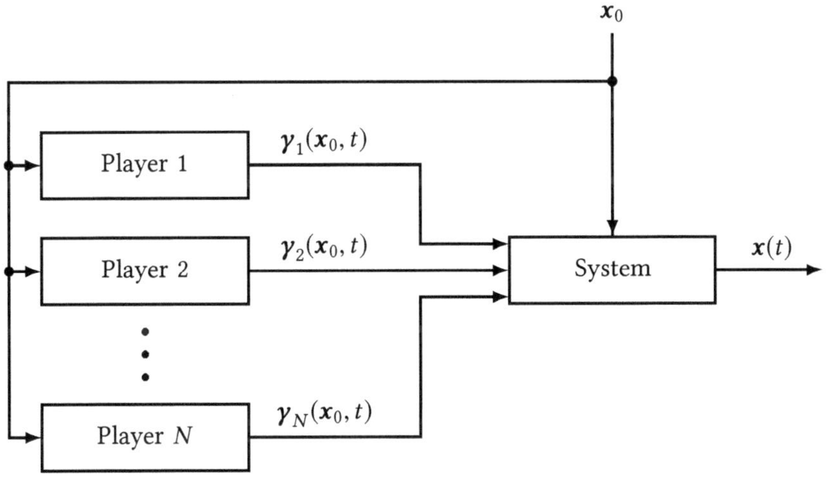

Figure 3.1: Differential game with an open-loop information structure.

In case of a **memoryless perfect state** pattern, the players have information of the initial state $\boldsymbol{x}_0$ and the current state $\boldsymbol{x}(t)$. The inclusion of the initial state becomes necessary for solving differential games where some of the players have an OL information pattern and others have access to the states $\boldsymbol{x}(t)$. In this thesis, the converse case—equal information patterns for all players—is considered such that a **feedback** information pattern can be used equivalently.[14] These last two information structures imply "closing the loop" in a control-theoretical sense. The resulting multiplayer control loop for a feedback information structure is exemplarily depicted in Figure 3.2.

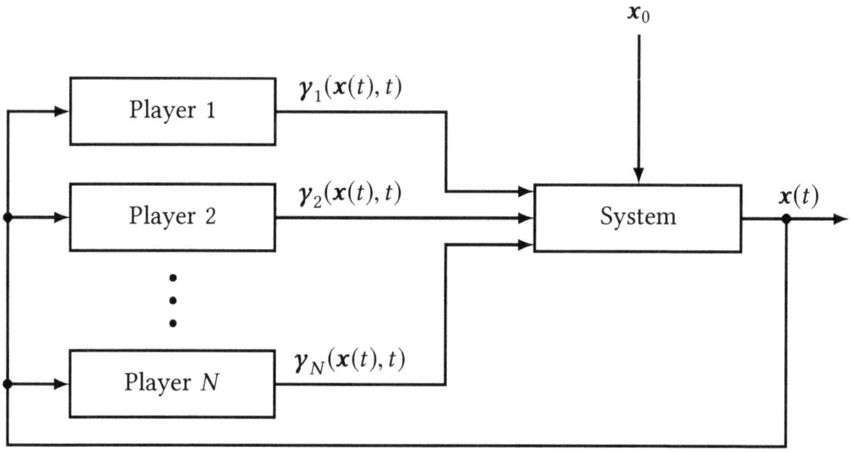

Figure 3.2: Differential game with a feedback information structure.

The different information patterns lead to various kinds of strategies selected by the players, each of which leads to a particular solution of the differential game, i.e. resulting state and control trajectories.

3.4 Strategies

As mentioned previously, the strategy defines the controls of the players based on the information available to them. Therefore, for each information structure defined above, we obtain a different class of strategy. The next definitions specify the corresponding strategy classes to the open-loop and the feedback information patterns.

[14] The later defined Nash equilibrium solution (cf. Section 3.5.1) is identical under both MPS and FB information patterns since the equilibrium dependence on x_0 is given only for the initial time $t = 0$. Therefore, these information patterns can be considered as equivalent in this sense [BO99, p. 278]. For this reason, in the following only the OL and the FB information patterns shall be considered.

Definition 3.5 (Open-Loop Strategy)
An open-loop strategy γ_i for player $i \in \mathcal{P}$ selects a control action according to

$$u_i(t) = \gamma_i(x_0, t), \quad \forall x_0 \in \mathbb{R}^n,\ \forall t \in [0, T], \tag{3.6}$$

where γ is a continuous function in t and defined for each possible initial state x_0. The set of all such possible strategies is denoted by Γ_i^{OL}.

Definition 3.6 (Feedback Strategy)
A feedback strategy γ_i for player $i \in \mathcal{P}$ selects a control action according to

$$u_i(t) = \gamma_i(x(t), t), \quad \forall t \in [0, T], \tag{3.7}$$

where γ is continuous in t and globally Lipschitz in x. The set of all such possible strategies is denoted by Γ_i^{FB}.

An **open-loop strategy** describes the situation where all players decide at $t = 0$ the control trajectories $u_i(t)$ to be applied for $t \in [0, T]$ based solely on the initial state value x_0 of the dynamic system. The control decision remains unchanged for the whole duration of the game, regardless of any possible disturbance on the states. The **feedback strategy** implies that the players define their actions based on the current state $x(t)$. Therefore, each player commits to a particular reaction to the information concerning the state of the system.

These strategy types are the basis for the solution of differential games. In the following, different solution concepts are presented.

3.5 Solution Concepts in Differential Games

A differential game may have different outcomes depending on its properties. The main difference arises from the cooperative or non-cooperative nature of the interacting players. In a non-cooperative game, all players act strictly rationally in order to minimize their own cost function, regardless of the detriment this may cause to other players. In this kind of game, the most common solution concepts are described as game-theoretical equilibria. These are the so-called **Nash equilibrium** [Nas51] and the **Stackelberg equilibrium** [Sta52]. In turn, in cooperative differential games, players are able to cooperate and make agreements such that they can (potentially better) achieve their objectives. In this kind of games, **Pareto efficient solutions** [Par14] are mostly sought.

3.5.1 Non-Cooperative Games

Nash Equilibrium

The Nash equilibrium is a solution concept in game theory which arises if (i) all players act simultaneously and optimally with respect to their own cost function and their beliefs of the other players' strategies and (ii) these beliefs are correct[15]. An alternative, equivalent definition is the following: For each player, there is no other feasible input strategy than the current, optimal one, that would minimize his own costs, taking into account all the other players with their optimal input strategy [Nas51]. In other words, it is not possible for all players to obtain a lower value of the cost function by solely altering their individual strategy. A formal definition is given in the following:

Definition 3.7 (Nash Equilibrium)
A Nash equilibrium is described by the N-tuple of strategies $\gamma^* := \left(\gamma_1^*, \ldots, \gamma_N^*\right)$, *with* $\gamma_i^* \in \Gamma_i^\diamond$, $i \in \mathcal{P}$, $\diamond \in \{\mathrm{OL}, \mathrm{FB}\}$, *which satisfies*

$$J_i\left(\gamma_i^*, \gamma_{\neg i}^*\right) \leq J_i\left(\gamma_i, \gamma_{\neg i}^*\right), \quad \forall i \in \mathcal{P},$$

i.e. $\gamma_i^* = \boldsymbol{u}_i^*(t)$, $t \in [0, T]$ *is the optimal input strategy for each player i considering optimal input strategies of all other players* $\gamma_{\neg i}^*$. *The resulting tuple of control trajectories* $\boldsymbol{u}^* := \left(\boldsymbol{u}_1^*(t), \ldots, \boldsymbol{u}_N^*(t)\right)$ *is called Nash equilibrium solution.*

Definition 3.7 describes either an **open-loop Nash equilibrium (OLNE)** or a **feedback Nash equilibrium (FNE)**, depending on the kind of strategy which is applied by each player, i.e. whether the strategy set Γ_i is given by Γ_i^{OL} or Γ_i^{FB}, respectively. The corresponding state trajectories $\boldsymbol{x}^*(t)$ are determined by solving the initial value problem (3.1) using the control trajectories $\left(\boldsymbol{u}_1^*(t), \ldots, \boldsymbol{u}_N^*(t)\right)$. The OLNE has the property of being a *weakly time consistent* solution. This means that the players do not have any incentive of deviating from their strategy during the game, i.e. at any time step $t_1 \in [0, T]$. On the other hand, the FNE is *strongly time consistent*[16], which means that their strategy γ_i^* is still an equilibrium strategy if it was applied from any time $t_1 \in [0, T]$ and starting from any arbitrarily chosen state $\boldsymbol{x}(t_1)$ off the original equilibrium path (which is reachable from $\boldsymbol{x}(0)$). This makes the feedback Nash equilibrium more robust towards any possible disturbances on the system state.

In a differential game, there may exist no Nash equilibria. Moreover, a single or multiple Nash equilibria may also exist. Furthermore, a Nash equilibrium cannot be uniquely associated to a set of cost functions $\mathcal{J}$. This fact is of particular importance for the inverse differential game problem and will be discussed in Section 4.1 of the next chapter.

[15] An example of this is a situation where all cost functions are made public to all players [OR94, p. 14].
[16] Also called *subgame perfect*, see. e.g. [Eng05, Definition 8.2].

Stackelberg Solutions

Previously, it was assumed that the players select their strategies simultaneously. A scenario, where the players select their strategies one after the other can lead to a different outcome of the game. Such a setting was first introduced by von Stackelberg in the context of a duopoly output game [Sta52]. In a general N-player situation, one of the players is selected as a leader such that he announces his selected control strategy. Afterwards, the next player uses this information to make a decision on his own strategy such that his cost function is minimized. This process continues until player N chooses its strategy based on the announcements of the other $N-1$ players' strategies. Stackelberg solutions are mostly considered in economic applications, e.g. market models, and are typically defined in a 2-player setting (cf. [CC72]).

Definition 3.8 (Stackelberg Strategy)
The strategy tuple $\boldsymbol{\gamma}^s = (\boldsymbol{\gamma}_1^s, \boldsymbol{\gamma}_2^s)$ is called a Stackelberg strategy with player 1 as leader and player 2 as follower if for all $\boldsymbol{\gamma}_1 \in \Gamma_1$

$$J_1(\boldsymbol{\gamma}_1^s, \boldsymbol{\gamma}_2^s) \leq J_1(\boldsymbol{\gamma}_1, \boldsymbol{\gamma}_2^o(\boldsymbol{\gamma}_1)) \tag{3.8}$$

where $\boldsymbol{\gamma}_2^o(\boldsymbol{\gamma}_1) \in \Gamma_2$ denotes the optimal response of player 2 to a fixed strategy of player 1, i.e.

$$J_2(\boldsymbol{\gamma}_1, \boldsymbol{\gamma}_2^o(\boldsymbol{\gamma}_1)) = \min_{\boldsymbol{\gamma}_2} J_2(\boldsymbol{\gamma}_1, \boldsymbol{\gamma}_2), \tag{3.9}$$

and $\boldsymbol{\gamma}_2^s = \boldsymbol{\gamma}_2^o(\boldsymbol{\gamma}_1^s)$.

The Stackelberg strategy is an attractive strategy when the information pattern is biased or asymmetric. This means that player 1 does not know the cost function of player 2, but player 2 has knowledge of both cost functions. This is the case in a market model where there is a dominant company. The leader has an advantage in terms of the possibility to obtain better results due to the fact that he is aware that the rest of the players will act optimally based on whatever strategy he may apply.

First derivations of Stackelberg solutions for dynamic games were given e.g. in [CC72, Med78]. For the (continuous-time) differential game case with N players, [Rub06, Proposition 2.3] states that the Stackelberg solution coincides with the feedback Nash equilibrium solution —provided it exists—if and only if (i) the running costs g_i depend solely on the state $\boldsymbol{x}$ and each player's controls $\boldsymbol{u}_i$, i.e.

$$g_i\left(\boldsymbol{x}(t), \boldsymbol{u}_i(t), \boldsymbol{u}_{\neg i}(t)\right) = g_i(\boldsymbol{x}(t), \boldsymbol{u}_i(t)), \tag{3.10}$$

and (ii) the dynamics of the state depend, at the most, linearly on each player's controls, i.e. the system dynamics have the control-affine form

$$\dot{\boldsymbol{x}}(t) = \boldsymbol{f}_{\boldsymbol{x}}(\boldsymbol{x}(t), t) + \sum_{i=1}^{N} G_i(\boldsymbol{x}, t)\boldsymbol{u}_i(t). \tag{3.11}$$

3.5.2 Cooperative Games

Contrary to the non-cooperative case, a cooperative game includes players which not only seek the optimization of their own objectives but also consider the objectives of the other players in the selection of the control actions. Hence, it is assumed that they cooperate in order to achieve their objectives.[17] However, no side-payments take place, which means that their cooperative behavior is not explicitely rewarded by introducing a cost-lowering term in the objective function. Consequently, depending on how the players decide to distribute their efforts, several possible minima exist for each particular player $i \in \mathcal{P}$.

In the field of cooperative games, the concept of *dominating* strategies plays an important role. A strategy tuple $\boldsymbol{\gamma}_{(a)}$ will dominate another strategy tuple $\boldsymbol{\gamma}_{(b)}$ if the application of $\boldsymbol{\gamma}_{(a)}$ leads to lower costs for all players compared to $\boldsymbol{\gamma}_{(b)}$. Therefore, dominating strategies lead to a better result for all players. This line of thought motivates considering only solutions that are such that they cannot be improved by all players simultaneously and leads to the concept of Pareto efficient solutions.

Pareto Efficient Solutions

A Pareto efficient solution is a combination of strategies such that it is not possible to obtain a better result in terms of the own cost function of each player without affecting the result of other players negatively. This means that, while it may be possible for individual players to improve their own result by changing their own action unilaterally, this would lead to a worse result for at least one of the other players. A Pareto efficient solution is defined as follows [Eng05, Definition 6.1]:

[17] Nevertheless, coalitional games, where several groups of players may build coalitions to act non-cooperatively with respect to other ones, are excluded in this thesis. See the definitions given in [ER11].

Definition 3.9 (Pareto Efficient Solution of a Differential Game)
An N-tuple of strategies $\boldsymbol{\gamma}^p = \left(\boldsymbol{\gamma}_1^p, \dots, \boldsymbol{\gamma}_N^p\right)$ *is a Pareto efficient solution (PES) of a differential game if no other feasible tuple* $\boldsymbol{\gamma} = \left(\boldsymbol{\gamma}_1, \dots, \boldsymbol{\gamma}_N\right)$ *exists for which*

$$J_j(\boldsymbol{\gamma}) < J_j(\boldsymbol{\gamma}^p) \tag{3.12}$$

for at least one $j \in \mathcal{P}$ *and*

$$J_i(\boldsymbol{\gamma}) \leq J_i(\boldsymbol{\gamma}^p), \quad \forall i \in \mathcal{P},\ i \neq j. \tag{3.13}$$

Definition 3.9 states that a PES is a combination of strategies such that it is not possible that any player obtains a lower value of his cost function by deviating from the strategy without affecting at least one other player negatively. Therefore, Pareto optima do not represent a stable solution of a non-cooperative game, since in such a game each player strives for minimization of their own cost function. A non-cooperative player will deviate from the Pareto strategy if this implies a lower value of his cost function, regardless of the resulting drawback for other players.

3.6 Calculation of Differential Game Solutions

This thesis focuses on the Nash equilibrium and on Pareto efficient solutions of differential games. Therefore, in the following, the relevant means for calculating these solutions are presented.

3.6.1 Open-Loop Nash Equilibrium

The basis of the calculation of Nash equilibria is Definition 3.7. The inequality implies that the optimal strategy $\boldsymbol{\gamma}_i^* \in \Gamma_i^{\mathrm{OL}}$ leads to a control trajectory $\boldsymbol{u}_i^*(t)$ which minimizes the cost function $J(\boldsymbol{u}_i(t), \boldsymbol{u}_{\neg i}^*(t))$ subject to the system dynamics

$$\dot{\boldsymbol{x}}(t) = \boldsymbol{f}(\boldsymbol{x}(t), \boldsymbol{u}_i(t), \boldsymbol{u}_{\neg i}^*(t), t), \tag{3.14}$$

i.e. the system dynamics with the optimal controls of the other players $j \in \mathcal{P}, j \neq i$. Therefore, we obtain an optimal control problem for player i since $\boldsymbol{u}_{\neg i}^*(t)$ does not depend on $\boldsymbol{u}_i(t)$. Hence, the tools of classical optimal control can be applied. In particular, Pontryagin's minimum principle (see e.g. [Nai03, Chapter 6]) can be used to determine a set of differential equations which represent necessary conditions for Nash equilibria. As in optimal control, the analysis of differential games is based on the Hamiltonian function

$$H_i(\boldsymbol{\psi}_i(t), \boldsymbol{x}(t), \boldsymbol{u}_i(t), \boldsymbol{u}_{\neg i}(t), t) = g_i\left(\boldsymbol{x}(t), \boldsymbol{u}_i(t), \boldsymbol{u}_{\neg i}(t), t\right) + \boldsymbol{\psi}_i^\top(t)\, \boldsymbol{f}\left(\boldsymbol{x}(t), \boldsymbol{u}_i(t), \boldsymbol{u}_{\neg i}(t), t\right) \quad (3.15)$$

for all $t \in [0, T]$ and all players $i \in \mathcal{P}$, where $\boldsymbol{\psi}_i : [0, T] \mapsto \mathbb{R}^n$ are so-called costate functions or Lagrangian multiplier functions. Given the case of an open-loop information structure and corresponding strategies as defined in Definition 3.5, the equilibrium is said to be an open-loop Nash equilibrium. The following theorem gives necessary conditions for such equilibria.

Theorem 3.1 (Necessary Conditions for Open-Loop Nash Equilibria)
For an N-player differential game of fixed duration $[0, T]$, let $\boldsymbol{f}(\boldsymbol{x}, \boldsymbol{u}_1, ..., \boldsymbol{u}_N, t)$, $g_i(\boldsymbol{x}, \boldsymbol{u}_1, ..., \boldsymbol{u}_N, t)$ and $h_i(\boldsymbol{x}(T), T)$ be continuously differentiable with respect to $\boldsymbol{x}$ for all $t \in [0, T]$, $i \in \mathcal{P}$.
Then, if $\boldsymbol{\gamma}^{\mathrm{OL}} = \left(\boldsymbol{\gamma}_1^*(\boldsymbol{x}_0, t), ..., \boldsymbol{\gamma}_N^*(\boldsymbol{x}_0, t)\right)$, where $\boldsymbol{\gamma}_i^* \in \Gamma_i^{\mathrm{OL}}$ and $\boldsymbol{\gamma}_i^*(\boldsymbol{x}_0, t) = \boldsymbol{u}_i^*(t)$, $i \in \mathcal{P}$, provides an open-loop Nash equilibrium (OLNE) solution with $\boldsymbol{x}^*(t)$ as the corresponding state trajectory, the trajectories of the N costate functions $\boldsymbol{\psi}_i(t)$, $i \in \mathcal{P}$, satisfy the relations:

$$\dot{\boldsymbol{x}}^*(t) = \boldsymbol{f}(\boldsymbol{x}^*(t), \boldsymbol{u}_1^*(t), ..., \boldsymbol{u}_N^*(t), t), \quad \boldsymbol{x}^*(0) = \boldsymbol{x}_0 \quad (3.16a)$$

$$\boldsymbol{u}_i^*(t) = \arg\min_{\boldsymbol{u}_i(t)} H_i\left(\boldsymbol{\psi}_i(t), \boldsymbol{x}^*(t), \boldsymbol{u}_i(t), \boldsymbol{u}_{\neg i}^*(t), t\right) \quad (3.16b)$$

$$\dot{\boldsymbol{\psi}}_i(t) = -\nabla_{\boldsymbol{x}} H_i\left(\boldsymbol{\psi}_i(t), \boldsymbol{x}^*(t), \boldsymbol{u}_i^*(t), \boldsymbol{u}_{\neg i}^*(t), t\right) \quad (3.16c)$$

$$\boldsymbol{\psi}_i(T) = \nabla_{\boldsymbol{x}} h_i(\boldsymbol{x}^*(T), t), \quad (3.16d)$$

where $\nabla_{\boldsymbol{x}}$ denotes the partial derivative with respect to the state variable $\boldsymbol{x}$.

Proof:
See the proof of Theorem 6.11 of [BO99]. □

The set of differential equations (3.16) have to be fulfilled for all open-loop Nash equilibria and is valid for the general case where $\boldsymbol{u}_i$ is constrained. In case the optimal controls lie strictly inside the set defining the constraints or if we have unconstrained controls $\boldsymbol{u}_i \in \mathbb{R}^{m_i}$ as considered in Definition 3.2, the control equation (3.16b) leads to

$$\boldsymbol{0} = \nabla_{\boldsymbol{u}_i} H_i(\boldsymbol{\psi}_i(t), \boldsymbol{x}^*(t), \boldsymbol{u}_i(t), \boldsymbol{u}_{\neg i}^*(t), t), \quad (3.17)$$

where $\nabla_{\boldsymbol{u}_i}$ denotes the partial derivative with respect to $\boldsymbol{u}_i$. Therefore, with the application of Theorem 3.1 we obtain a set of coupled differential equations. Under some further assumptions including the cost functions being decoupled with respect to each player's controls, i.e. (3.10) holds, and the system dynamics having the form (3.11), it is possible to formulate a two-point boundary value problem (TPBVP), generally consisting of $(N+1)n$ ODEs and $(N+1)n$ boundary conditions which can potentially be solved using numerical methods, e.g. shooting techniques [AMR95, Chapter 4]. Further details are given in Section B.3 of the Appendix. Note that the minimum principle of Pontryagin and therefore Theorem 3.1 represents only necessary conditions for Nash equilibria. It generates candidates for OLNE solutions but there is no guarantee that they are indeed a Nash equilibrium. However, under further assumptions, the minimum principle becomes a sufficient condition for optimality. Therefore, following [Doc00, Theorem 3.2], it can be stated that if $H_i\left(\boldsymbol{\psi}_i(t), \boldsymbol{x}(t), \boldsymbol{u}_i(t), \boldsymbol{u}_{\neg i}(t), t\right)$ is convex in $\boldsymbol{x}$ and also continuously differentiable in $\boldsymbol{x}$, and furthermore h_i is convex, then the controls $\boldsymbol{u}_i^*(t)$ are optimal with respect to each corresponding optimization problem and hence describe an OLNE.

In the following, an example is given to illustrate the procedure of calculating an OLNE by means of Theorem 3.1.

Example 3.1:

We consider a scenario consisting of two players controlling a system given by

$$\dot{x}(t) = -x(t) + u_1(t) + u_2(t). \tag{3.18}$$

Each player acts based on the cost function

$$J_i = \int_0^\infty \frac{1}{2}x^2(t) + \frac{1}{2}u_i^2(t)\,\mathrm{d}t, \quad i \in \{1, 2\}. \tag{3.19}$$

In the following, i and j are used to denote any player from the set $\mathcal{P} = \{1, 2\}$ such that $i \neq j$. Furthermore, time dependencies are omitted for brevity.

To determine the OLNE, we first determine the Hamiltonian of each player:

$$H_i = \frac{1}{2}x^2 + \frac{1}{2}u_i^2 + \psi_i\left(-x + u_i + u_j\right), \quad i, j \in \{1, 2\},\ i \neq j. \tag{3.20}$$

We now can utilize the necessary conditions for open-loop Nash equilibria given by Theorem 3.1. The control equation (3.16b) *leads to*

$$\frac{\partial H_i}{\partial u_i} = u_i + \psi_i = 0 \Leftrightarrow u_i = -\psi_i. \tag{3.21}$$

From (3.16c) *we obtain the differential equation*

$$\dot{\psi}_i = -\frac{\partial H_i}{\partial x} = -x + \psi_i. \tag{3.22}$$

Furthermore, the system dynamics equation (3.16a) *given by*

$$\dot{x} = -x + u_1 + u_2 \tag{3.23}$$

must hold as well.

By combining (3.21), (3.22) *and* (3.23) *we obtain the linear system of differential equations*

$$\begin{bmatrix} \dot{x} \\ \dot{\psi}_1 \\ \dot{\psi}_2 \end{bmatrix} = \begin{bmatrix} -1 & -1 & -1 \\ -1 & 1 & 0 \\ -1 & 0 & 1 \end{bmatrix} \begin{bmatrix} x \\ \psi_1 \\ \psi_2 \end{bmatrix}. \tag{3.24}$$

Given that optimal control and differential game problems usually specifiy initial conditions for the state vector and terminal conditions for the costates ψ_i, this system of differential equations represents a TPBVP. In this case, it can be solved both analytically and numerically. The general analytical solution can be determined e.g. by the eigenvalue and eigenvector method (see e.g. [HS14, Section 5.3]) and results in

$$x^*(t) = C_1(\sqrt{3} - 1)\exp(-\sqrt{3}t) + C_2(1 - \sqrt{3})\exp(\sqrt{3}t), \tag{3.25}$$

$$\psi_1^*(t) = C_1 \exp(-\sqrt{3}t) + C_2 \exp(\sqrt{3}t) - C_3 \exp(t), \tag{3.26}$$

$$\psi_2^*(t) = C_1 \exp(-\sqrt{3}t) + C_2 \exp(\sqrt{3}t) + C_3 \exp(t), \tag{3.27}$$

where the constants C_l, $l \in \{1, \ldots, 3\}$ are determined by using the aforementioned boundary conditions for states and costates. The OLNE solution results directly from the costate functions (3.26) *and* (3.27). *Finally, we recognize that in this example the conditions of Theorem 3.1 are both necessary and sufficient.*

3.6.2 Feedback Nash Equilibrium

Consider a differential game where the players apply a feedback strategy as in Definition 3.6. By applying the minimum principle, similar equations to the ones of Theorem 3.1 result. Nevertheless, instead of (3.16c), the equation

$$\dot{\boldsymbol{\psi}}_i(t) = -\nabla_{\boldsymbol{x}} H_i(\boldsymbol{\psi}_i(t), \boldsymbol{x}^*(t), \boldsymbol{u}_i^*(t), \boldsymbol{\gamma}_{\neg i}^*(\boldsymbol{x}^*, t), t) \tag{3.28}$$

holds. The time dependency of the state in the strategies $\boldsymbol{\gamma}_{\neg i}^*$ is dropped here and in the following for brevity. In this new costate equation, the controls $\boldsymbol{u}_{\neg i}^*(t) = \boldsymbol{\gamma}_{\neg i}^*(\boldsymbol{x}^*, t)$ have an influence on the partial derivative in (3.16c) since, contrary to the open-loop case, they now depend on the current value of $\boldsymbol{x}(t)$. Even though these new equations define a closed-loop no-memory Nash equilibrium, they are not computationally convenient [SH69a]. Furthermore, there is in general an uncountable number of solutions to the resulting differential equations, one of which is the open-loop solution determined in (3.16) [BO99, p. 277].

In order to eliminate this so-called "informational non-uniqueness", the concept of feedback Nash equilibria is introduced. This refinement states that if an N-tuple of strategies $\boldsymbol{\gamma}^* = (\boldsymbol{\gamma}_1^*, ..., \boldsymbol{\gamma}_N^*)$ constitutes a FNE solution of a differential game with duration $[0, T]$, then its restriction to the time interval $[t, T]$, for any $t \in [0, T]$, describes a FNE solution for the same differential game defined on this shorter time interval $[t, T]$. A consequence of this requirement is the strong time consistency of FNE solutions (cf. Section 3.5.1). Furthermore, any FNE also fufills the equations of Theorem 3.1 with the costate equation (3.28).

The core of the results concerning feedback Nash equilibria is given by N coupled Hamilton-Jacobi-Bellman (HJB) equations for which the value function, known from optimal control, is extended to the N-player case.

Definition 3.10 (Value Function)

Consider a player $i \in \mathcal{P}$. Let the optimal strategies of the other players $\boldsymbol{\gamma}_{\neg i}^*$ associated to an N-player non-cooperative differential game be given. The value function $V_i : \mathbb{R}^n \times [0, T] \mapsto \mathbb{R}$ of player i is defined by

$$V_i(\boldsymbol{x}, t) = \min_{\{\boldsymbol{\gamma}_i(\boldsymbol{x},s),\, t \leq s \leq T\}} \int_t^T g_i\left(\bar{\boldsymbol{x}}_i(s), \boldsymbol{\gamma}_i(\boldsymbol{x}, s), \boldsymbol{\gamma}_{\neg i}^*(\boldsymbol{x}, s), s\right) \mathrm{d}s + h_i(\boldsymbol{x}(T), T) \tag{3.29}$$

$$V_i(\boldsymbol{x}, t) = \int_t^T g_i(\boldsymbol{x}^*(s), \boldsymbol{\gamma}_i^*(\boldsymbol{x}, s), \boldsymbol{\gamma}_{\neg i}^*(\boldsymbol{x}, s), s)\, \mathrm{d}s \tag{3.30}$$

satisfying the boundary condition

$$V_i(\boldsymbol{x}, T) = h_i(\boldsymbol{x}, T), \tag{3.31}$$

and where

$$\dot{\bar{\boldsymbol{x}}}_i(s) = \boldsymbol{f}(\bar{\boldsymbol{x}}_i(s), \boldsymbol{\gamma}_i(\boldsymbol{x}, s), \boldsymbol{\gamma}_{\neg i}^*(\boldsymbol{x}, s)); \; \bar{\boldsymbol{x}}_i(t) = \boldsymbol{x}. \tag{3.32}$$

The value function V_i, $i \in \mathcal{P}$ represents the minimum cost-to-go from any initial state $\boldsymbol{x}$ and any initial time t which is attainable by player i, where the optimal strategies of the other $N - 1$ players are fixed. With this definition, the following theorem can be stated.

Theorem 3.2 (Sufficient Conditions for Feedback Nash Equilibria)
For an N-player differential game of prescribed fixed duration $[0,T]$, *an N-tuple of feedback strategies* $\boldsymbol{\gamma}^{\text{FB}} = (\gamma_1^*, ..., \gamma_N^*)$ *where* $\boldsymbol{\gamma}_i^* \in \Gamma_i^{\text{FB}}$ *and* $\boldsymbol{\gamma}_i^*(\boldsymbol{x}, t) = \boldsymbol{u}_i^*(t)$, $i \in \mathcal{P}$, *provides a feedback Nash equilibrium (FNE) solution if there exist continuous differentiable value functions* V_i *according to Defintion 3.10 which satisfy the partial differential equations*

$$\begin{aligned} -\frac{\partial V_i(\boldsymbol{x},t)}{\partial t} &= \min_{\boldsymbol{u}_i} \left[\nabla_{\boldsymbol{x}} V_i(\boldsymbol{x},t)\tilde{\boldsymbol{f}}_i^*(\boldsymbol{x}(t),\boldsymbol{u}_i(t),t) + \tilde{g}_i^*(\boldsymbol{x}(t),\boldsymbol{u}_i(t),t)\right] \\ &= \nabla_{\boldsymbol{x}} V_i(\boldsymbol{x},t)\tilde{\boldsymbol{f}}_i^*(\boldsymbol{x}(t),\boldsymbol{\gamma}_i^*(\boldsymbol{x},t),t) + \tilde{g}_i^*(\boldsymbol{x}(t),\boldsymbol{\gamma}_i^*(\boldsymbol{x},t),t), \\ V_i(\boldsymbol{x},T) &= h_i(\boldsymbol{x},T), \quad i \in \mathcal{P}, \end{aligned} \tag{3.33}$$

where

$$\begin{aligned} \tilde{\boldsymbol{f}}_i^*(\boldsymbol{x}(t),\boldsymbol{u}_i(t),t) &= \boldsymbol{f}(\boldsymbol{x}(t),\boldsymbol{\gamma}_{\neg i}^*(\boldsymbol{x},t),\boldsymbol{u}_i(t),t), \\ \tilde{g}_i^*(\boldsymbol{x}(t),\boldsymbol{u}_i(t),t) &= g_i(\boldsymbol{x}(t),\boldsymbol{\gamma}_{\neg i}^*(\boldsymbol{x},t),\boldsymbol{u}_i(t),t). \end{aligned} \tag{3.34}$$

The corresponding Nash equilibrium cost for player i is $V_i(\boldsymbol{x}_0, 0)$.

Proof:
See the proof of Theorem 6.16 of [BO99]. □

The following example illustrates the use of Theorem 3.2 to determine a FNE solution of a differential game.

Example 3.2:
Consider the differential game with 2 players from Example 3.1, where they control a system with dynamics (3.18) *and each of them chooses his actions such that his individual cost function* (3.19) *is minimized. However, contrary to last example, each of the players applies a feedback strategy according to Definition 3.6. Again, function dependencies are neglected for brevity, unless a variable dependence demands special attention.*

Given time-independent functions $g_i(\boldsymbol{x}, \boldsymbol{u}_i, \boldsymbol{u}_{\neg i})$ *and system dynamics as well as the infinite horizon (*$T \longrightarrow \infty$*), the value function also does not depend explicitely on time (cf. [HKZ12, Remark 7.5]), and therefore the HJB equation of each player results in*

$$0 = \min_{u_i} \left(\frac{1}{2}x^2 + \frac{1}{2}u_i^2 + \frac{\partial V_i}{\partial x}\left[-x + u_i + u_j\right]\right), \quad i \in \{1,2\},\ i \neq j. \tag{3.35}$$

Minimizing the expression at the right hand side leads to

$$u_i^* + \frac{\partial V_i}{\partial x} = 0 \Leftrightarrow u_i^* = -\frac{\partial V_i}{\partial x} \mathrel{\hat{=}} \gamma_i^*(x). \tag{3.36}$$

At this point, it is usually necessary to guess the structure of the value function V_i. Given the linear system dynamics and the quadratic cost function, we hypothesize a quadratic value function. Moreover, given the symmetric structure[18] of the game, we are interested in symmetrical equilibrium actions $u_i^* = u_j^*$ leading to identical value functions.

For any player $i \in \{1, 2\}$, we write the value function as

$$V_i(x) = \frac{A}{2}x^2 + Bx + C \Leftrightarrow \frac{\partial V_i}{\partial x} = Ax + B \tag{3.37}$$

with $A, B, C \in \mathbb{R}$. By using (3.36) and (3.37), the HJB equation (3.35) leads after some simplification to

$$0 = \left(-\frac{3}{2}A^2 - A + \frac{1}{2}\right)x^2 - (3AB + B)\,x - \frac{3}{2}B^2. \tag{3.38}$$

By comparing both equation sides we obtain $B = 0$ and two possible values $A_1 = -1$, $A_2 = 1/3$. Given the positive integrand in (3.19), the value function must be positive and therefore, A_1 is discarded. With (3.36) and (3.37) we obtain the optimal feedback strategy

$$\gamma_i^*(x) = -\frac{1}{3}x(t) \tag{3.39}$$

and the corresponding state trajectory

$$x^*(t) = C\exp\left(-\frac{5}{3}t\right), \tag{3.40}$$

where $C \in \mathbb{R}$ is determined by using an initial state condition $x(0) = x_0 \in \mathbb{R}$.

3.6.3 Pareto Efficient Solutions

In general, a dynamic game has various Pareto efficient solutions. The set of all of these solutions is called Pareto frontier. In the following, a theorem presenting necessary and sufficient conditions for Pareto efficient solutions is given.

[18] Here, the notion of symmetry of [Doc00, p. 106] is considered, meaning that all players (usually two) have the same cost function J_i and control space $\mathcal{U}_i$. Furthermore, the system dynamics are symmetric with respect to the players in the sense that the equation is unaffected if e.g. u_1 is interchanged with u_2.

Theorem 3.3 (Necessary and Sufficient Conditions for Pareto Efficient Solutions)
Let $\tau_i > 0$, for all $i \in \mathcal{P}$, satisfy

$$\sum_{i=1}^{N} \tau_i = 1. \tag{3.41}$$

Now consider an N-player differential game. If $\boldsymbol{\gamma}^p = \left(\gamma_1^p, ..., \gamma_N^p\right)$ is such that

$$\begin{aligned} \boldsymbol{\gamma}^p &= \arg\min_{\gamma} \sum_{i=1}^{N} \tau_i J_i(\boldsymbol{\gamma}) \\ &\text{w.r.t} \\ \dot{\boldsymbol{x}} &= \boldsymbol{f}(\boldsymbol{x}(t), \boldsymbol{u}_1(t), ...\boldsymbol{u}_N(t), t) \\ \boldsymbol{x}(0) &= \boldsymbol{x}_0 \end{aligned} \tag{3.42}$$

then $\boldsymbol{\gamma}^p$ is a Pareto efficient solution (PES). Moreover, if the strategy spaces Γ_i are convex and J_i are convex for all $i \in \mathcal{P}$, then for all Pareto-efficient solutions $\boldsymbol{\gamma}^p$ there exist τ_i such that $\boldsymbol{\gamma}^P$ solves the optimization problem (3.42).

Proof:
The theorem can be found in [Eng05, Theorem 6.4]. The sufficiency result is proved in [Eng05, Lemma 6.1] while the necessary part is proved in [Eng05, Lemma 6.3]. □

The formulation of Theorem 3.3 as a dynamic optimization problem allows the use of the minimum principle to solve for the PES. The solution can sometimes be given with τ_i as a degree of freedom. Weighting parameters which fulfill (3.41) can also be chosen to find a particular PES, e.g. with $\tau_i = 1/N$.

In the following, an example is given to illustrate the calculation of a PES.

Example 3.3:

Consider the differential game with two players from Example 3.1. In this example, we assume the players are able to build cooperative strategies such that their overall performance is increased.

We choose $\tau_1 = \tau$ and $\tau_2 = 1 - \tau$ and state the cost function

$$J_p = \tau J_1 + (1-\tau) J_2 \tag{3.43}$$

$$= \int_0^T \frac{1}{2}x^2 + \frac{\tau}{2}(u_1^2 - u_2^2) + \frac{1}{2}u_2^2 \, \mathrm{d}t. \tag{3.44}$$

We now can utilize the minimum principle to determine the solution. The Hamiltonian which corresponds to J_p is given by

$$H_p = \frac{1}{2}x^2 + \frac{\tau}{2}(u_1^2 - u_2^2) + \frac{u_2^2}{2} + \psi_p\,(-x + u_1 + u_2)\,. \tag{3.45}$$

Since there is a coordination between both players, we consider the vector $\boldsymbol{u}_p = \begin{bmatrix} u_1 & u_2 \end{bmatrix}^\top$ as the overall control vector. The control equation

$$\frac{\partial H_p}{\partial \boldsymbol{u}_p} = \begin{bmatrix} \tau u_1 + \psi_p \\ u_2(1-\tau) + \psi_p \end{bmatrix} = \boldsymbol{0} \tag{3.46}$$

of the minimum principle leads to

$$u_1 = -\frac{1}{\tau}\psi_p \quad \text{and} \quad u_2 = -\frac{1}{1-\tau}\psi_p\,. \tag{3.47}$$

Furthermore, the canonical differential equation of the costates

$$\dot{\psi}_p = -\frac{\partial H_p}{\partial x} = -x + \psi_p \tag{3.48}$$

and the system dynamics equation

$$\dot{x} = -x + u_1 + u_2 \tag{3.49}$$

must hold for the optimal solution.

Similar to Example 3.1, by inserting (3.47) *into* (3.49) *and using* (3.48), *we obtain a system of differential equations*

$$\dot{\begin{bmatrix} x \\ \psi_p \end{bmatrix}} = \begin{bmatrix} -1 & -\frac{1}{\tau(1-\tau)} \\ -1 & 1 \end{bmatrix} \begin{bmatrix} x \\ \psi_p \end{bmatrix} \tag{3.50}$$

which can be solved analytically using the eigenvalue and eigenvector method. The general solution is

$$x^*(t) = C_1(\lambda + 1)\exp(-\lambda t) - C_2(\lambda - 1)\exp(\lambda t) \tag{3.51}$$

$$\psi_p^*(t) = C_1 \exp(-\lambda t) + C_2 \exp(\lambda t), \tag{3.52}$$

with

$$\lambda = \sqrt{1 + \frac{1}{\tau(1-\tau)}} \tag{3.53}$$

and where $C_l \in \mathbb{R}$, $l \in \{1, 2\}$ are determined using initial and terminal conditions.

3.6.4 Comparison of Solution Concepts

In general, the OLNE and FNE are not equal since they are based on different assumptions concerning the available information to the players. Furthermore, while there are some cases where Nash equilibria and Pareto efficient solutions coincide, this is also generally not the case. In order to illustrate the difference between the solutions, the following example is presented.

Example 3.4:

Consider the same two-player differential game as in Examples 3.1, 3.2 and 3.3. In the three examples, the OLNE, FNE and the PES were calculated, respectively. In this example, the exact trajectories which follow from $\tau = 0.5$ *and the boundary conditions*

$$x(0) = 2, \quad \psi_1(T \to \infty) = 0, \quad \psi_2(T \to \infty) = 0 \quad \text{and} \quad \psi_p(T \to \infty) = 0 \tag{3.54}$$

were determined analytically using MATLAB's `dsolve`. *Figure 3.3 shows state and control trajectories of the differential game defined by* (3.18) *and* (3.19). *Only one control trajectory is shown for each solution concept since the symmetry of the game leads to equal controls for both players. While the OLNE and FNE are similar to each other, the PES differs considerably more.*

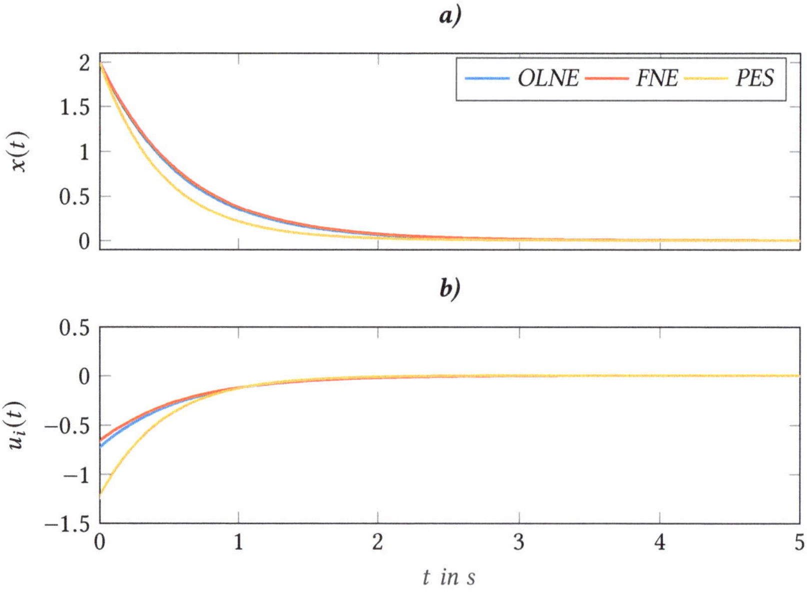

Figure 3.3: Open-loop Nash equilibrium, feedback Nash equilibrium and Pareto efficient solution of an example two-player differential game: **a)** State trajectories, **b)** Control trajectories.

Finally, in order to show that a cooperative differential game with a PES leads to a better outcome than a non-cooperative setting, we calculate the value of the objective function for each solution concept:

$$J^*_{i,\mathrm{OLNE}} = 0.655 \tag{3.55}$$

$$J^*_{i,\mathrm{FBNE}} = 0.667 \tag{3.56}$$

$$J_{i,\mathrm{PES}} = 0.618, \quad i \in \{1,2\}. \tag{3.57}$$

Hence, $J^*_{i,\mathrm{FBNE}} \geq J^*_{i,\mathrm{OLNE}} \geq J^*_{i,\mathrm{PES}}$ *holds. The lower costs of the PES demonstrates an advantage of acting cooperatively in this example.*

3.7 Tractable Differential Games

The solution of the coupled differential equations which arise from the necessary and sufficient conditions for Nash equilibria is in general not a trivial task, especially concerning the partial differential equations (HJB equations) which are needed to find an FNE. Indeed, finding Nash equilibria for general differential games is nontrivial and an object of current research. To find an FNE in nonlinear differential games, approximative or iterative solutions of the HJB equations are sought and therefore, the use of *reinforcement learning* or *adaptive dynamic programming* techniques are obtaining increased interest [KKD14, ZZWZ16, KVML18].

There are particular kinds of differential games which are similar to the examples presented in the previous subsections in the sense that the calculation of Nash equilibria is considerably simplified. These are therefore called *tractable differential games* [HKZ12, Section 7.6] and include

- linear-quadratic differential games
- linear-state differential games
- exponential differential games.

These kinds of differential games are treated e.g. in [DFJ85] and [Doc00, Chapter 7].

One of the structures considered in this thesis are linear-quadratic differential games, as it is an important and widespread class of differential games which has been used in several applications of automatic control including driver assistance systems [FFH17], collision avoidance [MSA17], control of mobile robots [Gu08] and control of energy grids [ZMSFZ16]. Therefore, the following section presents the most important results which are known for this particular class of games.

3.8 Linear-Quadratic Differential Games

A linear-quadratic (LQ) differential game is a class of differential games where the system the players control simultaneously has linear dynamics, i.e. the evolution of the states is governed by a system of linear differential equations. Furthermore, the players act based upon an individual quadratic cost function. This kind of games can therefore be seen as an extension of linear-quadratic optimal control to the N-player case. LQ differential games are considered a class of differential games which can be solved with reasonable effort. Their particular structure allows the derivation of necessary and sufficient conditions for Nash equilibria which are computationally tractable.

Definition 3.11 (Linear-Quadratic Differential Game)
A linear-quadratic (LQ) differential game is defined by the same elements as Definition 3.3. The system dynamics are linear, i.e. are defined by

$$\dot{\boldsymbol{x}}(t) = \boldsymbol{A}\boldsymbol{x}(t) + \sum_{i=1}^{N} \boldsymbol{B}_i \boldsymbol{u}_i(t), \tag{3.58}$$

where $\boldsymbol{x}(t) \in \mathbb{R}^n$, $\boldsymbol{u}(t) \in \mathbb{R}^{m_i}$ and $\boldsymbol{A}$ and $\boldsymbol{B}_i$, $i \in \mathcal{P}$, are the system and control matrices of appropriate dimensions, respectively, which form stabilizable matrix pairs $(\boldsymbol{A}, \boldsymbol{B}_i)$, $i \in \mathcal{P}$. Furthermore, the cost functions are quadratic, i.e.

$$J_i = \frac{1}{2}\boldsymbol{x}^\top(T)\boldsymbol{Q}_{i,T}\boldsymbol{x}(T) + \frac{1}{2}\int_0^T \boldsymbol{x}^\top(t)\boldsymbol{Q}_i\boldsymbol{x}(t) + \sum_{j=1}^{N} \boldsymbol{u}_j^\top(t)\boldsymbol{R}_{ij}\boldsymbol{u}_j(t)\,\mathrm{d}t, \tag{3.59}$$

where $\boldsymbol{Q}_{i,T}, \boldsymbol{Q}_i, \boldsymbol{R}_{ij}$ are symmetric matrices for all $i, j \in \mathcal{P}$ and $\boldsymbol{R}_{ii} \succ \boldsymbol{0}$.

The constraint of positive definiteness $\boldsymbol{R}_{ii} \succ \boldsymbol{0}$ is required in order to guarantee a meaningful minimization problem. Additional positive-semidefiniteness constraints are sometimes introduced, e.g. $\boldsymbol{Q}_{i,T}$, $\boldsymbol{Q}_i \succeq \boldsymbol{0}$. These are often convenient to obtain Nash equilibrium solutions but are not always strictly necessary, as will be discussed in the next subsection.[19] Furthermore, the stabilizable pairs $(\boldsymbol{A}, \boldsymbol{B}_i)$, $i \in \mathcal{P}$, imply that each player is able to stabilize the system on its own, a fact that is required for the following results on Nash equilibria in LQ differential games.

[19] A widespread case is given by a two-player differential game $N = 2$ where the players play in a stringent adversarial way. This is represented by cost function matrices $\boldsymbol{Q}_2 = -\boldsymbol{Q}_1$, $\boldsymbol{Q}_{2,T} = -\boldsymbol{Q}_{1,T}$, $\boldsymbol{R}_{12} = -\boldsymbol{R}_{22}$, $\boldsymbol{R}_{21} = -\boldsymbol{R}_{11}$ and is known as *zero-sum differential game* [SH69b].

3.8.1 Nash Equilibria in Open-Loop LQ Differential Games

Finite-Horizon

Consider a linear-quadratic differential game with finite horizon T. The calculation of open-loop Nash equilibria is based on the solution of coupled matrix Riccati differential equations (RDEs), which can be derived from Pontryagin's minimum principle. Therefore, applying Theorem 3.1 to LQ differential games leads to the following result.

Theorem 3.4 (Sufficient Conditions for OLNE solutions in Finite-Horizon LQ Differential Games)
Consider an N-player LQ differential game as in Definition 3.11 with the additional constraints $\boldsymbol{Q}_i, \boldsymbol{Q}_{i,T} \succeq \boldsymbol{0}$, $i \in \mathcal{P}$. Let there exist a set of matrix-valued functions $\boldsymbol{P}_i$, $i \in \mathcal{P}$, which satisfy the Riccati differential equations (RDEs)

$$\dot{\boldsymbol{P}}_i(t) = -\boldsymbol{P}_i(t)\boldsymbol{A} - \boldsymbol{A}^\top \boldsymbol{P}_i(t) + \sum_{j=1}^{N} \boldsymbol{P}_i(t)\boldsymbol{B}_j\boldsymbol{R}_{jj}^{-1}\boldsymbol{B}_j^\top \boldsymbol{P}_j(t) - \boldsymbol{Q}_i, \quad i \in \mathcal{P}, \tag{3.60}$$

with the transversality conditions

$$\boldsymbol{P}_i(T) = \boldsymbol{Q}_{i,T}, \quad i \in \mathcal{P}. \tag{3.61}$$

Then, the LQ differential game has a unique OLNE for every initial state $\boldsymbol{x}_0$. Moreover, the resulting N-tuple of equilibrium controls $\boldsymbol{u}^*$ is defined by the controls

$$\boldsymbol{u}_i^*(t) = \boldsymbol{\gamma}_i^*(\boldsymbol{x}_0, t) = -\boldsymbol{R}_{ii}^{-1}\boldsymbol{B}_i^\top \boldsymbol{P}_i(t)\boldsymbol{\Phi}(t,0)\boldsymbol{x}_0, \quad i \in \mathcal{P}. \tag{3.62}$$

Here, $\boldsymbol{\Phi}(t,0)$ satisfies the differential equation

$$\dot{\boldsymbol{\Phi}}(t,0) = \left(\boldsymbol{A} - \sum_{j=1}^{N} \boldsymbol{S}_j\boldsymbol{P}_j(t)\right)\boldsymbol{\Phi}(t,0), \quad \boldsymbol{\Phi}(t,t) = \boldsymbol{I}, \tag{3.63}$$

where

$$\boldsymbol{S}_j = \boldsymbol{B}_j\boldsymbol{R}_{jj}^{-1}\boldsymbol{B}_j^\top, \quad j \in \mathcal{P}. \tag{3.64}$$

Proof:
See Section B.1 of the Appendix. □

Theorem 3.4 gives an approach for calculating Nash equilibria by solving the RDEs (3.60) with the conditions (3.61). Nevertheless, cases exist where these do not have a solution, but the LQ differential game still has a solution [BO99, p. 314].

In case the system is not affected by any disturbance during the complete game duration, the controls can be formulated in the form of an optimal feedback law

$$\boldsymbol{\gamma}_i^*(\boldsymbol{x}, t) = -\boldsymbol{R}_{ii}^{-1}\boldsymbol{B}_i^\top \boldsymbol{P}_i(t)\boldsymbol{x}(t), \quad i \in \mathcal{P}. \tag{3.65}$$

Infinite-Horizon

In an infinite-horizon case, i.e. $T \to \infty$, the matrices $\boldsymbol{P}_i$ are constant ($\dot{\boldsymbol{P}}_i = \boldsymbol{0}$), resulting in coupled algebraic Riccati equations (ARE) and leading to the following result.

Theorem 3.5 (Sufficient Conditions for OLNE solutions in Infinite-Horizon LQ Differential Games)
Consider an N-player LQ differential game as in Definition 3.11 with $T \to \infty$ and with the additional constraints $\boldsymbol{Q}_i \succ \boldsymbol{0}$ and $\boldsymbol{Q}_{i,T} = \boldsymbol{0}$, $i \in \mathcal{P}$. Then, the LQ differential game has an OLNE for every initial state $\boldsymbol{x}_0$ if a set of matrices $\boldsymbol{P}_i$, $i \in \mathcal{P}$, exists which satisfies the algebraic Riccati equations (AREs)

$$\boldsymbol{0} = -\boldsymbol{P}_i\boldsymbol{A} - \boldsymbol{A}^\top\boldsymbol{P}_i + \sum_{j=1}^{N} \boldsymbol{P}_i\boldsymbol{B}_j\boldsymbol{R}_{jj}^{-1}\boldsymbol{B}_j^\top\boldsymbol{P}_j - \boldsymbol{Q}_i \quad , i \in \mathcal{P} \tag{3.66}$$

and additionally leads to a stable closed-loop system[20]

$$\boldsymbol{F} := \boldsymbol{A} - \sum_{j=1}^{N} \boldsymbol{S}_j\boldsymbol{P}_j, \tag{3.67}$$

i.e. the eigenvalues of $\boldsymbol{F}$ have a negative real part. The resulting N-tuple of Nash equilibrium controls $\boldsymbol{u}^*$ is defined by (3.62), *where $\boldsymbol{P}_i(t) = \boldsymbol{P}_i$, $i \in \mathcal{P}$.*

Proof:
See the proof of [BO99, Theorem 6.22]. □

According to [BO99, p. 336], the existence of OLNEs in an infinite-horizon LQ differential game does not imply the existence of an OLNE in the finite-horizon version of the game. Moreover, a unique solution of the RDEs in a finite-horizon differential game may converge

[20] Note that the stabilizability of $(\boldsymbol{A}, [\boldsymbol{B}_1, \ldots, \boldsymbol{B}_N])$ is necessary, a property which follows from the stabilizable pairs $(\boldsymbol{A}, \boldsymbol{B}_i)$, $i \in \mathcal{P}$, according to Definition 3.11.

for $T \longrightarrow \infty$ to a solution of the coupled AREs, but these are not necessarily stabilizing solutions and therefore would not constitute an OLNE of the infinite-horizon differential game.

3.8.2 Nash Equilibrium in Feedback LQ Differential Games

Finite Horizon

Consider a LQ differential game with finite horizon T. Similar to the open-loop case, the calculation of feedback Nash equilibria is based on the solution of coupled RDEs, which can be derived from Theorem 3.2. We shall now restrict our attention to linear feedback strategies belonging to the set

$$\Gamma_i^{\mathrm{FB}} = \{\boldsymbol{\gamma}_i \mid \boldsymbol{\gamma}_i(\boldsymbol{x}, t) = -\boldsymbol{K}_i(t)\boldsymbol{x}(t)\}. \tag{3.68}$$

This allows the formulation of the following theorem.

Theorem 3.6 (Necessary and Sufficient Conditions for FNE solutions in Finite-Horizon LQ Differential Games)
Consider an N-player LQ differential game as in Definition 3.11. The LQ differential game has a linear FNE for every initial state $\boldsymbol{x}_0$ if and only if a set of symmetric matrix-valued functions $\boldsymbol{P}_i$, $i \in \mathcal{P}$, exists which satisfy the Riccati differential equations (RDEs)

$$\begin{aligned}\dot{\boldsymbol{P}}_i(t) = &- \boldsymbol{Q}_i - \boldsymbol{P}_i(t)\boldsymbol{A} - \boldsymbol{A}^\top \boldsymbol{P}_i(t) + \sum_{j=1}^{N} \boldsymbol{P}_i(t)\boldsymbol{S}_j\boldsymbol{P}_j(t) + \ldots \\ &\ldots + \sum_{\substack{j=1 \\ j\neq i}}^{N} \boldsymbol{P}_j(t)\boldsymbol{S}_j\boldsymbol{P}_i(t) - \sum_{\substack{j=1 \\ j\neq i}}^{N} \boldsymbol{P}_j(t)\boldsymbol{S}_{ij}\boldsymbol{P}_j(t),\end{aligned} \tag{3.69}$$

where

$$\begin{aligned}\boldsymbol{S}_j &= \boldsymbol{B}_j\boldsymbol{R}_{jj}^{-1}\boldsymbol{B}_j^\top, && j \in \mathcal{P}, \\ \boldsymbol{S}_{ij} &= \boldsymbol{B}_j\boldsymbol{R}_{jj}^{-1}\boldsymbol{R}_{ij}\boldsymbol{R}_{jj}^{-1}\boldsymbol{B}_j^\top, && i, j \in \mathcal{P},\ i \neq j,\end{aligned} \tag{3.70}$$

and the transversality conditions

$$\boldsymbol{P}_i(T) = \boldsymbol{Q}_{i,T}, \quad i \in \mathcal{P}. \tag{3.71}$$

The resulting N-tuple of linear Nash equilibrium strategies $\boldsymbol{\gamma}^*$ is unique and defined by

$$\boldsymbol{\gamma}_i^*(\boldsymbol{x}, t) = -\boldsymbol{R}_{ii}^{-1}\boldsymbol{B}_i^\top \boldsymbol{P}_i(t)\boldsymbol{x}(t) =: -\boldsymbol{K}_i(t)\boldsymbol{x}(t), \quad i \in \mathcal{P}. \tag{3.72}$$

Proof:
See the proof of [Eng05, Theorem 8.3]. □

Generally speaking, the FNE arising from the solution of the coupled RDEs is not necessarily the only one. Basar reported in [Bas74] the existence of equilibrium strategies which are nonlinear functions of the state in discrete-time linear-quadratic dynamic games. Similarly, in [TM90] the authors present a specific LQ differential game example for which a nonlinear FNE exists. Therefore, Theorem 3.6 may not apply if the strategy space is enlarged as to include nonlinear strategies [Eng05, p. 365].

Infinite Horizon

As in the finite-horizon case, we restrict our attention to linear feedback strategies. Nevertheless, for infinite-horizon games, these are constant over time, i.e. they are defined by the set

$$\Gamma_i^{\mathrm{FB}} = \{\boldsymbol{\gamma}_i \mid \boldsymbol{\gamma}_i(\boldsymbol{x}, t) = -\boldsymbol{K}_i \boldsymbol{x}(t)\}. \tag{3.73}$$

Furthermore, these strategies (or alternatively, control laws) $\boldsymbol{K} = (\boldsymbol{K}_1, ..., \boldsymbol{K}_N)$ are assumed to belong to the set

$$\mathcal{F} = \left\{(\boldsymbol{K}_1, ..., \boldsymbol{K}_N) \mid \boldsymbol{F} \text{ is stable}\right\}, \tag{3.74}$$

which can be interpreted as a strive of the players for jointly stabilizing the system.[21] A necessary and sufficient condition for the non-emptiness of $\mathcal{F}$ is the stabilizability of the matrix pair $(\boldsymbol{A}, \begin{bmatrix}\boldsymbol{B}_1 & \cdots & \boldsymbol{B}_N\end{bmatrix})$ [EBS00]. With these conditions in mind, the following result is stated.

Theorem 3.7 (Necessary and Sufficient Conditions for FNE solutions in Infinite-Horizon LQ Differential Games)
Consider an N-player LQ differential game as in Definition 3.11 with $T \to \infty$. Let the matrices $\boldsymbol{P}_i$, $i \in \mathcal{P}$, be symmetric solutions to the ARE

$$\boldsymbol{0} = -\boldsymbol{Q}_i - \boldsymbol{P}_i\boldsymbol{A} - \boldsymbol{A}^\top\boldsymbol{P}_i + \sum_{j=1}^{N}\boldsymbol{P}_i\boldsymbol{S}_j\boldsymbol{P}_j + \sum_{\substack{j=1\\ j\neq i}}^{N}\boldsymbol{P}_j\boldsymbol{S}_j\boldsymbol{P}_i - \sum_{\substack{j=1\\ j\neq i}}^{N}\boldsymbol{P}_j\boldsymbol{S}_{ij}\boldsymbol{P}_j, \tag{3.75}$$

[21] According to [Eng05, p. 372], this corresponds to the supposition that both players have a first priority in stabilizing the system. Furthermore, for most games the equilibria without this stabilization constraint coincide with the ones corresponding to a game for which this constraint is included. Therefore, the stabilization constraint will not be active in most cases.

and additionally lead to a stable closed-loop system

$$F = A - \sum_{j=1}^{N} S_j P_j,$$

where

$$\begin{aligned} S_j &= B_j R_{jj}^{-1} B_j^\top, && j \in \mathcal{P}, \\ S_{ij} &= B_j R_{jj}^{-1} R_{ij} R_{jj}^{-1} B_j^\top, && i,j \in \mathcal{P},\ i \neq j. \end{aligned} \tag{3.76}$$

Then, there exists a linear FNE and the corresponding feedback strategies are defined by

$$u_i^*(t) = \gamma_i^*(x,t) = -R_{ii}^{-1} B_i^\top P_i x(t) = -K_i x(t). \tag{3.77}$$

Conversely, if a linear FNE exists and is defined by (3.77)*, then there exists a set of stabilizing matrices* P_i*,* $i \in \mathcal{P}$*, which solve the AREs* (3.75).

Proof:
In light of $(A, \begin{bmatrix} B_1 & \cdots & B_N \end{bmatrix})$ being stabilizable from the fact that the single pairs (A, B_i), $i \in \mathcal{P}$, are stabilizable according to Definition 3.11, the rest of the proof is stated in [Eng05, Theorem 8.5]. □

Theorem 3.7 was formulated with some freedom, as the results of the infinite-horizon case are established with the definition of a feedback Nash equilibrium specific for infinite-horizon LQ games which are based on the constant linear feedback strategies (3.73). Further details are given in Chapter 5, where the AREs are exploited to develop a method for inverse LQ dynamic games. In addition, it is worth noting that the solutions of the AREs (3.75) and therefore the FNE are generally not unique [Eng05, p. 381].

3.9 Summary

This chapter presented fundamentals of dynamic game theory needed for the understanding of the inverse dynamic game methods introduced in this thesis. The following chapters are all based on games with the basic properties presented in Definition 3.3 and with mainly the Nash equilibrium as a solution concept—Nevertheless, a possible application to dynamic games with Pareto efficient solutions shall additionally be mentioned. Inverse dynamic game problems depend on further characteristics of the game, e.g. the information structure and strategy types as well as the assumed class of dynamic systems and cost function structure. The following three chapters introduce different kinds of inverse dynamic games and corresponding methods for their solution.

4 Inverse Non-Cooperative Differential Games

This chapter presents results on the solution of inverse differential games.[22] As described in Chapter 2, the aim of an inverse differential game is to calculate the cost functions players minimized which gave rise to observed state and control trajectories. In the following, this problem is first formulated formally. Afterwards, the main contributions presented in this chapter are the proposal of an efficient method for solving inverse open-loop differential games and the formulation of sufficient conditions for the uniqueness of the solution. Furthermore, the applicability of the method for inverse differential games with feedback strategies is demonstrated.[23]

4.1 Problem Formulation

The theoretical framework of non-cooperative differential games describes N agents treated as entities controlling the system based on the minimization of their individual cost functions, as introduced in Chapter 3. The non-cooperative nature of the game means that no contracts or agreements between players are in place while attempting to minimize their individual costs. Within the inverse problem of differential games, the result of the interaction between all players, i.e. the state and control trajectories, are assumed as given. A further important characteristic of the inverse differential game is that the interaction led to a Nash equilibrium. Some work exists which investigates conditions under which Nash equilibria exist (see e.g. the results in [Luk71, Var70] and the discussions and references given in [BO99, Eng05]). However, these conditions are not general and not simple to formulate in terms of the system dynamics or the cost functions. Addressing the existence of Nash equilibria in general dynamic games is beyond the scope of this thesis and therefore, the following assumption will be made.

[22] In the remainder of this thesis, the term *inverse differential game* describes an inverse dynamic game in continuous time (cf. last paragraph of Section 3.1).

[23] The results of this chapter are based on the conference paper [RIK⁺17] and the author's contribution to the journal paper [MIF⁺20].

Assumption 4.1 (Nash Character of the Observed Trajectories)
The observed state trajectories $\tilde{\boldsymbol{x}}(t)$ and control trajectories $(\tilde{\boldsymbol{u}}_1(t), ..., \tilde{\boldsymbol{u}}_N(t))$ of all players are Nash equilibrium trajectories $\boldsymbol{x}^*(t)$ and $\boldsymbol{u}_i^*(t)$ generated by a non-cooperative differential game defined by a set of non-trivial cost functions $\mathcal{J}^* = \{J_1^*, ..., J_N^*\}$ and a dynamic system according to Definition 3.2.

With this assumption, the inverse differential game problem is defined as follows.

Definition 4.1 (Inverse Differential Game Problem)
Let Assumption 4.1 hold such that state trajectories $\boldsymbol{x}^*(t)$ and control trajectories $\boldsymbol{u}_i^*(t)$, $\forall i \in \mathcal{P}$, which correspond to a Nash equilibrium, are given. Find at least one set $\mathcal{J}$ such that J_i, $\forall i \in \mathcal{P}$, fulfill

$$\begin{aligned} \boldsymbol{u}_i^*(t) &= \arg\min_{\boldsymbol{u}_i(t)} \; J_i\left(\boldsymbol{x}^*(t), \boldsymbol{u}_i(t), \boldsymbol{u}_{\neg i}^*(t)\right) \\ &\text{w.r.t.} \\ \dot{\boldsymbol{x}}(t) &= \boldsymbol{f}\left(\boldsymbol{x}(t), \boldsymbol{u}_1(t), ..., \boldsymbol{u}_N(t), t\right) \\ \boldsymbol{x}(0) &= \boldsymbol{x}_0. \end{aligned} \tag{4.1}$$

The formulation of the inverse differential game problem implies determining the cost functions J_i, $i \in \mathcal{P}$, such that $\boldsymbol{u}_i^*(t)$ solves the optimal control problems (4.1) which follow from Definition 3.7. Definition 4.1 allows for several types of Nash equilibria which arise depending on the information structure of the game and the resulting strategy types. In particular, in this thesis open-loop and feedback Nash equilibria are considered. In addition, Definition 4.1 establishes the search of "at least one set" of cost functions in consequence of the ill-posedness nature of inverse problems in optimal control and dynamic games. This means that several sets of cost functions exist which are equivalent in the sense that all of them are able to explain the same state and control trajectories. The concept of equivalence of cost functions is formalized in Section B.2 of the Appendix.

The inverse differential game of Definition 4.1 is very general and represents a considerably complex task since there is an infinite number of possible cost functions varying in structure and parametrization which may potentially solve the inverse differential game problem. This issue is not unique to inverse dynamic or differential games as it also arises in the inverse problem of optimal control (single-player case). Therefore, parameters need to be introduced first. Two lines of research have been developed to achieve this.

- Approximation of non-linear cost function structures by means of Gaussian processes [LPK11, LHF14] or alternatively, artificial neural networks [WOP16].

- Setting the cost function structure as a linear combination of basis functions [MTL10, PJJB12, JAB13, AB14, MTFP16, PRBF18, JKL+19].

The first approach utilizes parameterized kernel functions which determine the structure of the Gaussian process. In this way, non-linear rewards can be learned by maximizing the likelihood function of the Gaussian process regression output and the kernel parameters under known observations of the state and control values. Nevertheless, finding these parameters is a computationally complex task which has only been solved succesfully in discretized state and control spaces (e.g. a grid world). On the other hand, the use of artificial neural networks usually demands large data sets and computation times.

Therefore, the second approach is followed and presented in the following subsection.

4.2 Basis Functions Approach

In this approach, the cost functions are given a structure specified with basis functions which are defined as follows.

Definition 4.2 (Basis Functions Vector)
The vector $\boldsymbol{\phi}_i \in \mathbb{R}^{M_i}$ *contains the non-trivial functions* $\phi_{i,(j)}(\boldsymbol{x}(t), \boldsymbol{u}_1(t), ..., \boldsymbol{u}_N(t), t)$, $j \in \{1, ..., M_i\}$ *which are called basis functions. Furthermore, the functions* $\phi_{i,(j)} : \mathbb{R}^n \times \mathbb{R}^{m_1} \times \ldots \times \mathbb{R}^{m_N} \times [0, T] \mapsto \mathbb{R}$ *are continuously differentiable in* $\boldsymbol{x}$ *and* $\boldsymbol{u}_1, ..., \boldsymbol{u}_N$ *for all* $j \in \{1, ..., M_i\}$.

The notation $a_{i,(j)}$ is used here and in the remainder of this thesis to represent the j-th entry of any vector $\boldsymbol{a}$ which corresponds to player $i \in \mathcal{P}$.

Based on Definition 4.2, cost functions which consist of a linear combination of the basis functions are introduced, i.e.

$$J_i(\boldsymbol{\phi}_i, \boldsymbol{\theta}_i) = \int_0^T \boldsymbol{\theta}_i^\top \boldsymbol{\phi}_i(\boldsymbol{x}(t), \boldsymbol{u}_1(t), ..., \boldsymbol{u}_N(t), t) \, \mathrm{d}t, \tag{4.2}$$

where $\boldsymbol{\theta}_i \in \Theta_i \subseteq \mathbb{R}^{M_i}$ are time-invariant parameters. The introduction of basis functions may appear stringent, yet it allows a wide variety of possible cost function structures.[24]

[24] Although the considered cost functions (4.2) have a so-called Lagrangian structure, i.e. cost functions with only integral costs, the methods and results of this chapter are also applicable to games with player cost functions with a Bolza structure, i.e. of the form (3.3). To do so, the terminal costs $h_i(\boldsymbol{x}(T), T)$ must be written as a linear combination of basis functions as well. Afterwards, the Bolza cost function can be transformed into a Lagrange cost function by means of the fundamental theorem of calculus (see e.g. [Nai03, Section 2.7.1]).

In order to define a well-posed inverse differential problem with the newly introduced basis functions, the dynamics f and basis functions $\boldsymbol{\phi}_i$ should be specified such that the observed states $\tilde{\boldsymbol{x}}(t)$ and controls $(\tilde{\boldsymbol{u}}_1(t), ..., \tilde{\boldsymbol{u}}_N(t))$ constitute a Nash equilibrium solution to the dynamic game for some (possibly non-unique) cost-functional parameters $\boldsymbol{\theta}_i \in \Theta_i$. Addressing the selection of suitable dynamics and basis functions is beyond the scope of this thesis. Therefore, the following assumption is introduced:

Assumption 4.2 (Nash Character of the Trajectories w.r.t. a Differential Game with Basis Functions)
The observed states $\tilde{\boldsymbol{x}}(t)$ and controls $(\tilde{\boldsymbol{u}}_1(t), ..., \tilde{\boldsymbol{u}}_N(t))$ constitute a Nash equilibrium solution to the differential game with system dynamics according to Definition 3.2 which are additionally continuously differentiable in $\boldsymbol{x}$ and $\boldsymbol{u}_1, ..., \boldsymbol{u}_N$, and cost functions of the form (4.2) consisting of basis functions $\boldsymbol{\phi}_i$ according to Definition 4.2 and the unknown cost function parameters $\boldsymbol{\theta}_i = \boldsymbol{\theta}_i^* \in \Theta_i$ for $i \in \mathcal{P}$.

Assumption 4.2 specifies Assumption 4.1 for the introduced cost function structure established with the basis functions of Definition 4.2. The assumption of continuous differentiability of the system dynamics f is standard and permits the consideration of Theorem 3.1 which shall be leveraged in the course of this chapter. With this introduced assumption, the inverse differential game problem regarded in this chapter is defined as follows.

Definition 4.3 (Inverse Differential Game with Basis Functions)
Let Assumption 4.2 be fulfilled such that state trajectories $\boldsymbol{x}^*(t)$ and control trajectories $\boldsymbol{u}_i^*(t)$, $i \in \mathcal{P}$, which correspond to a Nash equilibrium, are given. Determine at least one tuple of parameters $\boldsymbol{\theta} := (\boldsymbol{\theta}_1, ..., \boldsymbol{\theta}_N)$, with $\boldsymbol{\theta}_i \in \Theta_i$, $i \in \mathcal{P}$, such that

$$
\begin{aligned}
\boldsymbol{u}_i^*(t) = \arg\min_{\boldsymbol{u}_i(t)} \; & J_i\left(\boldsymbol{\phi}_i(\boldsymbol{x}^*(t), \boldsymbol{u}_i(t), \boldsymbol{u}_{\neg i}^*(t), t), \boldsymbol{\theta}_i\right) \\
\text{w.r.t.} \; & \\
\dot{\boldsymbol{x}}(t) = & \; f(\boldsymbol{x}(t), \boldsymbol{u}_1(t), ..., \boldsymbol{u}_N(t), t) \\
\boldsymbol{x}(0) = & \; \boldsymbol{x}_0
\end{aligned}
\tag{4.3}
$$

for all players $i \in \mathcal{P}$.

A consequence of the introduction of basis functions is the reduction of the general inverse differential game problem to a parameter identification problem. Despite this simplification, under Assumption 4.2, the inverse differential game problem will still have multiple solutions in general. One of the reasons is the following: if the trajectories $\boldsymbol{x}^*(t)$ and $\left(\boldsymbol{u}_1^*(t), \ldots, \boldsymbol{u}_N^*(t)\right)$

solve the dynamic optimization problems of Definition 4.3 with $\boldsymbol{\theta}_i = \boldsymbol{\theta}_i^* \in \Theta_i$, then the trajectories will also solve the dynamic optimization problems with $\boldsymbol{\theta}_i = c_i\boldsymbol{\theta}_i^*$ for all scaling factors $c_i > 0$. Furthermore, the zero vectors $\boldsymbol{\theta}_i = \mathbf{0}$ are trivial solutions to the inverse differential game problem. Therefore, without loss of generality, trivial solutions and ambiguous scaling shall be excluded by considering parameter sets of the form $\Theta_i = \{\boldsymbol{\theta}_i \in \mathbb{R}^{M_i} \mid \boldsymbol{\theta}_{i,(1)} = 1\}$ where $\boldsymbol{\theta}_{i,(1)}$ denotes the first element of $\boldsymbol{\theta}_i$. The choice of the fixed-element constraint $\boldsymbol{\theta}_{i,(1)} = 1$ is arbitrary and results analogous to those of this chapter will also hold with normalization constraints such as $||\boldsymbol{\theta}_i|| = 1$.[25]

4.3 Inverse Open-Loop Differential Games

The inverse differential games of Definitions 4.1 and 4.3 imply finding cost functions such the solution of the N optimal control problems correspond to the given controls $(\boldsymbol{u}_1^*(t), ..., \boldsymbol{u}_N^*(t))$. Since for a particular optimal control problem of player i, the other players' controls $\boldsymbol{u}_{\neg i}^*(t)$ are available, we can proceed to analyze these individual optimal control problems. For the forward problem of finding open-loop Nash equilibrium trajectories, the tools of optimal control theory, in particular the minimum principle of Pontryagin, are leveraged to obtain necessary conditions for open-loop Nash equilibria (cf. Section 3.6). Similarly, in this section, these conditions shall be exploited to find parameters $\boldsymbol{\theta}_i$ which solve the inverse differential game problem of Definition 4.3 in case of open-loop strategies.

4.3.1 Residual-Based Approach

The main idea consists of exploiting the fact that the observed trajectories correspond by assumption to a Nash equilibrium, i.e. $\tilde{\boldsymbol{x}}(t) = \boldsymbol{x}^*(t)$ and $\tilde{\boldsymbol{u}}_i(t) = \boldsymbol{u}_i^*(t)$. These must fulfill the equations of Theorem 3.1 as these represent necessary conditions for Nash equilibria. Consider any player $i \in \mathcal{P}$. Besides the system dynamics equation, the costate equation

$$\dot{\boldsymbol{\psi}}_i(t) = -\nabla_{\boldsymbol{x}} H_i\left(\boldsymbol{\psi}_i(t), \boldsymbol{x}^*(t), \boldsymbol{u}_i(t), \boldsymbol{u}_{\neg i}^*(t), t\right) \tag{4.4a}$$

$$\boldsymbol{\psi}_i(T) = 0, \tag{4.4b}$$

where (4.4b) follows from $h_i(\boldsymbol{x}(T), T) = 0$ due to the Lagrangian structure of the cost function (4.2), and the control equation

$$\boldsymbol{u}_i^*(t) = \arg\min_{\boldsymbol{u}_i(t)} H_i\left(\boldsymbol{\psi}_i(t), \boldsymbol{x}^*(t), \boldsymbol{u}_i(t), \boldsymbol{u}_{\neg i}^*(t), t\right) \tag{4.5}$$

25 Both fixed-element (e.g. [MTFP16]) and normalization-constraint parameter sets (e.g. [ARARU+11]) are popular in the related literature of inverse optimal control.

must be fulfilled. Since we consider no constraints on the control variables $\boldsymbol{u}_i(t)$, the control equation (4.5) results in the Hamiltonian gradient condition

$$\mathbf{0} = \nabla_{\boldsymbol{u}_i} H_i \left(\boldsymbol{\psi}_i(t), \boldsymbol{x}^*(t), \boldsymbol{u}_i(t), \boldsymbol{u}^*_{\neg i}(t), t\right). \tag{4.6}$$

With the Hamiltonian function of player i being given by

$$H_i = \boldsymbol{\theta}_i^\top \boldsymbol{\phi}_i \left(\boldsymbol{x}(t), \boldsymbol{u}_i(t), \boldsymbol{u}_{\neg i}(t), t\right) + \boldsymbol{\psi}_i^\top(t) \boldsymbol{f}\left(\boldsymbol{x}(t), \boldsymbol{u}_i(t), \boldsymbol{u}_{\neg i}(t), t\right) \tag{4.7}$$

as a result of the cost function structure (4.2), the following definition is introduced.

Definition 4.4 (Residuals)

The functions

$$r_{\mathrm{C}}(\boldsymbol{\theta}_i, \boldsymbol{\psi}_i, t) = \left\|\nabla_{\boldsymbol{u}_i} H_i \left(\boldsymbol{\psi}_i, \boldsymbol{\theta}_i, t\right)\right\|^2 \Big|_{\substack{\boldsymbol{u}_i(t) = \boldsymbol{u}_i^*(t) \\ \boldsymbol{x}(t) = \boldsymbol{x}^*(t)}} \tag{4.8}$$

and

$$r_{\mathrm{L}}(\boldsymbol{\theta}_i, \boldsymbol{\psi}_i, t) = \left\|\dot{\boldsymbol{\psi}}_i(t) + \nabla_{\boldsymbol{x}} H_i \left(\boldsymbol{\psi}_i, \boldsymbol{\theta}_i, t\right)\right\|^2 \Big|_{\substack{\boldsymbol{u}_i(t) = \boldsymbol{u}_i^*(t) \\ \boldsymbol{x}(t) = \boldsymbol{x}^*(t)}}, \tag{4.9}$$

where $\|\cdot\|$ denotes the Euclidean norm, are called ***residuals*** *of the control equation and the costate equation, respectively.*

The residuals of Definition 4.4 result from the insertion of the Hamiltonian (4.7) in (4.4a) and (4.6) and the subsequent insertion of the known optimal trajectories $\boldsymbol{x}^*(t)$ and $\boldsymbol{u}_i^*(t)$, which result in a dependence on the costate functions $\boldsymbol{\psi}_i$ and the parameters $\boldsymbol{\theta}_i$ only. Note that $r_{\mathrm{C}}(\boldsymbol{\theta}_i, \boldsymbol{\psi}_i)$ and $r_{\mathrm{L}}(\boldsymbol{\theta}_i, \boldsymbol{\psi}_i)$ are both equal to zero for $\boldsymbol{\theta}_i = \boldsymbol{\theta}_i^*$ and $\boldsymbol{\psi}_i(t) = \boldsymbol{\psi}_i^*(t)$. Therefore, in light of this formulation, the idea of the proposed **residual-based method** consists of the computation of $\hat{\boldsymbol{\theta}}_i \in \Theta_i$ and costate functions $\hat{\boldsymbol{\psi}}_i : [0, T] \mapsto \mathbb{R}^n$ for each player $i \in \mathcal{P}$ which solve the optimization problem

$$\begin{aligned} \min_{\boldsymbol{\psi}_i, \boldsymbol{\theta}_i} \quad & \int_0^T r_{\mathrm{C}}(\boldsymbol{\theta}_i, \boldsymbol{\psi}_i) + \rho\, r_{\mathrm{L}}(\boldsymbol{\theta}_i, \boldsymbol{\psi}_i)\, \mathrm{d}t \\ \text{s.t.} \quad & \boldsymbol{\theta}_i \in \Theta_i, \end{aligned} \tag{4.10}$$

where $\rho > 0$ is a specifiable weighting factor. The intuition behind (4.10) is the following: $\hat{\boldsymbol{\psi}}_i(t)$ and parameters $\hat{\boldsymbol{\theta}}_i$ are sought such that the costate condition (4.4a) and Hamiltonian gradient condition (4.6) hold for all $t \in [0, T]$.[26] Under Assumption 4.2, $\hat{\theta}_i = \theta_i^*$ will be a (possibly non-unique) solution to (4.10).

The solution of (4.10) is based on its reformulation as a quadratic program. For that purpose, it shall first rewritten as a LQ dynamic optimization problem. Let us define the matrix

$$L := \begin{bmatrix} I_{M_i} & 0_{M_i \times n} \end{bmatrix} \in \mathbb{R}^{M_i \times (M_i + n)} \tag{4.11}$$

where I_{M_i} denotes a square identity matrix with dimensions $M_i \times M_i$. Similarly, $0_{M_i \times n}$ denotes a zero matrix with dimensions $M_i \times n$. Furthermore, we define the matrices $R = I_n$, $B := \begin{bmatrix} 0_{n \times M_i} & I_n \end{bmatrix}^\top$ and the time-variant matrices

$$N_i(t) := \begin{bmatrix} \rho \nabla_x \phi_i(t) & \rho \nabla_x f(t) \end{bmatrix}^\top \tag{4.12}$$

$$Q_i(t) := \begin{bmatrix} \sqrt{\rho} \nabla_x \phi_i(t) & \sqrt{\rho} \nabla_x f(t) \\ \nabla_{u_i} \phi_i(t) & \nabla_{u_i} f(t) \end{bmatrix}^\top \begin{bmatrix} \sqrt{\rho} \nabla_x \phi_i(t) & \sqrt{\rho} \nabla_x f(t) \\ \nabla_{u_i} \phi_i(t) & \nabla_{u_i} f(t) \end{bmatrix} \tag{4.13}$$

where we use the shorthand $\nabla_x f(t) \in \mathbb{R}^{n \times n}$ and $\nabla_{u_i} f(t) \in \mathbb{R}^{m_i \times n}$ to denote the matrices of partial derivatives of f with respect to $x(t)$ and $u_i(t)$, respectively[27], and evaluated with θ_i, $x(t)$, and $u_i(t)$ for $i \in \mathcal{P}$. Similarly, we use $\nabla_x \phi_i(t) \in \mathbb{R}^{n \times M_i}$, and $\nabla_{u_i} \phi_i(t) \in \mathbb{R}^{m_i \times M_i}$ to denote the matrices of partial derivatives of ϕ_i evaluated with θ_i, $x(t)$, and $u_i(t)$ for $i \in \mathcal{P}$. The following lemma rewrites the problem (4.10) as a LQ dynamic optimization problem.

Lemma 4.1

Consider any player $i \in \mathcal{P}$. The optimization problem (4.10) over the costates ψ_i and parameters θ_i is equivalent to the LQ dynamic optimization problem

$$\begin{aligned} \min_{z_i, v_i} \int_0^T & z_i^\top(t) Q_i(t) z_i(t) + v_i^\top(t) \rho R v_i(t) + 2 z_i^\top(t) N_i(t) v_i(t) \, \mathrm{d}t \\ & \text{s.t.} \\ & \dot{z}_i(t) = B v_i(t), \; t \in [0, T] \\ & L z_i(t) \in \Theta_i, \; t \in [0, T] \end{aligned} \tag{4.14}$$

over the functions $z_i : [0, T] \mapsto \mathbb{R}^{M_i + n}$ and $v_i : [0, T] \mapsto \mathbb{R}^n$ with the variable substitutions

$$z_i(t) = \begin{bmatrix} \theta_i \\ \psi_i(t) \end{bmatrix} \quad \text{and} \quad v_i(t) = \dot{\psi}_i(t). \tag{4.15}$$

---

[26] If $N = 1$ and $\rho = 1$, this method recudes to the single-player approach presented in [JAB13].

[27] The partial derivatives $\nabla_x f(t)$ are defined here as the transposed Jacobian of f, i.e. $\nabla_x f(t) = \begin{bmatrix} \frac{\partial f(t)}{\partial x_1} & \cdots & \frac{\partial f(t)}{\partial x_n} \end{bmatrix}^\top$.

Proof:
We note that the integrand of the objective functional of (4.10) may be rewritten as

$$\begin{aligned}
&\left\|\nabla_{\boldsymbol{u}_i} H_i\left(t, \boldsymbol{\psi}_i(t), \boldsymbol{\theta}_i\right)\right\|^2 + \rho\left\|\dot{\boldsymbol{\psi}}_i(t) + \nabla_{\boldsymbol{x}} H_i\left(t, \boldsymbol{\psi}_i(t), \boldsymbol{\theta}_i\right)\right\|^2 \\
&= \left\|\begin{bmatrix} \sqrt{\rho}\dot{\boldsymbol{\psi}}_i(t) + \sqrt{\rho}\nabla_{\boldsymbol{x}} H_i\left(t, \boldsymbol{\psi}_i(t), \boldsymbol{\theta}_i\right) \\ \nabla_{\boldsymbol{u}_i} H_i\left(t, \boldsymbol{\psi}_i(t), \boldsymbol{\theta}_i\right) \end{bmatrix}\right\|^2 \\
&= \left\|\begin{bmatrix} \sqrt{\rho}\dot{\boldsymbol{\psi}}_i(t) + \sqrt{\rho}\nabla_{\boldsymbol{x}} \boldsymbol{\phi}_i(t)\boldsymbol{\theta}_i + \sqrt{\rho}\nabla_{\boldsymbol{x}} \boldsymbol{f}(t)\boldsymbol{\psi}_i(t) \\ \nabla_{\boldsymbol{u}_i} \boldsymbol{\phi}_i(t)\boldsymbol{\theta}_i + \nabla_{\boldsymbol{u}_i} \boldsymbol{f}(t)\boldsymbol{\psi}_i(t) \end{bmatrix}\right\|^2 \\
&= \left\|\begin{bmatrix} \sqrt{\rho}\nabla_{\boldsymbol{x}} \boldsymbol{\phi}_i(t) & \sqrt{\rho}\nabla_{\boldsymbol{x}} \boldsymbol{f}(t) \\ \nabla_{\boldsymbol{u}_i} \boldsymbol{\phi}_i(t) & \nabla_{\boldsymbol{u}_i} \boldsymbol{f}(t) \end{bmatrix}\begin{bmatrix} \boldsymbol{\theta}_i \\ \boldsymbol{\psi}_i(t) \end{bmatrix} + \begin{bmatrix} \sqrt{\rho}\boldsymbol{I}_n \\ \boldsymbol{0}_{m_i \times n} \end{bmatrix}\dot{\boldsymbol{\psi}}_i(t)\right\|^2 \\
&= \boldsymbol{z}_i^\top(t)\boldsymbol{Q}_i(t)\boldsymbol{z}_i(t) + \boldsymbol{v}_i^\top(t)\rho\boldsymbol{R}\boldsymbol{v}_i(t) + 2\boldsymbol{z}_i^\top(t)\boldsymbol{N}_i(t)\boldsymbol{v}_i(t)
\end{aligned}$$

where the second equality holds by recalling the definition of the player Hamiltonian (4.7), and the third and fourth equalities are obtained via matrix algebra by recalling the definitions of $\boldsymbol{Q}_i(t)$, $\boldsymbol{R}$, and $\boldsymbol{N}_i(t)$ together with the variable substitutions (4.15). We also note that the constraint $\boldsymbol{\theta}_i \in \Theta_i$ may be equivalently written as

$$\boldsymbol{L}\boldsymbol{z}_i(t) = \boldsymbol{\theta}_i \in \Theta_i$$

and the (implicit) constraint in (4.10) that $\boldsymbol{\theta}_i$ is time-invariant is equivalent to the constraint

$$\dot{\boldsymbol{z}}_i(t) = \begin{bmatrix} \dot{\boldsymbol{\theta}}_i \\ \dot{\boldsymbol{\psi}}_i(t) \end{bmatrix} = \begin{bmatrix} \boldsymbol{0}_{M_i \times n} \\ \dot{\boldsymbol{\psi}}_i(t) \end{bmatrix} = \boldsymbol{B}\boldsymbol{v}_i(t).$$

Minimization of the functional

$$\int_0^T \boldsymbol{z}_i^\top(t)\boldsymbol{Q}_i(t)\boldsymbol{z}_i(t) + \boldsymbol{v}_i^\top(t)\rho\boldsymbol{R}\boldsymbol{v}_i(t) + 2\boldsymbol{z}_i^\top(t)\boldsymbol{N}_i(t)\boldsymbol{v}_i(t)\,\mathrm{d}t$$

over $\boldsymbol{z}_i : [0,T] \mapsto \mathbb{R}^{M_i+n}$ and $\boldsymbol{v}_i : [0,T] \mapsto \mathbb{R}^n$ subject to the constraints $\dot{\boldsymbol{z}}_i(t) = \boldsymbol{B}\boldsymbol{v}_i(t)$ and $\boldsymbol{L}\boldsymbol{z}_i(t) \in \Theta_i$ for all $t \in [0,T]$ is therefore equivalent to the minimization of the objective functional of (4.10) over $\boldsymbol{\psi}_i(t)$ and $\boldsymbol{\theta}_i$ subject to $\boldsymbol{\theta}_i \in \Theta_i$ with the substitutions

$$\boldsymbol{z}_i(t) = \begin{bmatrix} \boldsymbol{\theta}_i \\ \boldsymbol{\psi}_i(t) \end{bmatrix} \text{ and } \boldsymbol{v}_i(t) = \dot{\boldsymbol{\psi}}_i(t).$$

The lemma result follows and the proof is complete. □

Lemma 4.1 establishes that (4.10) at the core of the proposed method can be rewritten as (4.14) with linear dynamic constraints, quadratic objective functional, and (partial) constraints on the function $\boldsymbol{L}\boldsymbol{z}_i(t)$. The following lemma shows that (4.14) can be solved as a LQ optimal control problem with an unknown initial state $\boldsymbol{z}_i(0)$ resulting in a quadratic program.

Lemma 4.2 (Quadratic Program Formulation)

Consider any player $i \in \mathcal{P}$ and suppose that $\rho > 0$ is selected such that the matrix $\boldsymbol{Q}_i(t) - \boldsymbol{N}_i(t)\rho^{-1}\boldsymbol{R}^{-1}\boldsymbol{N}_i^\top(t)$ is positive semidefinite for all $t \in [0, T]$. A pair of functions $\hat{\boldsymbol{z}}_i : [0, T] \mapsto \mathbb{R}^{M_i+n}$ and $\hat{\boldsymbol{v}}_i : [0, T] \mapsto \mathbb{R}^n$ solves the dynamic optimization problem (4.14) if and only if the initial value of $\hat{\boldsymbol{z}}_i(0) = \hat{\boldsymbol{\alpha}}_i \in \mathbb{R}^{M_i+n}$ solves the quadratic program

$$\begin{aligned} \min_{\boldsymbol{\alpha}_i} \quad & \boldsymbol{\alpha}_i^\top \boldsymbol{P}_i(0)\boldsymbol{\alpha}_i \\ \text{s.t.} \quad & \boldsymbol{L}\boldsymbol{\alpha}_i \in \Theta_i \end{aligned} \tag{4.16}$$

and the pair of functions satisfy the differential equation

$$\dot{\hat{\boldsymbol{z}}}_i(t) = \boldsymbol{B}\hat{\boldsymbol{v}}_i(t) = \boldsymbol{B}\boldsymbol{K}_i(t)\hat{\boldsymbol{z}}_i(t) \tag{4.17}$$

for all $t \in [0, T]$ where $\boldsymbol{K}_i(t) := -\rho^{-1}\left[\boldsymbol{B}^\top \boldsymbol{P}_i(t) + \boldsymbol{N}_i^\top(t)\right]$ and $\boldsymbol{P}_i : [0, T] \mapsto \mathbb{R}^{(M_i+n)\times(M_i+n)}$ is the unique symmetric positive semidefinite solution to the Riccati differential equation

$$\boldsymbol{0} = \dot{\boldsymbol{P}}_i(t) - \rho^{-1}(\boldsymbol{P}_i(t)\boldsymbol{B} + \boldsymbol{N}_i(t))(\boldsymbol{B}^\top \boldsymbol{P}_i^\top(t) + \boldsymbol{N}_i^\top(t)) + \boldsymbol{Q}_i(t) \tag{4.18}$$

for $t \in [0, T]$ with terminal boundary condition $\boldsymbol{P}_i(T) = \boldsymbol{0}$.

Proof:

Consider any player $i \in \mathcal{P}$. We first note that given a function $\boldsymbol{v}_i : [0, T] \mapsto \mathbb{R}^n$ together with an initial value $\boldsymbol{z}_i(0) = \boldsymbol{\alpha}_i \in \mathbb{R}^{M_i+n}$ with $\boldsymbol{L}\boldsymbol{\alpha}_i \in \Theta_i$, we may solve the differential equation $\dot{\boldsymbol{z}}_i(t) = \boldsymbol{B}\boldsymbol{v}_i(t)$ for the unique function $\boldsymbol{z}_i : [0, T] \mapsto \mathbb{R}^{M_i+n}$. The constraints in the dynamic optimization problem (4.14) from Lemma 4.1 therefore imply that the optimization in (4.14) may be rewritten as only over $\boldsymbol{z}_i(0)$ and $\boldsymbol{v}_i$. Namely, (4.14) is equivalent to the unknown initial state optimal control problem

$$\begin{aligned} \min_{\boldsymbol{\alpha}_i} \min_{\boldsymbol{v}_i} & \int_0^T \boldsymbol{z}_i^\top(t)\boldsymbol{Q}_i(t)\boldsymbol{z}_i(t) + \boldsymbol{v}_i^\top(t)\rho\boldsymbol{R}\boldsymbol{v}_i(t) + 2\boldsymbol{z}_i^\top(t)\boldsymbol{N}_i(t)\boldsymbol{v}_i(t)\, \mathrm{d}t \\ & \text{s.t.} \\ & \dot{\boldsymbol{z}}_i(t) = \boldsymbol{B}\boldsymbol{v}_i(t), \ t \in [0, T] \\ & \boldsymbol{z}_i(0) = \boldsymbol{\alpha}_i \\ & \boldsymbol{L}\boldsymbol{\alpha}_i \in \Theta_i. \end{aligned} \tag{4.19}$$

For any $\boldsymbol{\alpha}_i \in \mathbb{R}^{M_i+n}$, the inner optimization problem over the function $\boldsymbol{v}_i$ in (4.19) is a standard LQ optimal control problem with cross-product terms.

Under the positive definiteness of $\boldsymbol{R} = \boldsymbol{I}_n$ as well as $\rho > 0$ and the positive semidefiniteness of the expression $\boldsymbol{Q}_i(t) - \boldsymbol{N}_i(t)\rho^{-1}\boldsymbol{R}^{-1}\boldsymbol{N}_i^\top(t)$, Section 3.4 of [AM89] gives that for any $\boldsymbol{z}_i(0) = \boldsymbol{\alpha}_i \in \mathbb{R}^{M_i+n}$, the unique function solving the inner optimization problem over $\boldsymbol{v}_i$ in (4.19) is

$$\hat{\boldsymbol{v}}_i(t) = \boldsymbol{K}_i(t)\hat{\boldsymbol{z}}_i(t) \tag{4.20}$$

for all $t \in [0,T]$ where $\boldsymbol{K}_i(t) = -\rho^{-1}\left[\boldsymbol{B}^\top \boldsymbol{P}_i(t) + \boldsymbol{N}_i^\top(t)\right]$ and $\boldsymbol{P}_i : [0,T] \mapsto \mathbb{R}^{(M_i+n)\times(M_i+n)}$ is the unique symmetric positive semidefinite solution to the Riccati differential equation (4.18) with $\boldsymbol{P}_i(T) = \boldsymbol{0}$ (see also [Kuč73, Kal64]). Section 3.4 of [AM89] also gives that the minimum value of the inner optimization problem over $\boldsymbol{v}_i$ in (4.19) is

$$\boldsymbol{\alpha}_i^\top \boldsymbol{P}_i(0)\boldsymbol{\alpha}_i \tag{4.21}$$

for any initial state $\boldsymbol{z}_i(0) = \boldsymbol{\alpha}_i$. The function $\hat{\boldsymbol{z}}_i$ solving the inner optimization of (4.19) satisfies $\dot{\hat{\boldsymbol{z}}}_i(t) = \boldsymbol{B}\boldsymbol{K}_i(t)\hat{\boldsymbol{z}}_i(t)$ for any initial state $\boldsymbol{\alpha}_i$. Consequently, the unknown initial state optimal control problem (4.19) simplifies to the quadratic program (4.16). It follows that the pair of functions $(\hat{\boldsymbol{z}}_i, \hat{\boldsymbol{v}}_i)$ solves (4.14) if the functions satisfy the differential equation (4.17) and $\hat{\boldsymbol{z}}_i(0) = \hat{\boldsymbol{\alpha}}_i$ solves (4.16).

In the following, the "only if" part of the lemma assertion is proved — i.e., that if the pair of functions $(\hat{\boldsymbol{z}}_i, \hat{\boldsymbol{v}}_i)$ solves (4.14), then they satisfy the differential equation (4.17) and $\hat{\boldsymbol{z}}_i(0) = \hat{\boldsymbol{\alpha}}_i$ solves the quadratic program (4.16). We first note that the function $\hat{\boldsymbol{v}}_i$ solving the inner optimization problem over $\boldsymbol{v}_i$ in (4.19) is unique and given by (4.20) for any given $\boldsymbol{\alpha}_i \in \mathbb{R}^{M_i}$. Thus, if the pair of functions $(\hat{\boldsymbol{z}}_i, \hat{\boldsymbol{v}}_i)$ solves (4.14), then it must satisfy the differential equation (4.17). Since the unique form of $\hat{\boldsymbol{v}}_i$ implies that (4.19) reduces to the quadratic program (4.16), then $\hat{\boldsymbol{z}}_i(0) = \hat{\boldsymbol{\alpha}}_i$ if the pair of functions $(\hat{\boldsymbol{z}}_i, \hat{\boldsymbol{v}}_i)$ solves (4.14). The lemma result follows and the proof is complete. □

Lemma 4.2 allows us to solve the quadratic program (4.16) for the initial values $\hat{\boldsymbol{z}}_i(0) = \hat{\boldsymbol{\alpha}}_i$ instead of solving (4.14) for the functions $\hat{\boldsymbol{z}}_i$ over the entire interval $t \in [0,T]$. Recalling Lemma 4.1 and the definition of the vectors $\boldsymbol{z}_i(0)$, we note that the initial values $\hat{\boldsymbol{z}}_i(0) = \hat{\boldsymbol{\alpha}}_i$ correspond to the vector

$$\hat{\boldsymbol{\alpha}}_i = \begin{bmatrix} \hat{\boldsymbol{\theta}}_i^\top & \hat{\boldsymbol{\psi}}_i^\top(0) \end{bmatrix}^\top \tag{4.22}$$

where $\hat{\boldsymbol{\theta}}_i$ and $\hat{\boldsymbol{\psi}}_i$ are solutions to the residual-based method (4.10). Together, Lemmas 4.1 and 4.2 therefore allow us to sidestep the difficult problem of directly solving and analyzing the original optimization problem (4.10) and instead solve the quadratic program (4.16) for the parameters $\hat{\boldsymbol{\theta}}_i = \boldsymbol{L}\hat{\boldsymbol{\alpha}}_i$.

Remark 4.1:

The choice of $\rho = 1$ is always sufficient to ensure that the expression $\boldsymbol{Q}_i(t) - \boldsymbol{N}_i(t)\rho^{-1}\boldsymbol{R}^{-1}\boldsymbol{N}_i^\top(t)$ is positive semidefinite for all $t \in [0, T]$ since

$$\begin{aligned}
&\boldsymbol{Q}_i(t) - \boldsymbol{N}_i(t)\boldsymbol{R}^{-1}\boldsymbol{N}_i^\top(t) \\
&\quad = \boldsymbol{Q}_i(t) - \boldsymbol{N}_i(t)\boldsymbol{N}_i^\top(t) \\
&\quad = \begin{bmatrix} \nabla_{\boldsymbol{u}_i}\boldsymbol{\phi}_i^\top(t)\nabla_{\boldsymbol{u}_i}\boldsymbol{\phi}_i(t) & \nabla_{\boldsymbol{u}_i}\boldsymbol{\phi}_i^\top(t)\nabla_{\boldsymbol{u}_i}\boldsymbol{f}(t) \\ \nabla_{\boldsymbol{u}_i}\boldsymbol{f}^\top(t)\nabla_{\boldsymbol{u}_i}\boldsymbol{\phi}_i(t) & \nabla_{\boldsymbol{u}_i}\boldsymbol{f}^\top(t)\nabla_{\boldsymbol{u}_i}\boldsymbol{f}(t) \end{bmatrix} \\
&\quad = \begin{bmatrix} \nabla_{\boldsymbol{u}_i}\boldsymbol{\phi}_i(t) & \nabla_{\boldsymbol{u}_i}\boldsymbol{f}(t) \end{bmatrix}^\top \begin{bmatrix} \nabla_{\boldsymbol{u}_i}\boldsymbol{\phi}_i(t) & \nabla_{\boldsymbol{u}_i}\boldsymbol{f}(t) \end{bmatrix}
\end{aligned}$$

with the first equality holding due to the definition of $\boldsymbol{R}$, and the second and third equalities following by substituting the definitions $\boldsymbol{Q}_i(t)$ and $\boldsymbol{N}_i(t)$. Other values of ρ may result in $\boldsymbol{Q}_i(t) - \boldsymbol{N}_i(t)\rho^{-1}\boldsymbol{R}^{-1}\boldsymbol{N}_i^\top(t)$ not being positive semidefinite and thus leading to multiple solutions of (4.18).

In the following, the results of Lemmas 4.1 and 4.2 shall be used to establish novel explicit expressions for the parameters $\hat{\boldsymbol{\theta}}_i$ that solve the inverse differential game problem. Furthermore, sufficient conditions shall be presented under which the parameters $\hat{\boldsymbol{\theta}}_i$ are guaranteed to be unique and identical to the original parameters $\boldsymbol{\theta}_i^*$ up to a multiplying positive factor.

4.3.2 Sufficient Conditions for the Uniqueness of the Solution

To present the main result on the solution of the residual-based method (4.10), consider the matrix $\boldsymbol{P}_i(0)$ of the optimization problem (4.16) and define

$$\bar{\boldsymbol{P}}_i := \begin{bmatrix} P_{i,(2,2)}(0) & \dots & P_{i,(2,M_i+n)}(0) \\ P_{i,(3,2)}(0) & \dots & P_{i,(3,M_i+n)}(0) \\ \vdots & \ddots & \vdots \\ P_{i,(M_i+n,2)}(0) & \dots & P_{i,(M_i+n,M_i+n)}(0) \end{bmatrix} \tag{4.23}$$

as the principal submatrix of $\boldsymbol{P}_i(0)$ formed by deleting the first row and column of $\boldsymbol{P}_i(0)$, and

$$\bar{\boldsymbol{p}}_i := \begin{bmatrix} P_{i,(2,1)}(0) & P_{i,(3,1)}(0) & \dots & P_{i,(M_i+n,1)}(0) \end{bmatrix}^\top \tag{4.24}$$

which denotes the first column of $\boldsymbol{P}_i(0)$ with deleted first element. Furthermore, let

$$\bar{\boldsymbol{P}}_i = \boldsymbol{U}_i\Sigma_i^P\boldsymbol{U}_i^\top$$

be the singular value decomposition (SVD) of $\bar{\boldsymbol{P}}_i$ where $\Sigma_i^P \in \mathbb{R}^{(M_i+n-1)\times(M_i+n-1)}$ is a diagonal matrix, and

$$\boldsymbol{U}_i = \begin{bmatrix} \boldsymbol{U}_i^{11} & \boldsymbol{U}_i^{12} \\ \boldsymbol{U}_i^{21} & \boldsymbol{U}_i^{22} \end{bmatrix} \in \mathbb{R}^{(M_i+n-1)\times(M_i+n-1)} \tag{4.25}$$

is a block matrix with submatrices $\boldsymbol{U}_i^{11} \in \mathbb{R}^{(M_i-1)\times r_i^{\bar{P}}}$, $\boldsymbol{U}_i^{12} \in \mathbb{R}^{(M_i-1)\times(M_i+n-1-r_i^{\bar{P}})}$, $\boldsymbol{U}_i^{21} \in \mathbb{R}^{n\times r_i^{\bar{P}}}$ and $\boldsymbol{U}_i^{22} \in \mathbb{R}^{n\times(M_i+n-1-r_i^{\bar{P}})}$. Finally, $\bar{\boldsymbol{P}}_i^+$ and $r_i^{\bar{P}}$ represent the pseudoinverse and rank of the submatrix $\bar{\boldsymbol{P}}_i$, respectively. To present the main result, we recall the introduced parameter set

$$\Theta_i = \{\boldsymbol{\theta}_i \in \mathbb{R}^{M_i} \mid \boldsymbol{\theta}_{i,(1)} = 1\} \tag{4.26}$$

so as to exclude the trivial solution $\hat{\boldsymbol{\theta}}_i = \mathbf{0}$ and to exclude non-uniqueness due to scaling. As discussed in Section 4.2, there is no loss of generality with this parameter set since the ordering and scaling of the basis functions and cost function parameters is arbitrary.

Theorem 4.1 (General Solution of the Residual-Based Method)
Consider any player $i \in \mathcal{P}$, and let $\Theta_i = \{\boldsymbol{\theta}_i \in \mathbb{R}^{M_i} \mid \boldsymbol{\theta}_{i,(1)} = 1\}$. All of the parameter vectors $\hat{\boldsymbol{\theta}}_i \in \Theta_i$ corresponding to all solutions $(\hat{\boldsymbol{\psi}}_i, \hat{\boldsymbol{\theta}}_i)$ of the proposed method (4.10) *are of the form*

$$\hat{\boldsymbol{\theta}}_i = \boldsymbol{L}\hat{\boldsymbol{\alpha}}_i \tag{4.27}$$

where $\hat{\boldsymbol{\alpha}}_i = \begin{bmatrix} 1 & \hat{\bar{\boldsymbol{\alpha}}}_i^\top \end{bmatrix}^\top \in \mathbb{R}^{M_i+n}$ are (potentially non-unique) solutions to the quadratic program (4.16) *with $\hat{\bar{\boldsymbol{\alpha}}}_i \in \mathbb{R}^{M_i+n-1}$ given by*

$$\hat{\bar{\boldsymbol{\alpha}}}_i = -\bar{\boldsymbol{P}}_i^+\bar{\boldsymbol{p}}_i + \boldsymbol{U}_i \begin{bmatrix} \mathbf{0}_r \\ \boldsymbol{b} \end{bmatrix} \tag{4.28}$$

where $\mathbf{0}_r \in \mathbb{R}^{r_i^{\bar{P}}}$ and for any $\boldsymbol{b} \in \mathbb{R}^{M_i+n-1-r_i^{\bar{P}}}$. Furthermore, if either $\boldsymbol{U}_i^{12} = \mathbf{0}$ or $\bar{\boldsymbol{P}}_i$ has full rank, i.e. $r_i^{\bar{P}} = M_i + n - 1$, then all solutions $(\hat{\boldsymbol{\psi}}_i, \hat{\boldsymbol{\theta}}_i)$ to the proposed method (4.10) *correspond to the single unique parameter vector $\hat{\boldsymbol{\theta}}_i \in \Theta_i$ given by*

$$\hat{\boldsymbol{\theta}}_i = \boldsymbol{L}\begin{bmatrix} 1 \\ -\bar{\boldsymbol{P}}_i^+\bar{\boldsymbol{p}}_i \end{bmatrix}. \tag{4.29}$$

Proof:
Lemmas 4.1 and 4.2 together imply that all solutions to the original optimization problem (4.10) of the proposed residual-based method have parameter vectors given by $\hat{\boldsymbol{\theta}}_i = \boldsymbol{L}\hat{\boldsymbol{\alpha}}_i$ where $\hat{\boldsymbol{\alpha}}_i$ is a solution to the quadratic program (4.16). We thus proceed by analyzing (4.16).

For any $\boldsymbol{\alpha}_i \in \mathbb{R}^{M_i+n}$ with $\boldsymbol{L}\boldsymbol{\alpha}_i \in \Theta_i$ where $\Theta_i = \{\boldsymbol{\theta}_i \in \mathbb{R}^{M_i} \mid \boldsymbol{\theta}_{i,(1)} = 1\}$, we have that $\boldsymbol{\alpha}_i = \begin{bmatrix} 1 & \bar{\boldsymbol{\alpha}}_i^\top \end{bmatrix}^\top$ where $\bar{\boldsymbol{\alpha}}_i \in \mathbb{R}^{M_i+n-1}$ and so

$$\begin{aligned}\boldsymbol{\alpha}_i^\top \boldsymbol{P}_i(0)\boldsymbol{\alpha}_i &= \begin{bmatrix} 1 & \bar{\boldsymbol{\alpha}}_i^\top \end{bmatrix} \boldsymbol{P}_i(0) \begin{bmatrix} 1 \\ \bar{\boldsymbol{\alpha}}_i \end{bmatrix} \\ &= \boldsymbol{P}_{i,(1,1)}(0) + \bar{\boldsymbol{\alpha}}_i^\top \bar{\boldsymbol{P}}_i \bar{\boldsymbol{\alpha}}_i + 2\bar{\boldsymbol{\alpha}}_i^\top \bar{\boldsymbol{p}}_i\end{aligned}$$

where $\boldsymbol{P}_{i,(1,1)}(0)$ is the first element of $\boldsymbol{P}_i(0)$. All solutions $\hat{\boldsymbol{\alpha}}_i$ of the constrained quadratic program (4.16) with $\Theta_i = \{\boldsymbol{\theta}_i \in \mathbb{R}^{M_i} \mid \boldsymbol{\theta}_{i,(1)} = 1\}$ are therefore of the form $\hat{\boldsymbol{\alpha}}_i = \begin{bmatrix} 1 & \hat{\bar{\boldsymbol{\alpha}}}_i' \end{bmatrix}^\top$ where $\hat{\bar{\boldsymbol{\alpha}}}_i \in \mathbb{R}^{M_i+n-1}$ are solutions to the unconstrained quadratic program

$$\min_{\bar{\boldsymbol{\alpha}}_i} \left\{ \frac{1}{2} \bar{\boldsymbol{\alpha}}_i^\top \bar{\boldsymbol{P}}_i \bar{\boldsymbol{\alpha}}_i + \bar{\boldsymbol{\alpha}}_i^\top \bar{\boldsymbol{p}}_i \right\}.$$

We note that $\boldsymbol{P}_i(0)$ is symmetric positive semidefinite which guarantees the existence of a solution of (4.16). Furthermore, this leads to $\bar{\boldsymbol{P}}_i$ also being symmetric positive semidefinite. With these conditions fulfilled, [Gal11, Proposition 15.2] gives that the equivalent unconstrained quadratic program is solved by any $\hat{\bar{\boldsymbol{\alpha}}}_i$ satisfying

$$\hat{\bar{\boldsymbol{\alpha}}}_i = -\bar{\boldsymbol{P}}_i^+ \bar{\boldsymbol{p}}_i + \boldsymbol{U}_i \begin{bmatrix} \boldsymbol{0}_r \\ \boldsymbol{b} \end{bmatrix}$$

for any $\boldsymbol{b} \in \mathbb{R}^{M_i+n-1-r_i^{\bar{P}}}$. The first theorem assertion (4.27) follows.

Now, to prove the second theorem assertion we note that if $\boldsymbol{U}_i^{12} = \boldsymbol{0}$, then

$$\begin{aligned}\hat{\bar{\boldsymbol{\alpha}}}_i &= -\bar{\boldsymbol{P}}_i^+ \bar{\boldsymbol{p}}_i + \begin{bmatrix} \boldsymbol{U}_i^{11} & \boldsymbol{0} \\ \boldsymbol{U}_i^{21} & \boldsymbol{U}_i^{22} \end{bmatrix} \begin{bmatrix} \boldsymbol{0}_r \\ \boldsymbol{b} \end{bmatrix} \\ &= -\bar{\boldsymbol{P}}_i^+ \bar{\boldsymbol{p}}_i + \begin{bmatrix} \boldsymbol{0}_{M_i-1} \\ \boldsymbol{U}_i^{22}\boldsymbol{b} \end{bmatrix}\end{aligned}$$

for any $\boldsymbol{b} \in \mathbb{R}^{M_i+n-1-r_i^{\bar{P}}}$ where $\boldsymbol{U}_i^{22}\boldsymbol{b} \in \mathbb{R}^n$. Clearly, if $r_i^{\bar{P}} = M_i + n - 1$, then we also have that

$$\hat{\bar{\boldsymbol{\alpha}}}_i = -\bar{\boldsymbol{P}}_i^+ \bar{\boldsymbol{p}}_i.$$

Thus, if either $\boldsymbol{U}_i^{12} = \boldsymbol{0}$ or $r_i^{\bar{P}} = M_i + n - 1$, then the first $M_i - 1$ components of $\hat{\bar{\boldsymbol{\alpha}}}_i$ are invariant with respect to the free vector $\boldsymbol{b} \in \mathbb{R}^{M_i+n-1-r_i^{\bar{P}}}$, and so all solutions $\hat{\boldsymbol{\alpha}}_i = \begin{bmatrix} 1 & \hat{\bar{\boldsymbol{\alpha}}}_i^\top \end{bmatrix}^\top$ of the constrained quadratic program (4.16) satisfy

$$\boldsymbol{L}\hat{\boldsymbol{\alpha}}_i = \boldsymbol{L} \begin{bmatrix} 1 \\ -\bar{\boldsymbol{P}}_i^+ \bar{\boldsymbol{p}}_i \end{bmatrix}$$

due to the definition of $\boldsymbol{L}$ (cf. (4.11)). The second theorem assertion follows since $\hat{\boldsymbol{\theta}}_i = \boldsymbol{L}\hat{\boldsymbol{\alpha}}_i$ which completes the proof. □

Theorem 4.1 establishes that the conditions $\boldsymbol{U}_i^{12} = \boldsymbol{0}$ and $r_i^{\bar{\boldsymbol{P}}} = M_i + n - 1$ are both sufficient for ensuring the uniqueness of the player cost-functional parameters $\hat{\boldsymbol{\theta}}_i$ computed with the proposed method (4.10). These conditions will not hold when the inverse differential game problem is ill-posed — for example, on short time-horizons T, due to degenerate system dynamics, or when the trajectories are uninformative (e.g. when the trajectories $\boldsymbol{x}^*(t)$ and $(\boldsymbol{u}_1^*(t), \ldots, \boldsymbol{u}_N^*(t))$ correspond to a dynamic equilibrium of the dynamics in the sense that $\dot{\boldsymbol{x}}(t) = 0$ for all $t \in [0, T]$). The conditions $\boldsymbol{U}_i^{12} = \boldsymbol{0}$ and $r_i^{\bar{\boldsymbol{P}}} = M_i + n - 1$ may be interpreted as analogous conditions to the persistence of excitation conditions known from parameter estimation and adaptive control.

The following corollary establishes that, under the assumption that the ill-posedness of the inverse differential game problem is only due to an unknown scaling factor, then $\boldsymbol{U}_i^{12} = \boldsymbol{0}$ and $r_i^{\bar{\boldsymbol{P}}} = M_i + n - 1$ become sufficient conditions for ensuring that the residual-based method (4.10) yields unique player cost-functional parameters that only differ from the true player cost-functional parameters $\boldsymbol{\theta}_i^*$ by an unknown scaling factor $c_i > 0$ when Assumption 4.2 holds.

Corollary 4.1 (Uniqueness up to a Scaling Factor)
Suppose that Assumption 4.2 holds. Consider any player $i \in \mathcal{P}$, and let $\Theta_i = \{\boldsymbol{\theta}_i \in \mathbb{R}^{M_i} \mid \boldsymbol{\theta}_{i,(1)} = 1\}$. If either $\boldsymbol{U}_i^{12} = \boldsymbol{0}$ or $r_i^{\bar{\boldsymbol{P}}} = M_i + n - 1$, and if there exists a $c_i > 0$ such that $c_i\boldsymbol{\theta}_i^* \in \Theta_i$, then

$$\hat{\boldsymbol{\theta}}_i = \boldsymbol{L}\begin{bmatrix} 1 \\ -\bar{\boldsymbol{P}}_i^+\bar{\boldsymbol{p}}_i \end{bmatrix} = c_i\boldsymbol{\theta}_i^* \tag{4.30}$$

is the unique parameter vector corresponding to all optimal solutions $(\hat{\boldsymbol{\psi}}_i, \hat{\boldsymbol{\theta}}_i)$ of the residual-based method (4.10).

Proof:
The necessary conditions for open-loop Nash equilibria of Theorem 3.1, i.e. (4.4) and (4.6) imply that $(\boldsymbol{\psi}_i, c_i\boldsymbol{\theta}_i^*)$ (with $\boldsymbol{\psi}_i$ solving (4.4) under $\boldsymbol{\psi}_i(T) = 0$ and $\boldsymbol{\theta}_i = c_i\boldsymbol{\theta}_i^*$) is always a solution to the proposed method (4.10) under Assumption 4.2. Since the conditions of the corollary give that $c_i\boldsymbol{\theta}_i^*$ is in Θ_i, and since the second assertion of Theorem 4.1 implies the uniqueness of the parameter vector $\hat{\boldsymbol{\theta}}_i \in \Theta_i$ corresponding to all optimal solutions of the residual-based method (4.10) if either $\boldsymbol{U}_i^{12} = \boldsymbol{0}$ or $r_i^{\bar{\boldsymbol{P}}} = M_i + n - 1$, we must have that $\hat{\boldsymbol{\theta}}_i = c_i\boldsymbol{\theta}_i^*$ when either $\boldsymbol{U}_i^{12} = \boldsymbol{0}$ or $r_i^{\bar{\boldsymbol{P}}} = M_i + n - 1$ holds. The corollary assertion follows. □

In the following, the implications of each condition of Theorem 4.1 to the originally posed residual-based method (4.10) is analyzed.

Full-Rank Condition

In Corollary 4.1 and Theorem 4.1, if $r_i^{\bar{P}} = M_i + n - 1$ holds then both the player cost-functional parameters $\hat{\boldsymbol{\theta}}_i$ and costate functions $\hat{\boldsymbol{\psi}}_i$ solving (4.10) are unique. To see that a unique pair $(\hat{\boldsymbol{\psi}}_i, \hat{\boldsymbol{\theta}}_i)$ solves (4.10) when $r_i^{\bar{P}} = M_i + n - 1$, we note that the first assertion of Theorem 4.1, specifically (4.28), implies that the vectors $\hat{\boldsymbol{\alpha}}_i = \begin{bmatrix} 1 & \hat{\bar{\boldsymbol{\alpha}}}_i^\top \end{bmatrix}^\top$ are unique solutions to the quadratic program (4.16) if $r_i^{\bar{P}} = M_i + n - 1$ because the free vector $\boldsymbol{b}$ will be zero-dimensional. Now, since Lemmas 4.1 and 4.2 imply that the vectors $\hat{\boldsymbol{\alpha}}_i = \hat{\boldsymbol{z}}_i(0)$ correspond to $\begin{bmatrix} \hat{\boldsymbol{\theta}}_i^\top & \hat{\boldsymbol{\psi}}_i^\top(0) \end{bmatrix}^\top$, and since Lemma 4.2 implies a unique function $\hat{\boldsymbol{\psi}}_i$ for each initial condition $\hat{\boldsymbol{\psi}}_i(0)$, we have that the pair $(\hat{\boldsymbol{\psi}}_i, \hat{\boldsymbol{\theta}}_i)$ is indeed the unique solution to (4.10) when $r_i^{\bar{P}} = M_i + n - 1$.

SVD Matrix Condition

The condition $\boldsymbol{U}_i^{12} = \boldsymbol{0}$ can hold when $r_i^{\bar{P}} < M_i + n - 1$. If $\boldsymbol{U}_i^{12} = \boldsymbol{0}$ but $r_i^{\bar{P}} < M_i + n - 1$, then the second assertion of Theorem 4.1 implies that all pairs $(\hat{\boldsymbol{\psi}}_i, \hat{\boldsymbol{\theta}}_i)$ solving (4.10) will share the unique parameter vector $\hat{\boldsymbol{\theta}}_i$ given by (4.29) but may not share a common costate function $\hat{\boldsymbol{\psi}}_i(t)$. The condition $\boldsymbol{U}_i^{12} = \boldsymbol{0}$ prohibits the elements of $\hat{\boldsymbol{\alpha}}_i$ corresponding to $\hat{\boldsymbol{\theta}}_i$ (but not $\hat{\boldsymbol{\psi}}_i(0)$) from depending on the free vector $\boldsymbol{b}$ in (4.28).

4.3.3 Algorithm and Example

In light of Theorem 4.1 and the role of the conditions $\boldsymbol{U}_i^{12} = \boldsymbol{0}$ and $r_i^{\bar{P}} = M_i + n - 1$, the residual-based method (4.10) can be implemented for each player $i \in \mathcal{P}$ with the following algorithm:

Algorithm 1 Residual-based method for player i in an inverse OL differential game.

Input: State and control trajectories $\boldsymbol{x}^*(t)$ and $(\boldsymbol{u}_1^*(t), \dots, \boldsymbol{u}_N^*(t))$, dynamics $\boldsymbol{f}$, basis functions $\boldsymbol{\phi}_i$, and parameter constraint set $\Theta_i = \{\boldsymbol{\theta}_i \in \mathbb{R}^{M_i} \mid \theta_{i,(1)} = 1\}$.
Output: Computed Player i cost-functional parameters $\boldsymbol{\theta}_i$.
Compute $\boldsymbol{Q}_i(t)$ and $\boldsymbol{N}_i(t)$ from (4.12) and (4.13), $t \in [0, T]$.
Solve Riccati equation (4.18) with $\boldsymbol{P}_i(T) = \boldsymbol{0}$ for $\boldsymbol{P}_i(0)$.
Compute submatrix $\bar{\boldsymbol{P}}_i$ from (4.23) and vector $\bar{\boldsymbol{p}}_i$ from (4.24).
Compute rank $r_i^{\bar{P}}$ of $\bar{\boldsymbol{P}}_i$.
Compute pseudoinverse $\bar{\boldsymbol{P}}_i^+$ of $\bar{\boldsymbol{P}}_i$.

if $r_i^{\bar{P}} = M_i + n - 1$ **then**
  return Unique $\boldsymbol{\theta}_i = \hat{\boldsymbol{\theta}}_i$ given by (4.29).
else
  Compute $\boldsymbol{U}_i$ and $\boldsymbol{U}_i^{12}$ in (4.25) through SVD of $\bar{\boldsymbol{P}}_i$.
  if $\boldsymbol{U}_i^{12} = \boldsymbol{0}$ **then**
    return Unique $\boldsymbol{\theta}_i = \hat{\boldsymbol{\theta}}_i$ given by (4.29).
  else
    return Any $\boldsymbol{\theta}_i = \hat{\boldsymbol{\theta}}_i$ from (4.27) with any $\boldsymbol{b} \in \mathbb{R}^{M_i+n-1-r_i^{\bar{P}}}$.
  end if
end if

Hence, the core of the proposed residual-based method with Algorithm 1 is the solution of a RDE and thus we avoid the need to solve nested differential game or optimal control problems. Furthermore, we are also free to compute the cost function parameters of each player separately (rather than as part of the same optimization). Finally, the presented method gives conditions under which the computed parameters are unique in the parameter set Θ_i. These conditions hold for N-player inverse differential games and therefore valid for the special case of (single-player) inverse optimal control as well.

To conclude, an example illustrating the results of this section is presented.

Example 4.1:

Consider an optimal control problem, i.e. a one-player differential game, with system dynamics

$$\dot{x}(t) = u_1(t) \tag{4.31}$$

where $u_1(t) \in \mathbb{R}$ *and with an initial state value* $x_0 = 1$. *Let the cost function be of the form* (4.2) *with* $T = 3$ *and the basis functions*

$$\boldsymbol{\phi}_1(x(t), u_1(t), t) = \begin{bmatrix} u_1^2(t) & x^2(t) & x(t)u_1(t) \end{bmatrix}^\top \tag{4.32}$$

and cost function parameters

$$\boldsymbol{\theta}_1 = \boldsymbol{\theta}_1^* = \begin{bmatrix} 1 & 5 & 2 \end{bmatrix}^\top. \tag{4.33}$$

The optimal control problem is solved for the optimal state and control trajectories in Figure 4.1 by applying the minimum principle and solving the coupled differential equations numerically. These trajectories are unique solutions to the problem since $\boldsymbol{\theta}_1^*$ *satisfies the positive definite and positive semidefinite conditions of [AM89, Section 3.4]. To solve the inverse optimal control problem, Algorithm 1 is applied. The Riccati equation leads to the submatrix*

$$\overline{\boldsymbol{P}}_1 = \begin{bmatrix} 0.4614 & -0.6126 & -0.6126 \\ -0.6126 & 0.9951 & 0.9951 \\ -0.6126 & 0.9951 & 0.9951 \end{bmatrix} \tag{4.34}$$

which is rank deficient. Computing the SVD of $\overline{\boldsymbol{P}}_1$ *yields*

$$\boldsymbol{U}_1 = \begin{bmatrix} -0.4113 & -0.9115 & 0.0000 \\ 0.6445 & -0.2909 & -0.7071 \\ 0.6445 & -0.2909 & 0.7071 \end{bmatrix} \tag{4.35}$$

and therefore $\boldsymbol{U}_1^{12} = \begin{bmatrix} 0 & -0.7071 \end{bmatrix}^\top \neq \boldsymbol{0}$ *which implies that there are not unique parameters* $\boldsymbol{\theta}_1 \in \Theta_1$ *solving the inverse optimal control problem. Thus, the general solution is given by* (4.28). *By inspecting this solution, we observe that the first parameter of* $\boldsymbol{\alpha}_i(0)$ *which corresponds to* $\theta_{1,(2)}$ *can uniquely be recovered (cf. the first entry of* $\boldsymbol{U}_1^{12}$*). Nevertheless, the free parameter* $b \in \mathbb{R}$ *affects the parameter* $\theta_{1,(3)}$*, leading to the non-uniqueness. Using* (4.28), *the general solution of* $\boldsymbol{\theta}_1$ *can be formulated as*

$$\boldsymbol{\theta}_1 = \begin{bmatrix} 1 \\ 5 \\ 4.467 \end{bmatrix} + \begin{bmatrix} 0 \\ 0 \\ -0.7071 \end{bmatrix} b, \quad b \in \mathbb{R}. \tag{4.36}$$

Indeed, by solving the optimal control problem again with (4.36) *and any* $b \in \mathbb{R}$*, it is confirmed that the optimal trajectories* $x^*(t)$ *and* $u_1^*(t)$ *are unaffected by the choice of* b.

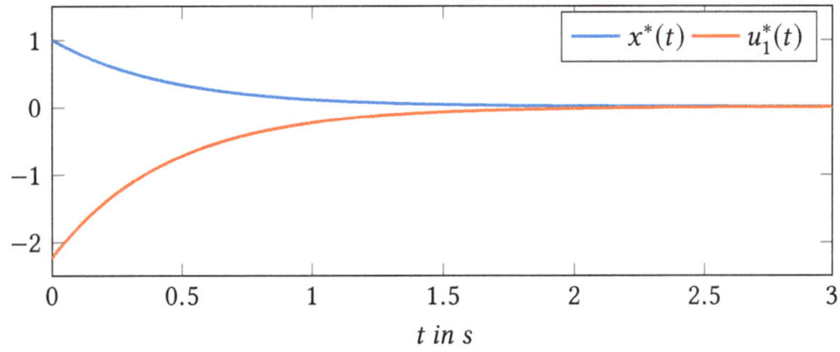

Figure 4.1: State and control trajectories solving the optimal control problem of Example 4.1

4.4 Inverse Feedback Differential Games

The inverse differential game problem assuming a feedback information structure consists in finding the cost function parameters of all players such that the observed trajectories correspond to a feedback Nash equilibrium.

As already noted in Section 3.6.2, the Nash solution of one player depends on the Nash controls of all other players. More specifically, the differential equation of the costate variables corresponding to player i depends on the other controls since, due to the closed-loop information structure, these depend on the state variables. In other words, the control $\boldsymbol{u}_i(t)$ is determined by a feedback strategy in the form of a control law $\boldsymbol{u}_i(t) = \boldsymbol{\gamma}_i(\boldsymbol{x}, t)$. As discussed in Section 3.6.2, the conditions presented in Theorem 3.1 now include the new costate equation

$$\dot{\boldsymbol{\psi}}_i(t) = -\nabla_{\boldsymbol{x}} H_i(\boldsymbol{\psi}_i(t), \boldsymbol{x}^*(t), \boldsymbol{u}_i^*(t), \boldsymbol{\gamma}_{\neg i}^*(\boldsymbol{x}^*, t), t)$$

in order to account for the other players' strategies dependency on the state variable.

4.4.1 Residual-Based Approach

In order to apply the residual-based method, the following assumption is introduced.

Assumption 4.3 (Control Laws)
The Nash equilibrium control laws $\boldsymbol{u}_i^*(t) = \boldsymbol{\gamma}_i^*(\boldsymbol{x}, t)$ are known for all players $i \in \mathcal{P}$.

Under Assumption 4.3, instead of (3.1), we have

$$\dot{\boldsymbol{x}}(t) = \boldsymbol{f}_i\left(\boldsymbol{x}(t), \boldsymbol{\gamma}_1^*(\boldsymbol{x}, t), \ldots, \boldsymbol{u}_i(t), \ldots, \boldsymbol{\gamma}_N^*(\boldsymbol{x}, t), t\right), \quad \boldsymbol{x}(0) = \boldsymbol{x}_0 \tag{4.37}$$

which represent system dynamics from the point of view of each player $i \in \mathcal{P}$. Furthermore, Assumption 4.3 leads to the basis functions

$$\boldsymbol{\phi}_i\left(\boldsymbol{x}(t), \boldsymbol{u}_i(t), \boldsymbol{\gamma}_{\neg i}^*(\boldsymbol{x}, t), t\right), \quad i \in \mathcal{P}. \tag{4.38}$$

Under Assumption 4.3 and the consequently introduced system dynamics (4.37) and basis functions (4.38), we obtain the Hamiltonian function of player i

$$H_i = \boldsymbol{\theta}_i^\top \boldsymbol{\phi}_i\left(\boldsymbol{x}, \boldsymbol{u}_i, \boldsymbol{\gamma}_{\neg i}, t\right) + \boldsymbol{\psi}_i^\top \boldsymbol{f}_i\left(\boldsymbol{x}, \boldsymbol{u}_i, \boldsymbol{\gamma}_{\neg i}^*, t\right), \tag{4.39}$$

where the implicit dependencies were omitted for brevity.

Residuals can be introduced analogously to Section 4.3 according to Definition 4.4. Thus, the inverse differential game with feedback strategies can be solved by applying the residual-based method (4.10). Using (4.37) and (4.38), we redefine the matrices

$$\boldsymbol{N}_i(t) := \begin{bmatrix} \rho \nabla_{\boldsymbol{x}} \boldsymbol{\phi}_i(t) & \rho \nabla_{\boldsymbol{x}} \boldsymbol{f}_i(t) \end{bmatrix}^\top \tag{4.40}$$

$$\boldsymbol{Q}_i(t) := \begin{bmatrix} \sqrt{\rho} \nabla_{\boldsymbol{x}} \boldsymbol{\phi}_i(t) & \sqrt{\rho} \nabla_{\boldsymbol{x}} \boldsymbol{f}_i(t) \\ \nabla_{\boldsymbol{u}_i} \boldsymbol{\phi}_i(t) & \nabla_{\boldsymbol{u}_i} \boldsymbol{f}_i(t) \end{bmatrix}^\top \begin{bmatrix} \sqrt{\rho} \nabla_{\boldsymbol{x}} \boldsymbol{\phi}_i(t) & \sqrt{\rho} \nabla_{\boldsymbol{x}} \boldsymbol{f}_i(t) \\ \nabla_{\boldsymbol{u}_i} \boldsymbol{\phi}_i(t) & \nabla_{\boldsymbol{u}_i} \boldsymbol{f}_i(t) \end{bmatrix} \tag{4.41}$$

and note that the differences with respect to the open-loop case arise from the influence of the new system dynamics f_i and basis functions on the partial derivatives. With these definitions, we can proceed analogously to the open-loop case, ultimately yielding Lemmas 4.1 and 4.2. Consequently, analogous results to Theorem 4.1 and Corollary 4.1 can be formulated. The formal theorem statements and proofs are omitted here.

4.4.2 Example

The following example illustrates the application of the residual-based method for inverse feedback differential games.

Example 4.2:

Consider a two-player differential game with system dynamics

$$\dot{x}(t) = -x(t) + u_1(t) + u_2(t) \tag{4.42}$$

where $u_i(t) \in \mathbb{R}$, $i \in \mathcal{P}$, *and with an initial state value* $x_0 = 5$. *Let the cost function be of the form* (4.2) *with* $T = 6$ *and the basis functions*

$$\boldsymbol{\phi}_i\left(\boldsymbol{x}(t), u_i(t), t\right) = \begin{bmatrix} u_i^2(t) & x_1^2(t) & u_j^2(t) \end{bmatrix}^\top, \quad i, j \in \{1, 2\},\ i \neq j \tag{4.43}$$

and cost function parameters

$$\boldsymbol{\theta}_1 = \boldsymbol{\theta}_1^* = \begin{bmatrix} 1 & 1 & 10 \end{bmatrix}^\top \tag{4.44}$$

$$\boldsymbol{\theta}_2 = \boldsymbol{\theta}_2^* = \begin{bmatrix} 1 & 2 & 1 \end{bmatrix}^\top. \tag{4.45}$$

These parameters are used to solve for the Nash equilibrium state and control trajectories depicted in Figure 4.2. Since a linear-quadratic differential game lies at hand, this was done by solving the coupled Riccati equations (3.69), *which also confirms the Nash character of the trajectories according to Theorem 3.6. This inverse differential problem is illustrated by recovering the cost function parameters of player* 1. *The feedback strategies of each player have the form* $u_i^*(t) = \gamma_i^*(x, t) = -k_i^*(t)x(t)$, *leading to the system dynamics*

$$\dot{x}(t) = -x(t) + u_1(t) - k_2^*(t)x(t) \tag{4.46}$$

and the basis functions

$$\boldsymbol{\phi}_1\left(x(t), u_i(t), t\right) = \begin{bmatrix} u_1^2(t) & x^2(t) & (-k_2^*(t)x(t))^2 \end{bmatrix}^\top \tag{4.47}$$

according to (4.37) and (4.38). The Riccati equation of the residual-based method leads to the submatrix

$$\overline{\boldsymbol{P}}_1 = \begin{bmatrix} 16.045 & 7.499 & -7.470 \\ 7.499 & 3.505 & -3.491 \\ -7.470 & -3.492 & 3.642 \end{bmatrix} \tag{4.48}$$

which has full rank equal to $M_i + n - 1 = 3$. Therefore, with the results of Theorem 4.1 and Corollary 4.1, we obtain the unique solution

$$\hat{\boldsymbol{\theta}}_1 = \begin{bmatrix} 1.000 & 1.000 & 10.000 \end{bmatrix}^\top = \boldsymbol{\theta}_1^*. \tag{4.49}$$

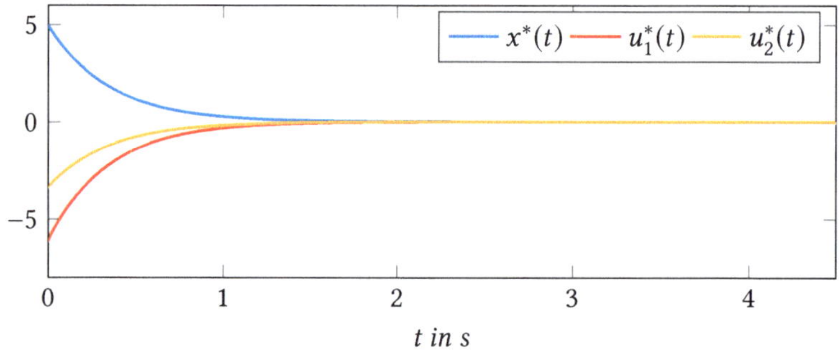

Figure 4.2: State and control trajectories solving the differential game in Example 4.2

The presented example illustrates Theorem 4.1 for inverse feedback differential games, allowing the identification of cost function parameters if the control laws of all players are known, according to Assumption 4.3. Interestingly, in Example 4.2, the cost function parameters of player 1 could be exactly recovered, even though the basis functions were partially redundant due to the fact that the control of player 2 depends on the state variable. However, since $k_2(t)$ was exactly known for all $t \in [0, T]$, the proportion of its influence on the state variable could be distinguished by the method.

4.5 Method Limitations

Before concluding this chapter, possible limitations of the presented methods shall be discussed. A first issue could emerge if only truncated trajectories are available, i.e. we only have access to the trajectories (and control laws, in the feedback case) for $t \in [0, \bar{T}]$ with $\bar{T} < T$. The method can still be applied, but the quality of identification depends on the extent up to which the available truncated trajectories represent the complete optimal trajectories.

A further issue arises if Assumption 4.2 does not hold. This assumption may be violated e.g. due to misspecified dynamics or basis functions, or imperfect trajectories.[28] Additionally, the violation might be even more severe if the trajectories do not even represent a Nash equilibrium, regardless of the basis functions or the system dynamics. In either case, by solving (4.10), parameters $\hat{\boldsymbol{\theta}}_i$ and functions $\hat{\boldsymbol{\psi}}_i(t)$ result such that (4.4a) and (4.6) hold approximately with their priority assigned via choice of ρ. Due to the fact that the approach is based on conditions for Nash equilibria which are generally only necessary, it cannot be always guaranteed that the resulting parameters can be used for determining Nash equilibrium trajectories.

Lastly, the exact knowledge of the feedback strategies as implied by Assumption 4.3 is a rather strict assumption. Nevertheless, given that the state $\boldsymbol{x}^*(t)$ and control trajectories $\boldsymbol{u}_i^*(t)$, $i \in \mathcal{P}$, are available, it is possible to at least determine an approximation using parameter estimation techniques. This will be examined in the next chapter in the context of inverse linear-quadratic differential games.

4.6 Conclusion

In this chapter, an inverse differential game method based on necessary conditions for Nash equilibria was presented. The main idea consisted in the formulation of residuals which represent the violation of the open-loop Nash equilibrium conditions if the parameters (and costate functions) do not correspond to a Nash equilibrium under the observations of the state and control trajectories. The minimization of the residuals lead to a dynamic optimization problem for each player i, the minimizers of which are given by the sought cost function parameters of that specific player. The method is substantially based on the solution of a Riccati differential equation and a static quadratic program, thus avoiding the expensive computation of Nash equilibrium trajectories in each iteration and allowing for the statement of sufficient conditions for the unique solution of the cost function parameters in an inverse open-loop differential game.

Moreover, an approach to solve inverse differential games with feedback strategies was presented. It was shown that it is possible to formulate a residual-based method for the feedback case by assuming the knowledge of the control laws. In this way, the sufficient conditions for the solution uniqueness are still valid. Nevertheless, in general, the control's dependence on the states may lead to redundant basis functions which potentially make the exact estimation of the cost function parameters difficult due to the ambiguity of the solution of the residual-based method.

This chapter presented results for finite-horizon inverse differential games. The following chapter deals with inverse problems for the class of infinite-horizon linear-quadratic differ-

[28] The latter two cases shall be examined in Chapter 7.

ential games and aims at gaining additional insight by exploiting the particular system and cost function structure.

5 Inverse Non-Cooperative Linear-Quadratic Differential Games

This chapter is devoted to the solution of inverse problems in non-cooperative linear-quadratic differential games. This particular class of inverse differential games arises if the dynamic system all players are controlling is linear and a quadratic structure of the player cost functions is given. Furthermore, the considered planning horizon is infinite, leading to constant linear feedback strategies of the players. Linear system dynamics and quadratic cost functions are ubiquitous in control theory and therefore, the properties of this kind of inverse differential games are thoroughly investigated. The techniques employed in this chapter are similar to the ones applied in Chapter 4 in the sense that control-theoretical conditions for Nash equilibria are leveraged, i.e. an inverse optimal control approach is applied. The main contribution presented in this chapter consists of the formulation of explicit solution sets describing all possible solutions of an inverse LQ differential game with an infinite horizon. The dimensions of this solution set depend on the characteristics of the differential game, e.g. number of states, controls and players. Furthermore, necessary and sufficient conditions are given for the uniqueness (up to a positive factor) of the inverse differential game solutions. Finally, on a more practical side, a quadratic program is formulated which allows the efficient computation of one solution (belonging to the whole solution set) and the corresponding algorithm for implementation is presented. The chapter ends with an illustrative example of the method and a conclusion.[29]

5.1 Problem Definition

Consider a continuous-time N-player noncooperative differential game of linear-quadratic type according to Definition 3.11. Therefore, the continuous-time state process of the game is described by the initial value problem

$$\dot{\boldsymbol{x}}(t) = \boldsymbol{A}\boldsymbol{x}(t) + \sum_{i=1}^{N} \boldsymbol{B}_i \boldsymbol{u}_i(t) \tag{5.1a}$$

$$\boldsymbol{x}(0) = \boldsymbol{x}_0 \tag{5.1b}$$

[29] The results of this chapter were partially previously published in the journal paper [IBM$^+$19].

where it is further assumed that $(\boldsymbol{A}, [\boldsymbol{B}_1 \quad \cdots \quad \boldsymbol{B}_N])$ is stabilizable. Following the explanations in Section 3.8.2, the results of this chapter shall be restricted to the consideration of constant linear feedback strategies, i.e. strategies $\boldsymbol{\gamma}_i$ belonging to the set (3.73). Therefore, the control trajectories are given by

$$\boldsymbol{u}_i(t) = -\boldsymbol{K}_i\boldsymbol{x}(t), \quad \forall\, i \in \mathcal{P}, \tag{5.2}$$

with the control laws $\boldsymbol{K} = (\boldsymbol{K}_1, \ldots, \boldsymbol{K}_N)$ (cf. (3.77)). In particular, these lead to a stable closed-loop system (cf. (3.67))

$$\boldsymbol{F} = \boldsymbol{A} - \sum_{j=1}^{N} \boldsymbol{B}_j\boldsymbol{K}_j, \tag{5.3}$$

i.e. they belong to the set of stabilizing control law tuples defined in (3.74).

In this chapter, a Lagrangian quadratic cost function

$$J_i(\boldsymbol{x}_0, \boldsymbol{K}, \boldsymbol{Q}_i, \boldsymbol{R}_{ij}) = \frac{1}{2}\int_0^{\infty} \boldsymbol{x}^\top \boldsymbol{Q}_i \boldsymbol{x} + \sum_{j=1}^{N} \boldsymbol{u}_j^\top \boldsymbol{R}_{ij} \boldsymbol{u}_j \, \mathrm{d}t, \tag{5.4}$$

is considered for each player $i \in \mathcal{P}$, where the same matrix assumptions as in Definition 3.11 are made, i.e. $\boldsymbol{Q}_i$, $\boldsymbol{R}_{ij}$ are symmetric for all $i, j \in \mathcal{P}$ and $\boldsymbol{R}_{ii} \succ \boldsymbol{0}$ for all $i \in \mathcal{P}$.[30] By posing (5.4), a particular structure of the cost functions of all players is defined, similar to the basis function approach considered in Section 4.2. Indeed, a cost function of the form (5.4) can be equivalently represented as a cost function with basis functions as introduced in (4.2).[31] The cost function J_i in (5.4) is written as a function of the N-tuple of feedback laws $\boldsymbol{K} = (\boldsymbol{K}_1, \ldots, \boldsymbol{K}_N)$ and the initial state $\boldsymbol{x}_0$ since together these generate the state and control trajectories $\boldsymbol{x}(t)$ and $\boldsymbol{u}_i(t)$ via (5.1) and (5.2). Finite cost function values are guaranteed by the restriction to strategies or feedback laws belonging to $\mathcal{F}$ as defined in (3.74).

In this chapter, feedback Nash equilibria are considered which are defined in the context of infinite-horizon LQ differential games as follows (cf. Definition 3.7).

[30] Note that no definiteness assumptions on $\boldsymbol{Q}_i$, $i \in \mathcal{P}$, are made since the control laws are restricted to the stabilizing set $\mathcal{F}$ (cf. [EBS00]).

[31] This follows directly from e.g. $\frac{1}{2}\boldsymbol{x}^\top \boldsymbol{Q}_i \boldsymbol{x} = \boldsymbol{\theta}_i^\top \boldsymbol{\phi}_i$ with $\boldsymbol{\theta}_i = \mathrm{vec}(\boldsymbol{Q}_i)$ and where $\boldsymbol{\phi}_i$ has the elements $\phi_{i,(j)} = \frac{1}{2}x_l x_p, \forall l, p \in \{1, \ldots, n\}$.

Definition 5.1 (Feedback Nash Equilibrium ([EBS00]))
An N*-tuple* $\boldsymbol{K}^* = (\boldsymbol{K}_1^*, ..., \boldsymbol{K}_N^*) \in \mathcal{F}$ *is called a stationary linear feedback Nash equilibrium if*

$$J_i(\boldsymbol{x}_0, \boldsymbol{K}^*, \boldsymbol{Q}_i, \boldsymbol{R}_{ij}) \leq J_i(\boldsymbol{x}_0, \boldsymbol{K}_{\neg i}^*(\boldsymbol{\beta}), \boldsymbol{Q}_i, \boldsymbol{R}_{ij}), \quad (5.5)$$

holds for all $i \in \mathcal{P}$*, all* $\boldsymbol{x}_0 \in \mathbb{R}^n$*, and all* $\boldsymbol{\beta}$ *such that* $\boldsymbol{K}_{\neg i}^*(\boldsymbol{\beta}) \in \mathcal{F}$*, where* $\boldsymbol{K}_{\neg i}^*(\boldsymbol{\beta}) = (\boldsymbol{K}_1^*, ..., \boldsymbol{K}_{i-1}^*, \boldsymbol{\beta}, \boldsymbol{K}_{i+1}^*, ..., \boldsymbol{K}_N^*)$.

The FNE is generally not unique (cf. Section 3.8.2), i.e. various tuples $\boldsymbol{K}^*$ corresponding to a particular infinite-horizon LQ differential game may exist. However, in the following, one specific FNE denoted by $\boldsymbol{K}^*$ shall be considered.

The following definition is introduced before formalizing the inverse LQ differential game problem.

Definition 5.2 (Canonical Parameter Set)
The canonical parameter set of the LQ differential game is the set Θ *which contains all possible cost function parameters of* (5.4)*, i.e. all possible matrices* $\boldsymbol{Q}_i$ *and* $\boldsymbol{R}_{ij}$*,* $\forall i, j \in \mathcal{P}$*, which lead to the Nash equilibrium given by* $\boldsymbol{K}^*$*, i.e.*

$$\Theta = \{\boldsymbol{\theta}_i \mid i \in \mathcal{P},\ \boldsymbol{K}^* = \boldsymbol{K}(\boldsymbol{\theta}_1, ..., \boldsymbol{\theta}_N)\ \text{fulfills}\ (5.5)\}, \quad (5.6)$$

where $\boldsymbol{\theta}_i$ *contains the elements of the matrices* $\boldsymbol{Q}_i$ *and* $\boldsymbol{R}_{ij}$*,* $i, j \in \mathcal{P}$*.*

This definition follows directly from the ill-posedness characteristic of inverse differential games. It allows for describing a general set of solutions of inverse differential game which do not necessarily differ in a constant parameter solely. Furthermore, the following assumption is introduced.

Assumption 5.1
The Nash equilibrium feedback matrices $\boldsymbol{K}^* \in \mathcal{F}$ *are known.*

With this assumption, which is similar to Assumption 4.1 made in the last chapter, the inverse infinite-horizon LQ differential game problem considered in this chapter is defined as follows.[32]

[32] In the remainder of this chapter, the considered inverse problem shall be referred to as *inverse linear-quadratic differential game problem*. The infinite-horizon property shall be omitted for the sake of brevity.

Definition 5.3 (Inverse Linear-Quadratic Differential Game Problem)
Consider an infinite-horizon LQ differential game consisting of system dynamics (5.1), *where* $\boldsymbol{A}, \boldsymbol{B}_i, \forall i \in \mathcal{P}$ *are given, and unknown cost functions* (5.4). *Furthermore, let Assumption 5.1 hold such that Nash equilibrium feedback matrices* $\boldsymbol{K}^*$ *are available. Determine the canonical parameter set* Θ *described in Definition 5.2*

While this problem definition is related to the problem in Definition 4.3, it is different in the sense that not only one single tuple of parameter vectors $\boldsymbol{\theta} = (\boldsymbol{\theta}_1, ..., \boldsymbol{\theta}_N)$ is sought, but the complete set of (equivalent) possible tuples of parameter vectors which lead to a given Nash equilibrium. Furthermore, instead of a Nash equilibrium described by the trajectories $\boldsymbol{x}^*(t)$ and $\boldsymbol{u}_i^*(t)$, $i \in \mathcal{P}$, the availability of a Nash equilibrium described by a tuple of control laws $\boldsymbol{K}^*$ is assumed.

Remark 5.1:
By solving the problem of Definition 5.3 we can also solve the related problem of finding Θ, *if instead of* $\boldsymbol{K}^*$, *trajectories* $\boldsymbol{x}^*(t)$ *and* $\boldsymbol{u}_i^*(t)$, $i \in \mathcal{P}$, *are given. This follows from the fact that* $\boldsymbol{K}_i^*$ *can be estimated via* (5.2). *Indeed, such an estimation is commonly performed in single-player inverse LQ optimal control, e.g. in [PCC+ 15] and [FMM+ 18], where the proposed methods also rely on the availability of a control law. Further details on the estimation of* $\boldsymbol{K}^*$ *are given in Section 5.4.2.*

5.2 Solution Sets for Inverse Linear-Quadratic Differential Games

This section presents general solution sets for inverse LQ differential games such that the problem of Definition 5.3 is solved. Similar to Chapter 4, available results on the conditions for feedback Nash equilibria shall be exploited. In the case of an infinite-horizon LQ differential game, the conditions are available in the form of coupled algebraic Riccati equations (ARE).

5.2.1 Coupled Algebraic Riccati Equations

The following theorem is introduced as a basis for the development of the results of this chapter.

Theorem 5.1 (Necessary and Sufficient Conditions for Feedback Nash Equilibria)
Let there exist an N-tuple of symmetric matrices P_i, $i \in \mathcal{P}$ satisfying the N matrix algebraic Riccati equations (ARE)

$$P_i F + F^\top P_i + \sum_{j \in \mathcal{P}} P_j B_j R_{jj}^{-1} R_{ij} R_{jj}^{-1} B_j^\top P_j + Q_i = 0 \tag{5.7}$$

such that F is stabilized. Furthermore, let K_i^* be defined as

$$K_i^* = R_{ii}^{-1} B_i^\top P_i. \tag{5.8}$$

Then, $K^* = (K_1^*, ..., K_N^*)$ is a FNE as in Definition 5.1 and $J_i(x_0, K^*, Q_i, R_{ij}) = x_0^\top P_i x_0$. Conversely, if K^* is a FNE then the set of ARE (5.7) has a stabilizing solution.

Proof:
See the proof of [EBS00, Theorem 4]. □

Remark 5.2:
The ARE given in (5.7) *are an alternative and equivalent formulation of the ARE given in* (3.75). *Both expressions are common in differential game theory.*

Theorem 5.1 represents a necessary and sufficient condition for feedback Nash equilibria. Hence, if the feedback matrices K^* are given, the cost function parameters in the matrices R_{ij} and Q_i, $i, j \in \mathcal{P}$, must fulfill (5.7). This fact shall be leveraged in order to develop a method to solve the inverse LQ differential game. Inspired by [JAK89] and [AKFIJ12], where numerical techniques for continuous-time Riccati equations and results on the properties of Sylvester and Lyapunov type algebraic equations were introduced, respectively, Kronecker products shall be applied to derive a reformulation of (5.7) which serves as a basis for the subsequent results.

Reformulation of the Algebraic Riccati Equations

Before presenting the reformulation, let us define a Kronecker sum [Bre78] as

$$X \oplus Y = \left(X \otimes I_q\right) + \left(I_r \otimes Y\right), \tag{5.9}$$

for squared matrices $X \in \mathbb{R}^{r \times r}$ and $Y \in \mathbb{R}^{q \times q}$, where I_q denotes a q-dimensional identity matrix and $\otimes$ is the Kronecker product. In order to develop a reformulation of (5.7), we require the following result.

Lemma 5.1 (Inverse Existence)

Define $F_\oplus := F^\top \oplus F^\top$ *where* F *is calculated by means of* (5.3) *with any tuple of feedback matrices* $K^* \in \mathcal{F}$ *(cf.* (3.74)*). The inverse* $F_\oplus^{-1}$ *exists.*

Proof:

$F_\oplus^{-1}$ exists if all eigenvalues $\lambda_l \in \sigma(F_\oplus)$, $l \in \{1, ..., n^2\}$ are different from zero. By using [Zha11, Theorem 4.8], we discern that $\lambda_l = \mu_j + \mu_k$, where $\mu_j, \mu_k \in \sigma(F)$, for $j, k \in \{1, ..., n\}$ such that l is associated to a particular combination of j and k, i.e. $j = \lceil \frac{l}{n} \rceil$ and $k = l - n(j-1)$. Since only stabilizing feedback matrices belonging to the set $\mathcal{F}$ in (3.74) are considered, F is a stable matrix and thus $\lambda_l < 0, \forall l \in \{1, ..., n^2\}$. The lemma assertion follows. □

Unless otherwise stated, the following calculations are with respect to a particular player $i \in \mathcal{P}$. With the results of Lemma 5.1, the matrices

$$Z_i := (I_n \otimes B_i^\top) F_\oplus^{-1} \in \mathbb{R}^{nm_i \times n^2} \tag{5.10}$$

and

$$K_i^\otimes := K_i^\top \otimes K_i^\top \in \mathbb{R}^{n^2 \times m_i^2} \tag{5.11}$$

are defined. Furthermore, K_i^* is written as K_i in (5.11) and in the following lemma for brevity.

Lemma 5.2 (Equivalent Formulation of the ARE)

Let the parameter $\bar{\theta}_i \in \mathbb{R}^L$ *denote the vectorized matrices of the cost function* (5.4), *i.e.*

$$\bar{\theta}_i = \begin{bmatrix} vec(Q_i)^\top & vec(R_{i1})^\top & \cdots & vec(R_{ii})^\top & \cdots & vec(R_{iN})^\top \end{bmatrix}^\top, \tag{5.12}$$

where $vec(X)$ *represents a column vectorization of a matrix* X*, leading to* $L = n^2 + \sum_{i=1}^{N} m_i^2$.
Then, the matrices Q_i, R_{ij}*,* $i, j \in \mathcal{P}$*, corresponding to* $\bar{\theta}_i$ *satisfy* (5.7) *if (and only if)* $\bar{\theta}_i$ *fulfills*

$$\bar{M}_i \bar{\theta}_i = 0 \tag{5.13}$$

where $\bar{M}_i \in \mathbb{R}^{n\, m_i \times L}$ *is given by*

$$\bar{M}_i := \begin{bmatrix} Z_i & Z_i K_1^\otimes & \cdots & Z_i K_{i-1}^\otimes & (Z_i K_i^\otimes + K_i \otimes I_p) & Z_i K_{i+1}^\otimes & \cdots & Z_i K_N^\otimes \end{bmatrix}. \tag{5.14}$$

Proof:
We rewrite (5.7) as

$$0 = \mathrm{vec}(P_i F) + \mathrm{vec}(F^\top P_i) + \sum_{j\in\mathcal{P}} \mathrm{vec}(P_j B_j R_{jj}^{-1} R_{ij} R_{jj}^{-1} B_j^\top P_j) + \mathrm{vec}(Q_i)$$

$$0 = \left[\left(F^\top \otimes I_n\right) + \left(I_n \otimes F^\top\right)\right] \mathrm{vec}(P_i) + \sum_{j\in\mathcal{P}} \left(K_j^\top \otimes K_j^\top\right) \mathrm{vec}(R_{ij}) + \mathrm{vec}(Q_i)$$

and thus

$$\mathrm{vec}(P_i) = -F_\oplus^{-1}\mathrm{vec}(Q_i) - \sum_{j\in\mathcal{P}} F_\oplus^{-1} K_j^\otimes \mathrm{vec}(R_{ij}). \tag{5.15}$$

The first equality follows from vectorizing (5.7), while for the second equality (5.8) was used and the following equivalence was applied:

$$\mathrm{vec}(XYZ) = \left(Z^\top \otimes X\right) \mathrm{vec}(Y). \tag{5.16}$$

This equivalence holds for any matrices X, Y and Z with suitable dimensions [Bre78]. The third equality (5.15) follows with the results of Lemma 5.1 and the definitions given in (5.11) and (5.9). Now we rewrite (5.8) as

$$\left(I_n \otimes B_i^\top\right)^{-1} \left(K_i^\top \otimes I_p\right) \mathrm{vec}(R_{ii}) = \mathrm{vec}(P_i) \tag{5.17}$$

using (5.16). Inserting (5.17) in (5.15) results in

$$Z_i \mathrm{vec}(Q_i) + \left(K_i^\top \otimes I_p\right) \mathrm{vec}(R_{ii}) + \sum_{j\in\mathcal{P}} Z_i K_j^\otimes \mathrm{vec}(R_{ij}) = 0 \tag{5.18}$$

and thus (5.13) follows immediately with (5.14) and (5.12). □

The parameters $\bar{\theta}_i$ for which (5.13) holds are valid solutions of (5.7) for a given K_i^*. Note that the feedback matrices $K^* = (K_1^*, \ldots, K_N^*)$ completely characterize the Nash equilibrium trajectories $x^*(t)$ and $u_i^*(t)$, $i \in \mathcal{P}$. This follows from (5.1) fulfilling all conditions for admitting a unique solution for any N-tuple of continuous controls (5.2) [BO99]. Thus, the parameters $\bar{\theta}_i$ are associated to a Nash equilibrium represented by either the feedback matrices or the state and control trajectories.

5.2.2 Canonical Parameter Set

The matrix Riccati equations (5.7) have multiple solutions which potentially represent different Nash equilibria [Wee01]. However, it is worth emphasizing that we are only interested in all parameter tuples $\bar{\theta}$ which represent a specific Nash equilibrium. Bearing this in mind, the following theorem gives the main result.

Theorem 5.2 (Canonical Parameter Set of Inverse LQ Differential Games)
Let a LQ differential game be given by (5.1) and (5.4). Furthermore, let Assumption 5.1 hold such that Nash equilibrium control laws K^* are given. Then, the canonical parameter set of the corresponding inverse LQ differential game is given by

$$\Theta = \bigcup_{i \in \mathcal{P}} \ker(\bar{M}_i), \tag{5.19}$$

with convex boundaries such that $R_{ii} > 0$, $\forall i \in \mathcal{P}$.

Proof:
By inspecting (5.13) from Lemma 5.2 we can recognize that all parameters which satisfy the ARE lie within the kernel of $\bar{M}_i$, which depends on K^*. Therefore, all possible cost function parameters of player i which lead to the known Nash equilibrium are given by $\mathrm{span}(\boldsymbol{v}_i^{(1)}, \ldots, \boldsymbol{v}_i^{(d_i)})$, where d_i represents the dimension of the kernel of $\bar{M}_i$ with basis vectors $\boldsymbol{v}_i$. The set including the cost function parameters of all players corresponding to the Nash equilibrium represented by K^* is thus given by (5.19). □

Note that the results of Lemma 5.2 together with Theorem 5.2 allow for a simple proof of the well-known invariance of the Nash equilibrium in case any cost function parameter $\bar{\theta}_i$ is multiplied by a positive constant.

Corollary 5.1
The trajectories constituting a Nash equilibrium under N cost functions $J_i(\bar{\theta}_i^*)$, $i \in \mathcal{P}$, of an infinite-horizon LQ differential game will constitute the same Nash equilibrium for $J_i(\bar{\theta}_i)$ with $\bar{\theta}_i = c_i \bar{\theta}_i^*, \forall c_i > 0$.

Proof:
This can be easily be seen from $\bar{M}_i c_i \bar{\theta}_i^* = c_i \bar{M}_i \bar{\theta}_i^* = \mathbf{0}$ which does not affect $\ker(\bar{M}_i)$ nor Θ. □

The results of Lemma 5.2 as well as Theorem 5.2 are derived with respect to the parameter definition in (5.12) which considers the most general case where no assumptions on the structure of the cost function matrices were made, e.g. symmetry. The characteristics of the differential game and in particular, the properties of the cost function matrices affect the dimensions of $\ker(\bar{M}_i)$ and consequently of the canonical parameter set Θ. Therefore, in the next section, some properties of inverse LQ differential games based on the possible structures of the cost function matrices are discussed.

5.3 Properties of Inverse Linear-Quadratic Differential Game Solution Sets

Cost function matrices in a quadratic cost function are largely assumed to be at least symmetric. Furthermore, in many applications, these are assumed to be diagonal. Since these matrix properties reduce the number of unknown parameters, inverse LQ differential games and their solution sets shall be analyzed considering all possible cases for the cost function matrices.

5.3.1 Preliminaries

Let us define the variable $M_i \in \mathbb{R}^+$ to denote the number of (non-redundant) parameters of a player's cost function. The specific value of M_i depends on whether the cost function matrices are symmetric or diagonal. We have

$$M_i = \begin{cases} \frac{n^2+n}{2} + \sum\limits_{i \in \mathcal{P}} \frac{m_i^2+m_i}{2}, & \text{symmetric matrices} \\ n + \sum\limits_{i \in \mathcal{P}} m_i, & \text{diagonal matrices} \\ L, & \text{else.} \end{cases} \tag{5.20}$$

Since $M_i \leq L$ holds, the analysis of inverse LQ differential games is based on the vectors $\boldsymbol{\theta}_i \in \mathbb{R}^{M_i}$ which have a potentially reduced dimension compared to the parameter vector of Lemma 5.2. The matrix $\boldsymbol{M}_i \in \mathbb{R}^{nm_i \times M_i}$ is introduced accordingly as a possible modification of the matrix $\bar{\boldsymbol{M}}_i$.

Remark 5.3:

The vector $\boldsymbol{\theta}_i \in \mathbb{R}^{M_i}$ and the modified matrix $\boldsymbol{M}_i \in \mathbb{R}^{nm_i \times M_i}$ comply with Lemma 5.2 in the sense that

$$\boldsymbol{M}_i \boldsymbol{\theta}_i = \boldsymbol{0} \tag{5.21}$$

holds. Consequently, the results of Theorem 5.2 and, obviously, Corollary 5.1 hold for these introduced variables as well.

In the following, an example illustrating the introduced modifications is presented.

Example 5.1:

Consider a 2-player LQ differential game with $n = 2$, $m_1 = m_2 = 1$, *where the cost functions are given by* (5.4). *By Lemma 5.2, we obtain* $M_i = L = 6$, *leading to the vector*

$$\bar{\boldsymbol{\theta}}_i = \begin{bmatrix} Q_{i,(1,1)} & Q_{i,(2,1)} & Q_{i,(1,2)} & Q_{i,(2,2)} & R_{i1} & R_{i2} \end{bmatrix}^\top, \tag{5.22}$$

where $Q_{i,(r,c)}$ *with* $r, c \in \{1,2\}$ *denotes the element of* Q *in the* r*-th row and* c*-th column. Furthermore, we have the matrix*

$$\bar{\boldsymbol{M}}_i = \begin{bmatrix} (\bar{\boldsymbol{m}}_i)_1 & (\bar{\boldsymbol{m}}_i)_2 & \cdots & (\bar{\boldsymbol{m}}_i)_6 \end{bmatrix}, \quad i \in \{1,2\}, \tag{5.23}$$

where $(\bar{\boldsymbol{m}}_i)_j$, $j \in \{1, 2, .., L\}$ *denotes the* j*-th column of* $\bar{\boldsymbol{M}}_i$.

Diagonal Matrices

In case of diagonal matrices, $Q_{i,(2,1)} = Q_{i,(1,2)} = 0$, $i \in \{1,2\}$. *Therefore, the reduced non-redundant parameter vector has the dimension* $M_i = 4$, $i \in \{1,2\}$, *and is given by*

$$\boldsymbol{\theta}_i = \begin{bmatrix} Q_{i,(1,1)} & Q_{i,(2,2)} & R_{i1} & R_{i2} \end{bmatrix}^\top. \tag{5.24}$$

Thus, we set

$$\boldsymbol{M}_i = \begin{bmatrix} (\bar{\boldsymbol{m}}_i)_1 & (\bar{\boldsymbol{m}}_i)_4 & (\bar{\boldsymbol{m}}_i)_5 & (\bar{\boldsymbol{m}}_i)_6 \end{bmatrix}, \quad i \in \{1,2\} \tag{5.25}$$

such that (5.21) *is fulfilled.*

Symmetric Matrices

In case of symmetric matrices, $Q_{i,(2,1)} = Q_{i,(1,2)}$, $i \in \{1,2\}$. *This leads to a reduced non-redundant parameter vector with the dimension* $M_i = 5$, $i \in \{1,2\}$, *and given by*

$$\boldsymbol{\theta}_i = \begin{bmatrix} Q_{i,(1,1)} & Q_{i,(1,2)} & Q_{i,(2,2)} & R_{i1} & R_{i2} \end{bmatrix}^\top. \tag{5.26}$$

Hence, we set

$$\boldsymbol{M}_i = \begin{bmatrix} (\bar{\boldsymbol{m}}_i)_1 & (\bar{\boldsymbol{m}}_i)_2 + (\bar{\boldsymbol{m}}_i)_3 & (\bar{\boldsymbol{m}}_i)_4 & (\bar{\boldsymbol{m}}_i)_5 & (\bar{\boldsymbol{m}}_i)_6 \end{bmatrix}, \quad i \in \{1,2\}, \tag{5.27}$$

such that (5.21) *is fulfilled.*

These modifications allow for the analysis of inverse LQ differential games and their solution sets in the case of symmetric or diagonal cost function matrices by means of the kernel of $\boldsymbol{M}_i$.

5.3.2 Sufficient Condition for Solution Sets

In the following, all possible parameters $\boldsymbol{\theta}_i$ which lead to the same Nash equilibrium, provided all other parameters $\boldsymbol{\theta}_{\neg i}$ are fixed, is denoted as the **solution set** of player $i \in \mathcal{P}$. This solution set is defined by the non-trivial solutions of (5.21). Therefore, one way to characterize these solutions is using the kernel of $\boldsymbol{M}_i$. Its dimension will depend on the number of linearly independent equations generated by the $n\,m_i$ rows of $\boldsymbol{M}_i$ compared to the number of unknown parameters M_i. Since $\operatorname{rank}(\boldsymbol{M}_i) \leq \min(M_i, n\,m_i)$, the number of players, states and controls of each player as well as the assumed properties of the cost function matrices are important for evaluating the existence of inverse differential game solutions.

Proposition 5.1:
The solution set of player i is at least one-dimensional if the number of rows of $\boldsymbol{M}_i$ is strictly less than the number of parameters in $\boldsymbol{\theta}_i$, i.e. $\ker(\boldsymbol{M}_i) \neq \emptyset$ if $n\,m_i < M_i$.

Proof:
The condition $n\,m_i < M_i$ implies $\operatorname{rank}(\boldsymbol{M}_i) < M_i$, leading to $\dim(\ker(\boldsymbol{M}_i)) > 0$. □

Proposition 5.1 gives a sufficient condition for the existence of vectors spanning the kernel of $\boldsymbol{M}_i$. The exact dimension of the kernel is defined by $\operatorname{rank}(\boldsymbol{M}_i)$. The following example illustrates the results of Theorem 5.2 and the solution set concept.

Example 5.2:
Consider an infinite-horizon LQ differential game where two players control a double-integrator system given by

$$\dot{\boldsymbol{x}}(t) = \begin{bmatrix} 0 & 1 \\ 0 & 0 \end{bmatrix} \boldsymbol{x}(t) + \begin{bmatrix} 0 \\ 1 \end{bmatrix} u_1(t) + \begin{bmatrix} 0 \\ 1 \end{bmatrix} u_2(t). \tag{5.28}$$

The cost functions of the two players are given by (5.4) *with* $\boldsymbol{Q}_1 = \operatorname{diag}(1, 2)$ *and* $\boldsymbol{Q}_2 = \operatorname{diag}(1, 0.7)$ *as well as* $R_{11} = 1$, $R_{12} = R_{21} = 0$ *and* $R_{22} = 1$. *The parameter vector of player i is given by*

$$\boldsymbol{\theta}_i = \begin{bmatrix} Q_{i,(1,1)} & Q_{i,(2,2)} & R_{ii} \end{bmatrix}, \quad i \in \{1, 2\}.$$

The game is solved by calculating the solution of the finite-horizon version of the game, i.e. solvin the corresponding RDEs (3.69), *and extracting the converged value of* $\boldsymbol{P}_i$ *afterwards. The resulting* $\boldsymbol{K}^*$ *represents a Nash equilibrium since the calculated* $\boldsymbol{P}_i$ *satisfies* (5.7) *for all players and the closed loop stability of the system dynamics was confirmed (cf. Theorem 5.1). The calculated Nash equilibrium is* $(\boldsymbol{K}_1^*, \boldsymbol{K}_2^*) = \left(\begin{bmatrix} 0.5773 & 1.2827 \end{bmatrix}, \begin{bmatrix} 0.5774 & 0.5882 \end{bmatrix}\right)$.

The kernels of the matrices $\boldsymbol{M}_i \in \mathbb{R}^{2\times 3}$ are defined by the span of the vectors

$$\boldsymbol{v}_1^{(1)} = [v_{1,(j)}^{(1)}]_{j=1,2,3} = \begin{bmatrix}0.4083 & 0.8165 & 0.4083\end{bmatrix}^\top \tag{5.29}$$

$$\boldsymbol{v}_2^{(1)} = [v_{2,(j)}^{(1)}]_{j=1,2,3} = \begin{bmatrix}0.6337 & 0.4437 & 0.6337\end{bmatrix}^\top \tag{5.30}$$

which result in the canonical parameter set

$$\Theta = \{\mu_i \hat{\boldsymbol{Q}}_i, \mu_i \hat{R}_{ii}\}_{i=1,2}, \quad \mu_i \in \mathbb{R}^+, \tag{5.31}$$

which consists of the solution sets of player 1 and 2 and where $\hat{\boldsymbol{Q}}_i = diag(v_{i,1}^{(1)}, v_{i,2}^{(1)})$ and $\hat{R}_{ii} = v_{i,3}^{(1)}$. This means that the cost function parameters are unique up to a constant parameter. In particular, $\mu_1 = 2.4494$ and $\mu_2 = 1.5779$ lead to the defined ground truth parameters.

As mentioned in the introduction of this section, the number of unknown parameters depend on the properties of the matrices, which in turn have an influence on the possible dimensions of each player's solution set for the inverse LQ differential game. This aspect is further examined in the following.

General Cost Function Matrices

In the case of arbitrary cost function matrices $M_i = L = n^2 + \sum_{j\in\mathcal{P}} m_j^2$ holds. Since $nm_i \leq 0.5(n^2 + m_i^2) < n^2 + \sum_{j\in\mathcal{P}} m_j^2$ for any choice of n, m_j, $\forall j \in \mathcal{P}$ and $N \in \mathbb{N}^+$, $\dim(\ker(\boldsymbol{M}_i)) > 0$ follows. The sufficient condition of Proposition 5.1 is fulfilled.

Symmetric Cost Function Matrices

If we assume symmetry of all cost function matrices, then $M_i = 0.5(n^2 + n + \sum_{j\in\mathcal{P}}(m_j^2 + m_j))$. Since

$$nm_i \leq 0.5(n^2 + m_i^2) < 0.5\left(n(n+1) + \sum_{j\in\mathcal{P}} m_j(m_j+1)\right) = M_i$$

for any choice of n, m_j, $\forall j \in \mathcal{P}$, and $N \in \mathbb{N}^+$, $\dim(\ker(\boldsymbol{M}_i)) > 0$ holds. The sufficient condition of Proposition 5.1 is fulfilled and the solution set of player i can be given in terms of the vectors $\boldsymbol{v}_i$ which span the kernel of $\boldsymbol{M}_i$.

Diagonal Cost Function Matrices

Only in the case of diagonal matrices, where $M_i = n + \sum_{j\in\mathcal{P}} m_j$, combinations of n, m_i, N exist such that $n\,m_i \geq M_i$, thus potentially leading to an empty solution set. Here we note that if $\mathrm{rank}(\boldsymbol{M}_i) = M_i - 1$, then the solution set of player i is one-dimensional and a

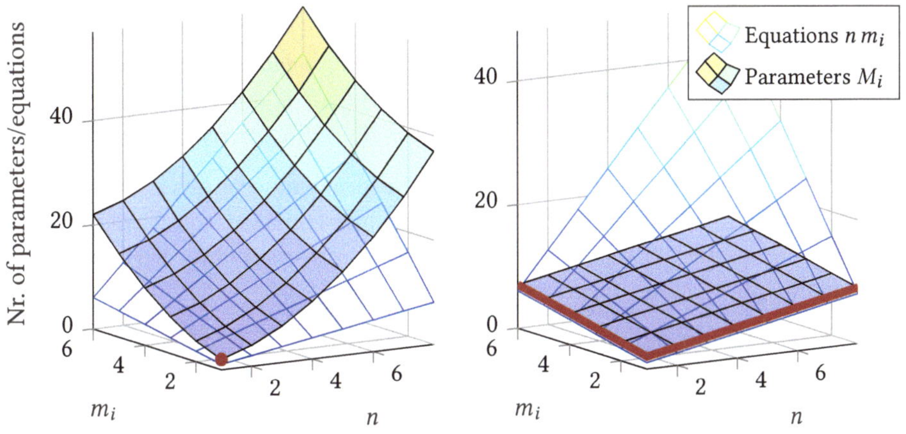

(a) Symmetrical cost function matrices

(b) Diagonal cost function matrices

Figure 5.1: Number of parameters and equations in the ILQDG problem depending on the number of states and controls in a one-player LQ differential game. The red thick line/dot denotes the cases where $n\,m_i = M_i - 1$.

unique algebraic solution for player i's parameters may be found by setting $\boldsymbol{\theta}_{i,(j)} = 1$ for one particular $j \in \{1, \ldots, M_i\}$ and proceeding analogously to [MZ18, Proposition 1], where the special case $N = 1$ is considered. This is possible e.g. if $n = 1$ and $m_1 = 1$ (besides $N = 1$).

The analysis of the sufficient condition for symmetric and diagonal cost function matrices is illustrated in Figure 5.1 for the case $N = 1$. The number of equations (rows of $\boldsymbol{M}_i$) and the number of parameters M_i are shown as a function of the number of states n and the number of controls m_i. In Figure 5.1(a), which depicts the case of symmetrical cost function matrices, the number of parameters M_i is always greater than the number of equations $n\,m_i$ such that the solution set of player 1 is at least one-dimensional. In Figure 5.1(b), which depicts the case of diagonal cost function matrices, we observe that there are combinations of n and m_i which lead to $n\,m_i \geq M_i$, thus not yet allowing for any conclusion concerning the solution set. In turn, the situations where the kernel of $\boldsymbol{M}_i$ is guaranteed to not be empty in this scenario are represented by the red thick line. It denotes the cases where $n\,m_i = M_i - 1 < M_i$ which fulfill the sufficient condition of Proposition 5.1.

These 3D maps are altered if $N > 1$ and in the general case where each player penalizes the other players' controls, i.e. $\boldsymbol{R}_{ij} \neq \boldsymbol{0}$ for $i \neq j$, $i \in \mathcal{P}$. The cases $N = 2$ and $N = 3$ are shown in the Appendix C for further illustration of how the properties of $\boldsymbol{M}_i$ are affected by the number of players, states and controls.

Remark 5.4:

The previous analysis shows the implications of $n\,m_i < M_i$ as a sufficient condition for the existence of a solution set for player i which is at least one-dimensional. The case $n\,m_i \geq M_i$

demands further attention, given that it potentially leads to an empty kernel of $\boldsymbol{M}_i$ *— this occurs if* $\operatorname{rank}(\boldsymbol{M}_i) \geq M_i$. *Nevertheless, this does not imply that a solution of the inverse differential game problem for player i does not exist. Indeed, the existence of a Nash equilibrium described by* $\boldsymbol{K}^*$ *implies the existence of at least one N-tuple* $\boldsymbol{\theta} = \boldsymbol{\theta}^*$ *which generated the equilibrium.*

In light of Remark 5.4, the next section presents a formulation of inverse LQ differential games which allows to find a solution of the inverse differential game problem regardless of the presented properties. In addition, it facilitates the derivation of further general results concerning the solution sets of each player.

5.4 Quadratic Programming Formulation for Inverse Linear-Quadratic Differential Games

The approach is based on the formulation of a residual function, analogously to Definition 4.4, which denotes the extent to which the necessary and sufficient conditions for Nash equilibria are violated. Since the conditions are represented by the coupled ARE (5.7) and its reformulation (5.21), where the matrix $\boldsymbol{M}_i$ depends on the given matrices $\boldsymbol{A}$, $\boldsymbol{B}_i$, $i \in \mathcal{P}$ and $\boldsymbol{K}^* = (\boldsymbol{K}_1^*, \ldots, \boldsymbol{K}_N^*)$, the following residual is introduced.

Definition 5.4 (Residual)

Let a function $\boldsymbol{r}_i : \mathbb{R}^{M_i} \mapsto \mathbb{R}^{n\, m_i}$, $i \in \mathcal{P}$, *be defined as*

$$\boldsymbol{r}_i(\boldsymbol{\theta}_i) = \boldsymbol{M}_i \boldsymbol{\theta}_i. \tag{5.32}$$

The function $\boldsymbol{r}_i$ *is called* ***residual*** *of the coupled ARE* (5.7).

The violation of the coupled ARE in terms of the residual function occurs if the parameters $\boldsymbol{\theta}_i$ do not represent a Nash equilibrium for given feedback control laws $\boldsymbol{K}_i^*$ and system dynamics matrices $\boldsymbol{A}$ and $\boldsymbol{B}_i$, $i \in \mathcal{P}$. While it would be possible to pose an optimization problem such that $||\boldsymbol{r}_i||$ is minimized, it is computationally more convenient to consider a quadratic residual function. The following lemma relates the quadratic residual function to the AREs.

Lemma 5.3

Let a LQ differential game be given by (5.1) *and* (5.4). *Furthermore, let Assumption 5.1 hold. The ARE* (5.7) *is fulfilled if and only if* $||\boldsymbol{M}_i \boldsymbol{\theta}_i||^2 = 0$.

Proof:
The proof is trivial given that the norm of a vector is zero if and only if the vector itself is a zero-vector. □

In light of Lemma 5.3, the optimization problem

$$\begin{aligned} \min_{\boldsymbol{\theta}_i} ||\boldsymbol{r}_i(\boldsymbol{\theta}_i)||_2^2 &= \min_{\boldsymbol{\theta}_i} \frac{1}{2}\boldsymbol{\theta}_i^\top \boldsymbol{H}_i \boldsymbol{\theta}_i, \\ &\text{s.t.} \\ &\theta_{i,(j)} > 0, \quad \forall j \in \{1, \dots, M_i\}, \\ &\boldsymbol{R}_{ii} \succ \boldsymbol{0} \end{aligned} \tag{5.33}$$

is posed, where $\boldsymbol{H}_i = 2(\boldsymbol{M}_i^\top \boldsymbol{M}_i) \in \mathbb{R}^{M_i \times M_i}$. Analogously to the residual-based approach in Section 4.3.1, the aim of the optimization problem (5.33) is to minimize the quadratic residual to obtain parameters $\boldsymbol{\theta}_i$ which fulfill the ARE.

Remark 5.5:
The constraints $\theta_{i,(j)} > 0$, $\forall j \in \{1, \dots, M_i\}$, in (5.33) are introduced in order to avoid trivial solutions. Literature in inverse optimal control and inverse games often introduce the constraint $\theta_{i,(j)} = 1$ for any $j \in \{1, \dots, M_i\}$ (see note 25 in page 51). Analogous results concerning the properties of (5.33) can easily be proved with this (additional) constraint. Also note that, in case of diagonal cost function matrices, $\theta_{i,(j)} > 0$, $\forall j \in \{1, \dots, M_i\}$, ensures $\boldsymbol{R}_{ii} \succ \boldsymbol{0}$.

5.4.1 Necessary and Sufficient Conditions for One-Dimensional Solution Sets

In the following, the quadratic program (5.33) is leveraged to obtain insights on inverse LQ differential games. The properties of the quadratic program (5.33) differ considerably depending on whether $\text{rank}(\boldsymbol{M}_i)$ is less, equal or greater than the number of parameters M_i. By considering the case $n\,m_i < M_i$, which leads to $\text{rank}(\boldsymbol{M}_i) < M_i$, the following proposition can be stated.

Proposition 5.2:
Let a LQ differential game be given by (5.1) and (5.4) such that $n\,m_i < M_i$. Then, the quadratic program (5.33) is convex and a solution is guaranteed to exist.

Proof:
It is clear that both the constraint set defined by $\theta_{i,(j)} > 0$, $\forall j \in \{1, ..., M_i\}$, and $\boldsymbol{R}_{ii} \succ \boldsymbol{0}$ are convex and therefore, their intersection is also convex. Under the conditions $n\,m_i < M_i$ we obtain $\text{rank}(\boldsymbol{H}_i) = \text{rank}(\boldsymbol{M}_i^\top \boldsymbol{M}_i) \leq \min(n\,m_i, M_i) = n\,m_i < M_i$, leading to a convex—since $\boldsymbol{M}_i^\top \boldsymbol{M}_i \succeq \boldsymbol{0}$—but not strictly convex objective function. Hence, the quadratic program is convex and therefore always has a solution. □

The results of Proposition 5.2 are not surprising for the case where Assumption 5.1 holds, since this guarantees that at least one solution for the parameters $\boldsymbol{\theta}_i$ of a particular player $i \in \mathcal{P}$ (and the ones generated by a multiplying positive constant) must exist. Note that solving the optimization problem (5.33) leads to one of the solutions belonging to $\ker(\boldsymbol{M}_i)$ (cf. Proposition 5.1 and Theorem 5.2), but it does not give any information on the dimensions of each player's solution set.

The following theorem is stated as the main result regarding the canonical parameter set of inverse LQ differential games.

Theorem 5.3 (Necessary and Sufficient Conditions for Uniqueness up to a Positive Factor)
Let a LQ differential game be given by (5.1) *and* (5.4). *Furthermore, let Assumption 5.1 hold. The inverse LQ differential game has a canonical parameter set of the form*

$$\Theta = \{c_i \boldsymbol{\theta}_i;\ c_i > 0,\ i \in \mathcal{P}\}, \tag{5.34}$$

if and only if $n\,m_i \geq M_i - 1$ *and additionally* $\text{rank}(\boldsymbol{M}_i) = M_i - 1$.

Proof:
We first state that $n\,m_i \geq M_i - 1$ is a necessary condition for unique solutions since $n\,m_i < M_i - 1$ leads to a solution set of a dimension greater than 1 (cf. Proposition 5.1). By the results of Lemma 5.3, (5.7) is fulfilled if and only if $||\boldsymbol{M}_i \boldsymbol{\theta}_i||^2 = 0$. We therefore proceed to analyze the quadratic program (5.33). Under the theorem condition $\text{rank}(\boldsymbol{M}_i) = M_i - 1$ we have $\dim(\ker(\boldsymbol{M}_i)) = 1$ which implies a one-dimensional solution set for each player $i \in \mathcal{P}$ of the form (5.34).

The case $\text{rank}(\boldsymbol{M}_i) < M_i - 1$ leads to solution sets with a dimension greater that 1 and is therefore excluded. Therefore, only the case $\text{rank}(\boldsymbol{M}_i) = M_i$ remains which we analyze using (5.33). If $\text{rank}(\boldsymbol{M}_i) = M_i$, which is only possible if $n\,m \geq q$, then we obtain $\boldsymbol{H}_i \succ \boldsymbol{0}$ and thus (5.33) is strictly convex. Strict convexity leads to a unique solution of (5.33) and therefore to a unique solution of the ARE (5.7). But the latter contradicts Corollary 5.1, from where we conclude that $\text{rank}(\boldsymbol{M}_i) = M_i - 1$ is also necessary. □

Theorem 5.3 gives necessary and sufficient conditions for the solution set of each player i to be one-dimensional, i.e. each player's parameters $\boldsymbol{\theta}_i$ are unique up to a real positive factor c_i.

Summarizing the results of this subsection, if the canonical parameter set has the form (5.34), then a particular $\boldsymbol{\theta}_i$ belonging to the corresponding solution set each player i can be computed by means of the quadratic program (5.33). If the conditions of Theorem 5.3 are not fulfilled, then with the results of Proposition 5.2, (5.33) yields any solution from the canonical parameter set with non-unique parameters for each player $i \in \mathcal{P}$.

5.4.2 Identification of Feedback Matrices

The optimization problem (5.33) always yields a solution which is associated with a given Nash equilibrium represented by $\boldsymbol{K}^*$. If only observed Nash equilibrium control and state trajectories are available, then it becomes necessary to estimate the control laws $\boldsymbol{K}^*$. For the N-player inverse differential game at hand, a least-squares identification based on (5.2) is proposed. For this purpose, let us introduce a finite sequence of sampling times

$$\mathcal{T}_i := \{t_k \in [0,T] : 1 \leq k \leq K_i \wedge 0 \leq t_1 < \ldots < t_{K_i} \leq T\} \tag{5.35}$$

for each player $i \in \mathcal{P}$, where $[0,T]$ is the time interval for which $\boldsymbol{x}^*(t)$ and $\boldsymbol{u}_i^*(t)$ are available. Let the value of the state and control trajectories at t_k be denoted by $\boldsymbol{x}^{[k]}$ and $\boldsymbol{u}_i^{[k]}$, respectively. Then, the feedback matrix can be estimated by means of

$$\hat{\boldsymbol{K}}_i = \arg\min_{\boldsymbol{K}_i} \sum_{k=1}^{K_i} ||\boldsymbol{K}_i \boldsymbol{x}^{[k]} + \boldsymbol{u}_i^{[k]}||^2, \tag{5.36}$$

where $||\cdot||$ denotes the Euclidean norm. Least-square estimation theory states that the parameters (in this case the entries of $\boldsymbol{K}_i$) can be recovered if persistence of excitation (PE) conditions are fulfilled [ÅW95, Section 2.4]. These conditions demand that the trajectories of $\boldsymbol{x}$ and $\boldsymbol{u}_i$ are "informative" enough and are e.g. not identical to zero. Furthermore, if the least-square estimation is considered from a stochastic point of view, i.e.

$$\boldsymbol{u}_i^{[k]} = -\boldsymbol{K}_i \boldsymbol{x}^{[k]} + \boldsymbol{\epsilon}_i, \tag{5.37}$$

where $\boldsymbol{\epsilon}_i \in \mathbb{R}^{m_i}$ denotes a vector of zero-mean Gaussian white noise, then the estimation is biasfree if $\boldsymbol{\epsilon}_i(t)$ is independent of the state $\boldsymbol{x}(t)$ [ÅW95, P. 47]. The conditions for a bias-free estimation are usually not given. For example, the state $\boldsymbol{x}(t)$ depends on the controls $\boldsymbol{u}_i(t)$ due to the system dynamics and is therefore not independent of the additive gaussian noise. Nevertheless, the LS estimation works well in practice, as shown later in Chapter 7.

5.4.3 Algorithm and Example

The inverse LQ differential game method for determining a particular solution parameter vector $\boldsymbol{\theta}_i$ of player i based on (5.33) can be implemented with the following algorithm.

Algorithm 2 IOC based method for player i in an inverse feedback LQ differential game.

Input: State and control trajectories $\boldsymbol{x}(t)$ and $(\boldsymbol{u}_1(t), \ldots, \boldsymbol{u}_N(t))$, system matrix $\boldsymbol{A}$, input matrices $\boldsymbol{B}_i$, $\forall i \in \mathcal{P}$.

Output: Computed player i cost function parameters $\boldsymbol{\theta}_i$.

1: Estimate $\boldsymbol{K}_i^*$ for all $i \in \mathcal{P}$ with (5.36) and determine the corresponding closed-loop system matrix $\boldsymbol{F}$ with (3.67).
2: Compute matrices $\boldsymbol{Z}_i$ with (5.10) and $\boldsymbol{K}_i^{\otimes}$ with (5.11).
3: Compute matrix $\boldsymbol{M}_i$ with (5.14).
4: Solve the quadratic optimization problem (5.33).

Note that, similar to the methods presented in Chapter 4, Algorithm 2 may be used for cost function parameter identification of any player $i \in \mathcal{P}$ in an N-player infinite-horizon LQ differential game. Furthermore, the method may also be applied for the special case of a single player, i.e. an inverse LQ optimal control problem. The core of the presented approach is the quadratic program which can be solved very efficiently with state-of-the-art methods, e.g. active-set and interior point methods [NW06, Chapter 16].

This section ends by the presentation of an example to illustrate Theorem 5.3 and the use of Algorithm 2 to identify cost function parameters in an inverse LQ differential game.

Example 5.3:

Consider an infinite-horizon LQ differential game where 2 players control a stabilizable linear system defined by the differential equation

$$\dot{\boldsymbol{x}}(t) = \begin{bmatrix} 1 & -1 \\ 1 & 0 \end{bmatrix} \boldsymbol{x}(t) + \begin{bmatrix} 1 & 0 \\ 0 & 1 \end{bmatrix} \boldsymbol{u}_1(t) + \begin{bmatrix} 1 & 0 \\ 0 & 1 \end{bmatrix} \boldsymbol{u}_2(t) \tag{5.38}$$

and select their feedback strategies according to a cost function of the form (5.4) *with cost function matrices*

$$\begin{aligned} \boldsymbol{Q}_1 &= \begin{bmatrix} 1 & 0 \\ 0 & 1 \end{bmatrix}, & \boldsymbol{Q}_2 &= \begin{bmatrix} 1 & 0 \\ 0 & 10 \end{bmatrix}, \\ \boldsymbol{R}_{11} &= \begin{bmatrix} 1 & 0 \\ 0 & 1 \end{bmatrix}, & \boldsymbol{R}_{22} &= \begin{bmatrix} 2 & 0 \\ 0 & 1 \end{bmatrix}, \\ \boldsymbol{R}_{12} &= \boldsymbol{0}, & \boldsymbol{R}_{21} &= \boldsymbol{0}. \end{aligned} \tag{5.39}$$

The vectorization of the cost function matrices according to (5.12) *leads to a parameter vector of dimension* $M_i = 4$ *given by*

$$\boldsymbol{\theta}_i = \begin{bmatrix} Q_{i,(1,1)} & Q_{i,(2,2)} & R_{ii,(1,1)} & R_{ii,(2,2)} \end{bmatrix}, \quad i \in \{1,2\},$$

where $Q_{i,(j,j)}$ *and* $R_{ii,(j,j)}$ *denote the* j*-th diagonal entry of* $\boldsymbol{Q}_i$ *and* $\boldsymbol{R}_{ii}$*, respectively. Analogously to the last example, the infinite-horizon LQ differential game was solved by calculating the solution of the corresponding RDEs* (3.69) *and extracting the converged value of* $\boldsymbol{P}_i$*. The resulting state and control trajectories* $\boldsymbol{x}^*(t)$ *and* $\boldsymbol{u}_i^*(t)$ *were confirmed to correspond to a stable system and hence, to a Nash equilibrium.*

In this example, the inverse method is given the resulting state and control trajectories $\boldsymbol{x}^*(t)$ *and* $\boldsymbol{u}_i^*(t)$ *instead of the Nash equilibrium feedback matrices* $\boldsymbol{K}^*$*. Following Algorithm 2, these trajectories were used to estimate the feedback matrices with the LS approach given in* (5.36)*, where a set* $\mathcal{T}_i$ *with* $T = 10$ *and* $K_i = 501$ *was selected according to* (5.35)*. The Nash equilibrium can be exactly estimated with deviations* $||\hat{\boldsymbol{K}}_i - \boldsymbol{K}_i^*|| < 10^{-14}$ *for all* $i = \{1,2\}$*. With* $\boldsymbol{K}^* = (\boldsymbol{K}_1^*, \boldsymbol{K}_2^*)$*, we obtain the matrices*

$$\boldsymbol{M}_1 = \begin{bmatrix} -0.436 & -0.026 & 0.466 & -0.004 \\ 0.100 & -0.027 & 0.053 & -0.126 \\ 0.100 & -0.027 & -0.078 & 0.006 \\ -0.032 & -0.153 & -0.020 & 0.204 \end{bmatrix} \tag{5.40}$$

and

$$\boldsymbol{M}_2 = \begin{bmatrix} -0.436 & -0.026 & 0.530 & -0.365 \\ 0.100 & -0.027 & 0.144 & -0.114 \\ 0.100 & -0.027 & 0.264 & -0.353 \\ -0.032 & -0.153 & -0.048 & 1.655 \end{bmatrix}. \tag{5.41}$$

We find that $\operatorname{rank}(\boldsymbol{M}_i) = M_i$ *holds for* $i = \{1,2\}$*, which indicates a one-dimensional solution set for each player* i *according to Theorem 5.3. By solving the quadratic program* (5.33) *we obtain the parameters*

$$\begin{aligned} \hat{\boldsymbol{\theta}}_1 &= \begin{bmatrix} 1.000 & 1.000 & 1.000 & 1.000 \end{bmatrix} \\ \hat{\boldsymbol{\theta}}_2 &= \begin{bmatrix} 0.602 & 6.024 & 1.204 & 0.602 \end{bmatrix}. \end{aligned} \tag{5.42}$$

The parameters $\boldsymbol{\theta}_1^*$ *were exactly identified, while for the second player, the parameters are equal up to a multiplying constant. In particular, we have* $\hat{\boldsymbol{\theta}}_2 = 0.6024\, \boldsymbol{\theta}_2^*$*.*

5.5 Method Limitations

Prior to this chapter's conclusion, possible limitations of the method are discussed. The first issue is given, similar to last chapter, if e.g. only noise-corrupted measurements of the state and control trajectories are available. Nevertheless, since the method relies on the feedback control laws and these are estimated by the LS method, it can be conjectured that the method has a considerable robustness to noise in the trajectories. This case shall be further examined in Section 7.5. In addition, truncated trajectories do not represent a problem as long as these fulfill the PE condition mentioned in Section 5.4.2. Informative trajectories can potentially fulfill this condition even with a small number of values.

A further issue arises if an $i \in \mathcal{P}$ exists such that $\boldsymbol{K}_i$ does not constitute a Nash equilibrium feedback law with respect to any set of cost function matrices $\boldsymbol{Q}_i$, $\boldsymbol{R}_{ii}$, and $\boldsymbol{R}_{ij}$ of the assumed structure, e.g. symmetric. More generally, $\boldsymbol{K}_i$ might not be a Nash equilibrium for any set of cost function matrices, regardless of their structure. This can occur e.g. if $\boldsymbol{K}_i$ is identified from trajectories $\boldsymbol{x}(t)$ and $\boldsymbol{u}_i(t)$ which do not represent a Nash equilibrium. However, by the results of Proposition 5.2, the existence of a solution to the quadratic program (5.33) is guaranteed, independently of the Nash character of the control laws. Since the presented quadratic programming approach is based on the coupled ARE which are necessary and sufficient conditions for feedback Nash equilibria, the identification results yield parameters which lead to the Nash equilibrium feedback law which is the closest to the original observed feedback law. The distance is measured in terms of the violation of the coupled AREs (cf. the discussion of the experimental results in Section 8.8). However, this distance may not be proportional to or correlate with the error between observed and identified trajectories.

5.6 Conclusion

In this chapter, the inverse problem of infinite-horizon LQ differential games was considered, where a feedback Nash equilibrium is given and cost function parameters are sought which explain this resulting equilibrium. The parameters correspond to the elements of the matrices of the quadratic cost functions of the players and the Nash equilibrium is assumed to be given in the form of an N-tuple of player feedback matrices. The solution of the inverse LQ differential game was given in the form of an explicit set—the *canonical parameter set*—which describes all possible cost function parameter vectors or matrices which lead to the same Nash equilibrium, and was achieved by a reformulation of the necessary and sufficient conditions for Nash equilibria. Importantly, sufficient conditions for the possibility of stating such explicit sets were given. In addition, these results were applied to formulate a quadratic program which allows an efficient computation of the cost function parameters. Moreover, the analysis of the resulting quadratic program allowed for the statement of necessary and sufficient conditions for the uniqueness of the solution set of a particular player up to a multiplying positive constant. Finally, it was demonstrated that the feedback matrices of all players can be estimated out of Nash equilibrium state and control trajectories by using a least-squares approach. Consequently, all of the results developed in this chapter can be applied if, instead of the player feedback matrices, observations of Nash equilibrium state and control trajectories are available.

The results of this chapter represent solutions related to one of the questions Kalman stated: *"What optimization problems lead to a constant, linear control law?"* (Problem A in [Kal64]). This problem was recently considered in [MZ18] for single-player infinite-horizon problems; these results have been generalized for N-player differential games in this chapter.

6 Inverse Dynamic Games Based on Inverse Reinforcement Learning

This chapter presents inverse dynamic game solutions such that cost functions which explain observed behavior of several players can be found. The methods presented in this chapter are based on inverse reinforcement learning techniques and on a discrete-time formulation of the infinite dynamic game. Therefore, the methods in this chapter represent an alternative approach to the IOC-based methods of the previous two chapters. Nevertheless, there is a similarity to the results of these aforementioned chapters, namely the development of an inverse dynamic game method which does not rely on a repeated solution of the forward problem, i.e. the repeated computation of Nash equilibrium state and control trajectories. After a short introduction to the principle of maximum entropy, which represents the basis of the methods, the main contribution of this chapter is shown, namely the derivation of a probabilistic method for inverse dynamic games based on Maximum Entropy Inverse Reinforcement Learning (MaxEnt IRL). The cases where the players' behavior corresponds to an open-loop and a feedback Nash equilibrium are considered. In addition, results on the unbiasedness of the estimation of cost function parameters are presented. After providing further details which are important for the practical implementation of these methods, examples for the solution of inverse linear-quadratic dynamic games are given. The chapter ends with conclusions on all presented results.[33]

6.1 Introduction to the Probabilistic Approach and Maximum Entropy

In this thesis, the aim is the development of IRL methods for inverse dynamic games which allow for continuous-valued control and state spaces, such that comparable methods to the ones presented in Chapters 4 and 5 based on inverse optimal control can be obtained. The inverse dynamic game problem is regarded in this chapter from a probabilistic perspective which is introduced in the following.

[33] Preliminary versions of the results of this chapter have been published in the conference paper [KIR+17]. The chapter's contents are based on the article [IBKH20].

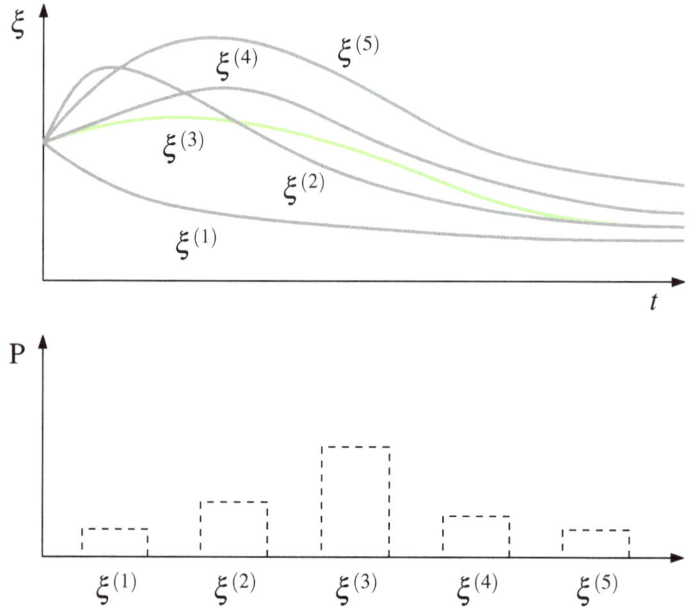

Figure 6.1: Example of a probability function for trajectories

For a simplified presentation, consider the results of a dynamic game as a single trajectory $\tilde{\xi}(t)$ which is assumed to stem from a probability function $\mathrm{P}(\xi)$ defined over a finite and discrete set of (in this case five) possible trajectories $\xi(t)$. This scenario is illustrated in Figure 6.1, where the observed trajectory $\tilde{\xi}(t) = \xi^{(3)}$ is colored green. In this example, a probability value is assigned to each of the five possible trajectories. Transferring this line of thought to an inverse problem in dynamic games leads to the fact that one or several trajectories ξ are observed, but their probabilities are unknown. The choice of a probability function which explains these observed trajectory is not unique, even if some constraints are introduced. This problem becomes even more complex if the trajectories originate from a probability density function $\mathrm{p}(\xi)$ instead of the previously presented probability mass function $\mathrm{P}(\xi)$ since this implies a potentially infinite number of possible trajectories. In order to resolve the ambiguity in this kind of problem, the principle of maximum entropy can be applied. This was introduced by Jaynes in [Jay57] as a means to infer probability distributions which are consistent with experimental data.[34] According to Jaynes, this method leads to the "least biased estimate possible on the given information". This is illustrated e.g. by the fact that the distribution which maximizes the entropy with the constraints of fixed and known expectation and variance is the Gaussian distribution. Similarly, the maximum entropy distribution where no constraints are introduced is the uniform distribution [CT06, Section 12.2].

[34] Jaynes' objective was to present a potential application of information theory results—obtained by Shannon ([Sha48])— to the field of statistical mechanics. The interested reader is also referred to [PGLD13] for a historical review.

This introduced probabilistic perspective of dynamic games constitutes the basis of the definition of the problem. Likewise, the principle of maximum entropy shall be leveraged for the development of inverse dynamic game solutions presented in the next sections.

6.2 Problem Definition

Consider an infinite dynamic game in discrete time[35], where N players simultaneously control a system with (potentially time-variant) dynamics of the form (see also Definition A.1)

$$\boldsymbol{x}^{(k+1)} = \boldsymbol{f}_D^{(k)}\left(\boldsymbol{x}^{(k)}, \boldsymbol{u}_1^{(k)}, \ldots, \boldsymbol{u}_N^{(k)}\right) \tag{6.1a}$$

$$\boldsymbol{x}^{(1)} = \boldsymbol{x}_1. \tag{6.1b}$$

The goal of each player $i \in \mathcal{P}$ is to minimize its individual cost function by applying a control strategy. The cost functions' structure is assumed to be defined by a linear combination of $M_i \in \mathbb{N}$ known features[36], i.e.

$$J_i = -\sum_{k=1}^{k_E} \boldsymbol{\theta}_i^\top \boldsymbol{\phi}_i\left(\boldsymbol{x}^{(k)}, \boldsymbol{u}_1^{(k)}, \ldots, \boldsymbol{u}_N^{(k)}\right), \tag{6.2}$$

where $k_E \in \mathbb{N}_{>0}$, $\boldsymbol{\phi}_i \in \mathbb{R}^{M_i}$ contains all features of player i defined analogously to Definition 4.2 and $\boldsymbol{\theta}_i \in \mathbb{R}^{M_i}$ represents the vector of player i's individual feature weights, i.e. the cost function parameters.

A main element of inverse problems in optimal control and dynamic games are the observed state and control trajectories. Generally speaking, a trajectory consists of a sequence of values according to the discrete-time formulation of the game. Therefore, the following definition is introduced.

---

[35] The discrete-time formulation is chosen following the line of a vast number of previous studies on single-player IRL (cf. Section 2.1.3). The results of this chapter are based on definitions analogous to the ones in Chapter 3. These discrete-time dynamic game definitions are given in Appendix A.

[36] In this chapter, the term *features* is used instead of *basis functions* in order to be consistent to IRL literature. Furthermore, in the following it is assumed that the feature functions in $\boldsymbol{\phi}_i$ are independent of k. Their corresponding values are still stage-dependent through the values of the states and the controls. In addition, note that the cost function has been multiplied with a factor of -1. This is done in order to be congruent with IRL literature which assumes a reward function to be maximized instead of a cost function to be minimized.

Definition 6.1 (Stacked State and Control Values)

Let

$$\underline{\boldsymbol{x}} = \left[\left(\boldsymbol{x}^{(1)}\right)^\top \quad \dots \quad \left(\boldsymbol{x}^{(k_E)}\right)^\top\right]^\top \in \mathbb{R}^{nk_E}, \tag{6.3}$$

$$\underline{\boldsymbol{u}}_i = \left[\left(\boldsymbol{u}_i^{(1)}\right)^\top \quad \dots \quad \left(\boldsymbol{u}_i^{(k_E)}\right)^\top\right]^\top \in \mathbb{R}^{m_i k_E}, \tag{6.4}$$

$\forall\, i \in \mathcal{P}$, *be vectors containing all values of the system state* $\boldsymbol{x}^{(k)}$ *and the control values* $\boldsymbol{u}_i^{(k)}$ *of player* $i \in \mathcal{P}$ *for all time steps* $k \in \mathcal{K}$*, respectively.*

Furthermore, the following notation is introduced for a set of trajectories in accordance with the system dynamics (6.1) which will facilitate a more compact representation of the results of this chapter.

Definition 6.2 (Trajectory Set)

A trajectory $\zeta := \{\underline{\boldsymbol{x}}, \underline{\boldsymbol{u}}_1, \dots, \underline{\boldsymbol{u}}_N\}$ *is defined as a set containing the stacked values of the system state* $\underline{\boldsymbol{x}}$ *and the stacked control values* $\underline{\boldsymbol{u}}_i$ *of all players* $i \in \mathcal{P}$*, which is feasible with respect to the system dynamics given by* (6.1).

The estimation of the cost function parameters $\boldsymbol{\theta}_i$ is based on an observed set of trajectories denoted by $\tilde{\zeta} := \{\tilde{\underline{\boldsymbol{x}}}, \tilde{\underline{\boldsymbol{u}}}_1, \dots, \tilde{\underline{\boldsymbol{u}}}_N\}$ which, following the probabilistic approach presented in Section 6.1, is assumed to be sampled by a probability density function $\mathrm{p}\left(\zeta \,|\, \boldsymbol{\theta}_1^*, \dots, \boldsymbol{\theta}_N^*\right)$ with unknown parameters $\boldsymbol{\theta}_1^*, \dots, \boldsymbol{\theta}_N^*$.

A further key value in IRL problems is the feature count (cf. [AN04, RBZ06, ZMBD08] in the single-player case) which is introduced in the following.

Definition 6.3 (Feature Count)

The feature count $\boldsymbol{\mu}_i(\zeta) \in \mathbb{R}^{M_i}$ *of a player* $i \in \mathcal{P}$ *along a trajectory* ζ *is defined as a vector containing the accumulated values of the features along that trajectory, i.e.*

$$\boldsymbol{\mu}_i(\zeta) = \sum_{k=1}^{k_E} \boldsymbol{\phi}_i\left(\boldsymbol{x}^{(k)}, \boldsymbol{u}_1^{(k)}, \dots, \boldsymbol{u}_N^{(k)}\right), \tag{6.5}$$

with $\boldsymbol{x}^{(k)}, \boldsymbol{u}_i^{(k)} \in \zeta$, $\forall\, i \in \mathcal{P}, k \in \mathcal{K}$.

Using the feature counts $\boldsymbol{\mu}_i(\zeta)$ and (6.2), the costs along a trajectory ζ for any player $i \in \mathcal{P}$ can be rewritten as

$$J_i(\zeta, \boldsymbol{\theta}_i) = -\boldsymbol{\theta}_i^\top \boldsymbol{\mu}_i(\zeta). \tag{6.6}$$

In the following and with some abuse of notation in favor of better readability, $\mathrm{p}(\zeta|\,\boldsymbol{\theta}_{1:N})$ represents the probability density of a trajectory ζ as a function of parameters $\boldsymbol{\theta}_1, \ldots, \boldsymbol{\theta}_N$ corresponding to the cost functions J_i, $\forall i \in \mathcal{P}$.

Having introduced these basic definitions, the inverse dynamic game problem considered in this chapter is defined as follows.

Definition 6.4 (Inverse Dynamic Game Based on IRL)
Find parameters $\hat{\boldsymbol{\theta}}_i$, $\forall\, i \in \mathcal{P}$, such that the expected costs of a trajectory sampled from the probability density $\mathrm{p}\left(\zeta|\,\hat{\boldsymbol{\theta}}_{1:N}\right)$ resulting from the identified parameters corresponds for each player $i \in \mathcal{P}$ to the expected costs of the observed trajectory sampled from the probability density $\mathrm{p}\left(\zeta|\,\boldsymbol{\theta}^*_{1:N}\right)$, i.e.

$$\mathbb{E}_{\mathrm{p}\left(\zeta|\hat{\boldsymbol{\theta}}_{1:N}\right)}\left\{J_i\left(\zeta, \boldsymbol{\theta}_i^*\right)\right\} \stackrel{!}{=} \mathbb{E}_{\mathrm{p}\left(\zeta|\boldsymbol{\theta}^*_{1:N}\right)}\left\{J_i\left(\zeta, \boldsymbol{\theta}_i^*\right)\right\}, \ \forall i \in \mathcal{P}. \tag{6.7}$$

The requirement (6.7) arises from the demand of obtaining for each player a cost function that results in an individual performance as good as the observed one, where the performance is measured with respect to each player's unknown true cost function $J_i\left(\zeta, \boldsymbol{\theta}_i^*\right)$.[37] Similar to the inverse differential game problem of Definition 4.3, Definition 6.4 implies that we are interested in finding one parameter vector $\boldsymbol{\theta}_i$ for each player $i \in \mathcal{P}$ such that (6.7) holds, i.e. the dynamic game with identified cost function parameters is able to explain the observed trajectories. This differs to the problem investigated in Section 5.2 where the complete solution set for each player $i \in \mathcal{P}$ is sought, since inverse problems in optimal control and dynamic games are naturally ill-posed.

6.3 Maximum Entropy Distribution of Trajectories in Dynamic Games

The principle of maximum entropy provides a means to resolve the ill-posedness issue such that parameters can be found which solve the problem given in Definition 6.4. In this section, we transfer the maximum entropy approach to inverse dynamic games with N players. The aim is to find a probability density function $\mathrm{p}(\zeta|\,\boldsymbol{\theta}_{1:N})$ which represents the probability of trajectories ζ as a function of the parameters $\boldsymbol{\theta}_1, \ldots, \boldsymbol{\theta}_N$, yet considering (6.7) as only constraint or a-priori knowledge. Finding an expression for $\mathrm{p}(\zeta|\,\boldsymbol{\theta}_{1:N})$ shall provide a useful result on our way towards the solution of inverse dynamic games with IRL.

[37] Similar objectives have been frequently defined in single-player IRL methods, see e.g. the seminal papers [NR00] and [AN04].

In order to state a relationship between observed trajectories $\tilde{\zeta}$ and the probability distribution $\mathrm{p}\left(\zeta \mid \boldsymbol{\theta}^*_{1:N}\right)$ which generated them, the following assumption is made:

Assumption 6.1

The feature count of player i along the trajectory $\tilde{\zeta}$ (denoted as $\tilde{\boldsymbol{\mu}}_i$ for all players $i \in \mathcal{P}$) represents the expectation of the feature count $\mathbb{E}_{\mathrm{p}(\zeta \mid \boldsymbol{\theta}^*_{1:N})}\left\{\boldsymbol{\mu}_i(\zeta)\right\}$ based on the probability density function $\mathrm{p}\left(\zeta \mid \boldsymbol{\theta}^*_{1:N}\right)$ which results from the parameters $\boldsymbol{\theta}^*_1, \ldots, \boldsymbol{\theta}^*_N$, i.e.

$$\mathbb{E}_{\mathrm{p}(\zeta \mid \boldsymbol{\theta}^*_{1:N})}\left\{\boldsymbol{\mu}_i(\zeta)\right\} = \tilde{\boldsymbol{\mu}}_i, \quad \forall i \in \mathcal{P}. \tag{6.8}$$

Assumption 6.1 means that each observation $\tilde{\zeta}_l$ is representative[38]. As no further information is available, the sample mean is used as an estimate for the expectation of the feature count. Furthermore, note that Assumption 6.1 implies that if $n_t \in \mathbb{N}_{>0}$ observed trajectories are given, i.e. a set of trajectories $\mathcal{D} = \{\tilde{\zeta}_1, \ldots, \tilde{\zeta}_{n_t}\}$, the expectation of the feature count of player i is given by

$$\mathbb{E}_{\mathrm{p}(\zeta \mid \boldsymbol{\theta}^*_{1:N})}\left\{\boldsymbol{\mu}_i(\zeta)\right\} = \frac{1}{n_t}\sum_{l=1}^{n_t} \boldsymbol{\mu}_i\left(\tilde{\zeta}_l\right), \tag{6.9}$$

where $\boldsymbol{\mu}_i\left(\tilde{\zeta}_l\right)$ denotes the feature count of the observed trajectory $\tilde{\zeta}_l$ with $l \in \{1, \ldots, n_t\}$.

Lemma 6.1 (Path Feature Count Equivalence to Costs)

Let the expectation of the feature count be equal for both the probability density $\mathrm{p}\left(\zeta \mid \hat{\boldsymbol{\theta}}_{1:N}\right)$ resulting from the identified parameters and the probability function $\mathrm{p}\left(\zeta \mid \boldsymbol{\theta}^*_{1:N}\right)$ with original parameters $\boldsymbol{\theta}^*_1, \ldots, \boldsymbol{\theta}^*_N$, i.e.

$$\mathbb{E}_{\mathrm{p}\left(\zeta \mid \hat{\boldsymbol{\theta}}_{1:N}\right)}\left\{\boldsymbol{\mu}_i(\zeta)\right\} = \mathbb{E}_{\mathrm{p}(\zeta \mid \boldsymbol{\theta}^*_{1:N})}\left\{\boldsymbol{\mu}_i(\zeta)\right\} \tag{6.10}$$

for each player $i \in \mathcal{P}$. Then, for any parameters where $\left\|\boldsymbol{\theta}^*_i\right\|_2 < \infty$, (6.7) is fulfilled.

38 A representative sample is a typical sample of a population [Mar91]. The latter means in this context all possible trajectories which can be generated from the assumed probability density function $\mathrm{p}\left(\zeta \mid \boldsymbol{\theta}^*_{1:N}\right)$.

Proof:
By rewriting (6.7), we can state the following relations:

$$0 \leq \left| \mathbb{E}_{p\left(\zeta | \hat{\boldsymbol{\theta}}_{1:N}\right)} \left\{ J_i^* \left(\zeta, \boldsymbol{\theta}_i^*\right) \right\} - \mathbb{E}_{p\left(\zeta | \boldsymbol{\theta}_{1:N}^*\right)} \left\{ J_i^* \left(\zeta, \boldsymbol{\theta}_i^*\right) \right\} \right| \tag{6.11}$$

$$= \left| \mathbb{E}_{p\left(\zeta | \hat{\boldsymbol{\theta}}_{1:N}\right)} \left\{ \boldsymbol{\theta}_i^{*\top} \boldsymbol{\mu}_i (\zeta) \right\} - \mathbb{E}_{p\left(\zeta | \boldsymbol{\theta}_{1:N}^*\right)} \left\{ \boldsymbol{\theta}_i^{*\top} \boldsymbol{\mu}_i (\zeta) \right\} \right| \tag{6.12}$$

$$\leq \left\| \boldsymbol{\theta}_i^* \right\|_2 \left\| \mathbb{E}_{p\left(\zeta | \hat{\boldsymbol{\theta}}_{1:N}\right)} \left\{ \boldsymbol{\mu}_i (\zeta) \right\} - \mathbb{E}_{p\left(\zeta | \boldsymbol{\theta}_{1:N}^*\right)} \left\{ \boldsymbol{\mu}_i (\zeta) \right\} \right\|_2 \tag{6.13}$$

Therefore, if (6.10) holds, then the right side of (6.13) is equal to zero and hence, together with the inequality in (6.11), this implies that (6.7) holds as well. □

Lemma 6.1 represents the principle of matching feature expectations for all players. This principle was introduced in [AN04] for $N = 1$ and used as a basis for numerous single-player IRL methods.

Since the inverse dynamic game problem defined in Definition 6.4 demands the fulfillment of (6.7), by the results of Lemma 6.1 and using Assumption 6.1 we require

$$\mathbb{E}_{p(\zeta | \boldsymbol{\theta}_{1:N})} \left\{ \boldsymbol{\mu}_i (\zeta) \right\} = \tilde{\boldsymbol{\mu}}_i, \tag{6.14}$$

for each player $i \in \mathcal{P}$. Moreover, for a density function,

$$\int_{\forall \zeta} p(\zeta | \boldsymbol{\theta}_{1:N}) \, d\zeta = 1 \tag{6.15}$$

must apply. Since the conditions (6.14) and (6.15) do not lead to a unique solution for the probability density function, the principle of maximum entropy is applied. For a continuous density function the entropy corresponding to a probability density function is given by [CT06, Section 8.1]

$$h(p(\zeta | \boldsymbol{\theta}_{1:N})) = - \int_{\forall \zeta} p(\zeta | \boldsymbol{\theta}_{1:N}) \ln (p(\zeta | \boldsymbol{\theta}_{1:N})) \, d\zeta. \tag{6.16}$$

In order to determine a probability density function $p(\zeta | \boldsymbol{\theta}_{1:N})$ which only takes the information of (6.14) and (6.15) into consideration, the differential entropy (6.16) is maximized with the requirements (6.14) and (6.15) as optimization constraints. The density function which leads to maximum entropy in dynamic games is presented in the following lemma.

Lemma 6.2 (Maximum Entropy Probability Distribution in Inverse Dynamic Games)

The maximum entropy distribution under the constraints defined by (6.14) *and* (6.15) *is given by*

$$\begin{aligned} \mathrm{p}(\zeta \,|\, \boldsymbol{\theta}_{1:N}) &= \frac{\exp\left(\sum_{i=1}^{N} \boldsymbol{\theta}_i^\top \boldsymbol{\mu}_i(\zeta)\right)}{\int_{\forall \zeta} \exp\left(\sum_{i=1}^{N} \boldsymbol{\theta}_i^\top \boldsymbol{\mu}_i(\zeta)\right) \mathrm{d}\zeta} \\ &= \frac{\exp\left(\sum_{i=1}^{N} -J_i(\zeta, \boldsymbol{\theta}_i)\right)}{\int_{\forall \zeta} \exp\left(\sum_{i=1}^{N} -J_i(\zeta, \boldsymbol{\theta}_i)\right) \mathrm{d}\zeta}, \end{aligned} \tag{6.17}$$

where the alternative representation given in the last equation follows from (6.6).

Proof:

A calculus-based approach is followed as suggested in [CT06, Section 12.1]. To maximize the differential entropy (6.16) under the constraints given by (6.14) and (6.15), we introduce Lagrange multipliers $\psi \in \mathbb{R}$ and $\boldsymbol{\theta}_i \in \mathbb{R}^{M_i \times 1}$, $\forall\, i \in \mathcal{P}$, and set up the objective function

$$\begin{aligned} &\Lambda(\mathrm{p}(\zeta \,|\, \boldsymbol{\theta}_{1:N}), \psi, \boldsymbol{\theta}_{1:N}) = \\ &\quad -\int_{\forall \zeta} \mathrm{p}(\zeta \,|\, \boldsymbol{\theta}_{1:N}) \ln(\mathrm{p}(\zeta \,|\, \boldsymbol{\theta}_{1:N}))\, \mathrm{d}\zeta + \psi\left(\int_{\forall \zeta} \mathrm{p}(\zeta \,|\, \boldsymbol{\theta}_{1:N})\, \mathrm{d}\zeta - 1\right) \\ &\quad + \boldsymbol{\theta}_1^\top\left(\int_{\forall \zeta} \mathrm{p}(\zeta \,|\, \boldsymbol{\theta}_{1:N})\, \boldsymbol{\mu}_1(\zeta)\, \mathrm{d}\zeta - \tilde{\boldsymbol{\mu}}_1\right) + \dots \\ &\quad \vdots \\ &\quad + \boldsymbol{\theta}_N^\top\left(\int_{\forall \zeta} \mathrm{p}(\zeta \,|\, \boldsymbol{\theta}_{1:N})\, \boldsymbol{\mu}_N(\zeta)\, \mathrm{d}\zeta - \tilde{\boldsymbol{\mu}}_N\right). \end{aligned} \tag{6.18}$$

In this way, the expression

$$\begin{aligned} \frac{\partial \Lambda}{\partial \mathrm{p}(\zeta \,|\, \boldsymbol{\theta}_{1:N})} &= -\int_{\forall \zeta} \ln(\mathrm{p}(\zeta \,|\, \boldsymbol{\theta}_{1:N}))\, \mathrm{d}\zeta - \int_{\forall \zeta} \frac{\mathrm{p}(\zeta \,|\, \boldsymbol{\theta}_{1:N})}{\mathrm{p}(\zeta \,|\, \boldsymbol{\theta}_{1:N})} \mathrm{d}\zeta \\ &\quad + \psi \int_{\forall \zeta} 1\, \mathrm{d}\zeta + \boldsymbol{\theta}_1^\top \int_{\forall \zeta} \boldsymbol{\mu}_1(\zeta)\, \mathrm{d}\zeta + \dots + \boldsymbol{\theta}_N^\top \int_{\forall \zeta} \boldsymbol{\mu}_N(\zeta)\, \mathrm{d}\zeta \\ &= \int_{\forall \zeta} \left(-\ln(\mathrm{p}(\zeta \,|\, \boldsymbol{\theta}_{1:N})) - 1 + \psi + \sum_{i=1}^{N} \boldsymbol{\theta}_i^\top \boldsymbol{\mu}_i(\zeta)\right) \mathrm{d}\zeta \\ &\overset{!}{=} 0 \end{aligned} \tag{6.19}$$

gives a necessary condition for the sought probability density function. By inspecting (6.19) we see that this condition is fulfilled if

$$-\ln(\mathrm{p}(\zeta|\,\boldsymbol{\theta}_{1:N})) - 1 + \psi + \sum_{i=1}^{N} \boldsymbol{\theta}_i^\top \boldsymbol{\mu}_i(\zeta) = 0. \tag{6.20}$$

By reformulating (6.20), we obtain the probability density function of a trajectory ζ, i.e.

$$\mathrm{p}(\zeta|\,\boldsymbol{\theta}_{1:N}) = \exp(-1 + \psi)\exp\left(\sum_{i=1}^{N} \boldsymbol{\theta}_i^\top \boldsymbol{\mu}_i(\zeta)\right). \tag{6.21}$$

Using (6.21), (6.15) is rewritten as

$$\begin{aligned} 1 &= \int_{\forall\zeta} \mathrm{p}(\zeta|\,\boldsymbol{\theta}_{1:N})\,\mathrm{d}\zeta \\ &= \exp(-1+\psi)\int_{\forall\zeta} \exp\left(\sum_{i=1}^{N} \boldsymbol{\theta}_i^\top \boldsymbol{\mu}_i(\zeta)\right)\mathrm{d}\zeta \\ \Leftrightarrow\ \exp(-1+\psi) &= \frac{1}{\displaystyle\int_{\zeta} \exp\left(\sum_{i=1}^{N} \boldsymbol{\theta}_i^\top \boldsymbol{\mu}_i(\zeta)\right)\mathrm{d}\zeta}. \end{aligned} \tag{6.22}$$

Inserting (6.22) in (6.21) leads to the probability density function (6.17). The entropy is maximized since

$$\frac{\partial^2 \Lambda}{\partial \mathrm{p}(\zeta|\,\boldsymbol{\theta}_{1:N})^2} = -\int_{\forall\zeta} \frac{1}{\mathrm{p}(\zeta|\,\boldsymbol{\theta}_{1:N})}\mathrm{d}\zeta < 0 \tag{6.23}$$

for all $\mathrm{p}(\zeta|\,\boldsymbol{\theta}_{1:N}) \neq 0$.

□

In order to obtain an estimate of the cost function parameters $\hat{\boldsymbol{\theta}}_i$, $i \in \mathcal{P}$, it may appear suitable to maximize the probability density function (6.17), analogously to similar 1-player IRL methods [ZMBD08, LK12]. However, given the dependence of (6.17) on the cost function parameters of all players, it is not possible to solve for a particular $\boldsymbol{\theta}_i$. Nevertheless, if ζ corresponds to a Pareto efficient solution according to Definition 3.9, then (6.17) can be used to identify corresponding parameters $\hat{\boldsymbol{\theta}}_i$ which explain the observations. This approach is presented in Appendix D.

The following sections present approaches to identify cost function parameters which explain observed Nash equilibrium trajectories.

6.4 Open-Loop Case

In this section, we shall consider inverse dynamic games where each player applies an open-loop strategy (cf. Definition A.4) and an open-loop Nash equilibrium (OLNE) arises from their interaction.

A suitable probability density function $\mathrm{p}(\zeta)$ is sought which allows for the estimation of cost function parameters.

6.4.1 Probability Density Function

The non-cooperative character of the dynamic game implies that each player only considers his own cost function and strives for its minimization by means of the selected open-loop strategy. From Theorem A.1 we see that the open-loop Nash equilibrium involves the solution of a set of differential equations which includes derivatives of the system dynamics and the features (which constitute the Hamiltonian) with respect to the system state $\boldsymbol{x}^{(k)}$ and player i's controls $\boldsymbol{u}_i^{(k)}$. The other players' controls do not depend on either of these, and therefore, they do not have any influence on player i's actions.

Consequently, the following probability function

$$\begin{aligned}\mathrm{p}(\zeta|\,\boldsymbol{\theta}_i) &= \frac{\exp\left(-J_i\left(\zeta\right)\right)}{\int_{\tilde{\zeta}} \exp\left(-J_i\left(\tilde{\zeta}\right)\right)\mathrm{d}\tilde{\zeta}} \\ &= \frac{\exp\left(\boldsymbol{\theta}_i^\top \boldsymbol{\mu}_i(\zeta)\right)}{\int_{\tilde{\zeta}} \exp\left(\boldsymbol{\theta}_i^\top \boldsymbol{\mu}_i(\tilde{\zeta})\right)\mathrm{d}\tilde{\zeta}}\end{aligned} \tag{6.24}$$

is defined, which represents the probability (density) of a particular trajectory from the point of view of player i. This density implies that the probability of a particular trajectory is inversely proportional to the costs generated by player i's own individual cost function J_i defined by player i's cost function parameter set $\boldsymbol{\theta}_i$. This simplifies the probability density function $\mathrm{p}\left(\zeta|\,\boldsymbol{\theta}_{1:N}\right)$ in such a way that N probability density functions $\mathrm{p}\left(\zeta|\,\boldsymbol{\theta}_i\right)$ which depend on each player's cost function parameters $\boldsymbol{\theta}_i$ are considered instead of one single probability density function which depends on all parameters.

Considering a possible total number of n_t demonstrations, the following likelihood function is defined based on the introduced probability density function.

Definition 6.5 (Likelihood Function)

Let a set of n_t trajectories denoted by $\mathcal{D} = \{\tilde{\zeta}_1, \ldots, \tilde{\zeta}_{n_t}\}$ be given. Then the likelihood of the data given a parameter vector $\boldsymbol{\theta}_i$ is defined as

$$\mathcal{L}(\boldsymbol{\theta}_i \mid \mathcal{D}) = \prod_{l=1}^{n_t} \mathrm{p}\left(\tilde{\zeta}_l \mid \boldsymbol{\theta}_i\right), \tag{6.25}$$

where $\mathrm{p}\left(\tilde{\zeta}_l \mid \boldsymbol{\theta}_i\right)$ is obtained by evaluating (6.24) at $\tilde{\zeta}_l$, $l \in \{1, \ldots, n_t\}$.

The likelihood describes the probability density of the trajectories when the parameters are set. Moreover, it is a function of $\boldsymbol{\theta}_i$. With this function, the foundation for a maximum likelihood estimation (MLE) of the cost function parameters is given. In order to show that maximizing the likelihood leads to an unbiased estimation of the cost function parameters, the following assumption adapts Assumption 6.1 (and (6.9)) to probability density functions depending only on the parameters $\boldsymbol{\theta}_i$ of one player $i \in \mathcal{P}$ as defined in (6.24).

Assumption 6.2 (Expectation and Mean Equivalence)

The mean of the feature count of the n_t observed trajectories is equal to the expectation of the feature count of the trajectories resulting from the probability density function with original parameters $\boldsymbol{\theta}_i^*$, i.e.

$$\mathbb{E}_{\mathrm{p}(\zeta \mid \boldsymbol{\theta}_i^*)}\left\{\boldsymbol{\mu}_j(\zeta)\right\} = \frac{1}{n_t}\sum_{l=1}^{n_t} \boldsymbol{\mu}_j\left(\tilde{\zeta}_l\right), \quad \forall\, i, j \in \mathcal{P}. \tag{6.26}$$

6.4.2 Cost Function Estimation and Unbiasedness Results

Before presenting the unbiasedness of the MLE as the main result for inverse non-cooperative dynamic games of this chapter, an alternative definition of the cost functions which will be convenient for the proof of the main theorem.

Definition 6.6 (Extended Features, Feature Count and Parameter Vector)
Let $\bar{\boldsymbol{\phi}}$ denote an extended feature vector which includes all features $\phi_{i,(q)}$, $i \in \mathcal{P}$, $q \in \{1, \ldots, M_i\}$ of all N players such that $\bar{\phi}_{(r)} \neq \bar{\phi}_{(s)}$ for all $r, s \in \{1, \ldots, \dim(\bar{\boldsymbol{\phi}})\}$ and $r \neq s$. In other words, the extended feature vector $\bar{\boldsymbol{\phi}}$ consists of the feature vectors $\boldsymbol{\phi}_i$ of all players such that no feature is included more than once and all features are linearly independent of each other. The extended feature count $\bar{\boldsymbol{\mu}}(\zeta)$ is defined analogously according to Definition 6.3. Furthermore, let the extended parameter vector $\bar{\boldsymbol{\theta}}_i$ be defined such that

$$J_i(\zeta) = \boldsymbol{\theta}_i^\top \boldsymbol{\mu}_i(\zeta) = \bar{\boldsymbol{\theta}}_i^\top \bar{\boldsymbol{\mu}}(\zeta), \quad i \in \mathcal{P}. \tag{6.27}$$

Remark 6.1:
For (6.27) to hold, $\bar{\boldsymbol{\theta}}_i$ has to include zeros in the positions corresponding to the elements of $\bar{\boldsymbol{\phi}}$ representing features which were not in $\boldsymbol{\phi}_i$ previously.

Remark 6.2:
Assumption 6.2 leads to

$$\mathbb{E}_{\mathrm{p}(\zeta|\boldsymbol{\theta}_i^*)}\{\bar{\boldsymbol{\mu}}(\zeta)\} = \frac{1}{n_t}\sum_{l=1}^{n_t} \bar{\boldsymbol{\mu}}\left(\tilde{\zeta}_l\right), \quad \forall\, i \in \mathcal{P}, \tag{6.28}$$

for the extended feature count $\bar{\boldsymbol{\mu}}(\zeta)$.

The following theorem presents the method for estimating cost function parameters from open-loop Nash equilibrium trajectories and states the unbiasedness of the estimation.

Theorem 6.1 (Unbiasedness of the Estimation)
Let a set of trajectories $\mathcal{D} = \{\tilde{\zeta}_1, \ldots, \tilde{\zeta}_{n_t}\}$ for which Assumption 6.2 is fulfilled be given. Then, the MLE with respect to the observed trajectories, i.e.

$$\hat{\boldsymbol{\theta}}_i = \arg\max_{\boldsymbol{\theta}_i} \ln \mathcal{L}\{\boldsymbol{\theta}_i | \mathcal{D}\}, \tag{6.29}$$

where $\mathcal{L}\{\boldsymbol{\theta}_i | \mathcal{D}\}$ is obtained by evaluating the likelihood function of Definition 6.5 at $\tilde{\zeta}_l$, $l \in \{1, \ldots, n_t\}$, leads to parameters $\hat{\boldsymbol{\theta}}_i$ such that $\mathrm{p}\left(\zeta | \hat{\boldsymbol{\theta}}_i\right)$ results in an expectation of the cost function values $J_j\left(\zeta, \boldsymbol{\theta}_j^*\right)$, $\forall j \in \mathcal{P}$ which is equal to the one corresponding to $\mathrm{p}\left(\zeta | \boldsymbol{\theta}_i^*\right)$, i.e.

$$\mathbb{E}_{\mathrm{p}\left(\zeta|\hat{\boldsymbol{\theta}}_i\right)}\left\{J_j\left(\zeta, \boldsymbol{\theta}_j^*\right)\right\} = \mathbb{E}_{\mathrm{p}(\zeta|\boldsymbol{\theta}_i^*)}\left\{J_j\left(\zeta, \boldsymbol{\theta}_j^*\right)\right\}, \tag{6.30}$$

holds for all $i, j \in \mathcal{P}$.

Proof:

Using the extended parameter vector $\bar{\boldsymbol{\theta}}_i$ (cf. Definition 6.6), (6.30) can be rewritten as

$$\mathbb{E}_{\mathrm{p}\left(\zeta\,|\,\hat{\bar{\boldsymbol{\theta}}}_i\right)}\left\{J_j\left(\zeta,\bar{\boldsymbol{\theta}}_j^*\right)\right\} = \mathbb{E}_{\mathrm{p}(\zeta\,|\,\bar{\boldsymbol{\theta}}_i^*)}\left\{J_j\left(\zeta,\bar{\boldsymbol{\theta}}_j^*\right)\right\} \tag{6.31}$$

for all $i, j \in \mathcal{P}$. Therefore, (6.31) shall be proved in the following.

The maximization of the log-likelihood function (6.29) implies

$$\mathbf{0} \overset{!}{=} \frac{\partial}{\partial\bar{\boldsymbol{\theta}}_i}\sum_{l=1}^{n_t}\ln\left(\frac{\exp\left(\bar{\boldsymbol{\theta}}_i^\top\bar{\boldsymbol{\mu}}(\tilde{\zeta}_l)\right)}{\int_\zeta \exp\left(\bar{\boldsymbol{\theta}}_i^\top\bar{\boldsymbol{\mu}}(\zeta)\right)\mathrm{d}\zeta}\right)\Bigg|_{\bar{\boldsymbol{\theta}}_i=\hat{\bar{\boldsymbol{\theta}}}_i} \tag{6.32}$$

$$= \sum_{l=1}^{n_t}\frac{\partial}{\partial\bar{\boldsymbol{\theta}}_i}\left(-\ln\left(\int_\zeta \exp\left(\bar{\boldsymbol{\theta}}_i^\top\bar{\boldsymbol{\mu}}(\zeta)\right)\mathrm{d}\zeta\right) + \bar{\boldsymbol{\theta}}_i^\top\bar{\boldsymbol{\mu}}(\tilde{\zeta}_l)\right)\Bigg|_{\bar{\boldsymbol{\theta}}_i=\hat{\bar{\boldsymbol{\theta}}}_i} \tag{6.33}$$

$$= \sum_{l=1}^{n_t}\left(\frac{\int_\zeta -\exp\left(\bar{\boldsymbol{\theta}}_i^\top\bar{\boldsymbol{\mu}}(\zeta)\right)\bar{\boldsymbol{\mu}}(\zeta)\,\mathrm{d}\zeta}{\int_{\check{\zeta}} \exp\left(\bar{\boldsymbol{\theta}}_i^\top\bar{\boldsymbol{\mu}}(\check{\zeta})\right)\mathrm{d}\check{\zeta}} + \bar{\boldsymbol{\mu}}(\tilde{\zeta}_l)\right)\Bigg|_{\bar{\boldsymbol{\theta}}_i=\hat{\bar{\boldsymbol{\theta}}}_i}. \tag{6.34}$$

Since the integrals in the numerator and the denominator in (6.34) are independent of each other, (6.34) can be rewritten as

$$\mathbf{0} \overset{!}{=} \sum_{l=1}^{n_t}\left(\int_\zeta \frac{-\exp\left(\bar{\boldsymbol{\theta}}_i^\top\bar{\boldsymbol{\mu}}(\zeta)\right)\bar{\boldsymbol{\mu}}(\zeta)}{\int_{\check{\zeta}} \exp\left(\bar{\boldsymbol{\theta}}_i^\top\bar{\boldsymbol{\mu}}(\check{\zeta})\right)\mathrm{d}\check{\zeta}}\mathrm{d}\zeta + \bar{\boldsymbol{\mu}}(\tilde{\zeta}_l)\right)\Bigg|_{\bar{\boldsymbol{\theta}}_i=\hat{\bar{\boldsymbol{\theta}}}_i}. \tag{6.35}$$

Using the defined probability density function (6.24), we obtain

$$\begin{aligned}\mathbf{0} &\overset{!}{=} \sum_{l=1}^{n_t}\left(-\int_\zeta \mathrm{p}\left(\zeta\,|\,\bar{\boldsymbol{\theta}}_i\right)\bar{\boldsymbol{\mu}}(\zeta)\,\mathrm{d}\zeta + \bar{\boldsymbol{\mu}}(\tilde{\zeta}_l)\right)\Bigg|_{\bar{\boldsymbol{\theta}}_i=\hat{\bar{\boldsymbol{\theta}}}_i}\\ &= \sum_{l=1}^{n_t}\left(-\mathbb{E}_{\mathrm{p}\left(\zeta\,|\,\hat{\bar{\boldsymbol{\theta}}}_i\right)}\left\{\bar{\boldsymbol{\mu}}(\zeta)\right\} + \bar{\boldsymbol{\mu}}(\tilde{\zeta}_l)\right).\end{aligned} \tag{6.36}$$

By rewriting (6.36) and considering Assumption 6.2 and Remark 6.2,

$$\mathbb{E}_{\mathrm{p}\left(\zeta\,|\,\hat{\bar{\boldsymbol{\theta}}}_i\right)}\left\{\bar{\boldsymbol{\mu}}(\zeta)\right\} = \frac{1}{n_t}\sum_{l=1}^{n_t}\bar{\boldsymbol{\mu}}(\tilde{\zeta}_l) = \mathbb{E}_{\mathrm{p}(\zeta\,|\,\bar{\boldsymbol{\theta}}_i^*)}\left\{\bar{\boldsymbol{\mu}}(\zeta)\right\} \tag{6.37}$$

results. Therefore, the expectations of the feature count $\bar{\mu}$ are equal for both probability density functions. By applying the results of Lemma 6.1 (which also hold for a probability density function $\mathrm{p}(\zeta | \theta_i)$) we conclude that (6.37) leads to (6.31) which is equivalent to (6.30). □

The results of Theorem 6.1 guarantee (6.30), which at first glance differs from the requirement (6.7) posed in the inverse dynamic game problem in Definition 6.4. However, for inverse open-loop dynamic games, it was proposed to consider N probability density functions $\mathrm{p}\left(\zeta | \theta_i^*\right)$ instead of a single one given by $\mathrm{p}\left(\zeta | \theta_{1:N}^*\right)$. Therefore, instead of the equivalence of expected costs with respect to this initially assumed probability density function $\mathrm{p}\left(\zeta | \theta_{1:N}^*\right)$, we obtain the equivalence of expected costs for all players $j \in \mathcal{P}$ with respect to each of the N probability density functions $\mathrm{p}\left(\zeta | \theta_i^*\right)$ as stated in (6.30). Consequently, the estimated parameters $\hat{\theta}_i$ solve the inverse dynamic game problem for an open-loop information structure.

Remark 6.3:

Solving the optimization problem (6.29) *demands the possibility of evaluating the likelihood function $\mathcal{L}\{\theta_i | \mathcal{D}\}$ and therefore the probability density function* (6.24) *at the trajectories ζ_l^*. The denominator in* (6.24) *includes an integral over all trajectories ζ which are feasible with respect to the system dynamics and an initial state. Calculating this integral is intractable given the continuous-valued control and action spaces. Therefore, approximations are usually applied. This will be tackled in Section 6.6.*

6.5 Feedback Case

In this section, solutions for inverse dynamic games with the feedback Nash equilibrium (FNE) as a solution concept are presented. Therefore, the MPS and feedback information structures according to Definition A.3 are considered. The resulting strategies are given by[39]

$$u_i^{(k)} = \gamma_i^{(k)}(x^{(k)}). \tag{6.38}$$

The following assumption is needed for the results of this section.

> **Assumption 6.3 (Control Laws)**
> *The Nash equilibrium control laws $\gamma_i^{(k)*}(x^{(k)})$, $k \in \mathcal{K}$ are known for all players $i \in \mathcal{P}$.*

[39] According to [BO99, p. 278], the feedback Nash equilibrium solution under the MPS information pattern solely depends on $x^{(k)}$ at the time step k. The dependency on $x^{(1)}$ is given only for $k = 1$. Therefore, we have feedback strategies as in Definition A.5 for both MPS and FB information structures.

For the case of a finite-horizon dynamic game, i.e. $k_E \in \mathbb{N}$, Assumption 6.3 demands the knowledge of the exact (time-dependent) function $\boldsymbol{\gamma}_i^{(k)*}(\boldsymbol{x}^{(k)})$. This case is analogous to Assumption 4.3 for inverse feedback differential games. In case of an infinite-horizon ($k_E \longrightarrow \infty$) dynamic game, Assumption 6.3 implies that the time-independent functional relationship of $\boldsymbol{\gamma}_i^{(k)}$ to $\boldsymbol{x}^{(k)}$ is known.

Remark 6.4:

Assumption 6.3 is rather restrictive for general nonlinear feedback Nash equilibria. However, not only the estimation of the control law is non-trivial, but also the calculation of the equilibria themselves which implies the solution of coupled partial differential equations (see Theorem 3.2) or coupled Bellman equations (see Theorem A.2). On the other hand, Assumption 6.3 is not restrictive for infinite-horizon linear-quadratic dynamic games, since the Nash equilibrium controls are given by

$$\boldsymbol{\gamma}_i^{(k)*}(\boldsymbol{x}^{(k)}) = \boldsymbol{K}_i^* \boldsymbol{x}^{(k)}, \tag{6.39}$$

with $\boldsymbol{K}_i^* \in \mathbb{R}^{m_i \times n}$ [Eng05, Section 8.3]. As mentioned in Section 5.4.2, the estimation of $\boldsymbol{K}_i^*$ can easily be performed by means of a least-squares approach.

If Assumption 6.3 holds, the control laws of the players $j \in \mathcal{P}, j \neq i$ can replace $\boldsymbol{u}_j^{(k)*}$ in (6.1), leading to system dynamics from player i's perspective defined as

$$\begin{aligned} \boldsymbol{x}^{(k+1)} &= \boldsymbol{f}^{(k)}\left(\boldsymbol{x}^{(k)}, \boldsymbol{u}_i^{(k)}, \boldsymbol{\gamma}_{\neg i}^{(k)*}\left(\boldsymbol{x}^{(k)}\right)\right) \\ &= \boldsymbol{f}_i^{(k)}\left(\boldsymbol{x}^{(k)}, \boldsymbol{u}_i^{(k)}\right). \end{aligned} \tag{6.40}$$

In this way, it is possible for player i to represent the system dynamics as a function of the system state $\boldsymbol{x}$ and his own control variable $\boldsymbol{u}_i$. The effect of the other players' controls are considered due to the implied knowledge of the control laws and the system state in every time step. Analogously, the features $\boldsymbol{\phi}_i$ of player i's cost function can be rewritten as a function of the state $\boldsymbol{x}$ and the control variables $\boldsymbol{u}_i$, i.e.

$$\begin{aligned} \boldsymbol{\phi}_i &= \boldsymbol{\phi}_i(\boldsymbol{x}^{(k)}, \boldsymbol{u}_1^{(k)}, \ldots, \boldsymbol{u}_N^{(k)}) \\ &= \boldsymbol{\phi}_i\left(\boldsymbol{x}^{(k)}, \boldsymbol{u}_i^{(k)}, \boldsymbol{\gamma}_{\neg i}^{(k)*}\left(\boldsymbol{x}^{(k)}\right)\right) \\ &= \boldsymbol{\phi}_i\left(\boldsymbol{x}^{(k)}, \boldsymbol{u}_i^{(k)}\right), \end{aligned} \tag{6.41}$$

where the same vector $\boldsymbol{\phi}_i$ is used with some mathematical freedom in favor of a simplified presentation. Based on the system dynamics from player i's perspective (6.40) and the rewritten features (6.41), the following theorem is presented which describes the method for an unbiased maximum likelihood estimation of cost function parameters in an inverse feedback Nash dynamic game.

Theorem 6.2 (Unbiasedness of the Estimation)
Let a set of trajectories $\mathcal{D} = \{\tilde{\zeta}_1, \ldots, \tilde{\zeta}_{n_t}\}$ *be given such that Assumption 6.2 is fulfilled. Furthermore, let Assumption 6.3 hold such that the feedback Nash control laws* $\boldsymbol{\gamma}_i^{(k)*}$ *are known for all* $i \in \mathcal{P}$*. Then, the MLE with respect to the observed trajectories, i.e.*

$$\hat{\boldsymbol{\theta}}_i = \arg\max_{\boldsymbol{\theta}_i} \ln \mathcal{L}\{\boldsymbol{\theta}_i | \mathcal{D}\} \tag{6.42}$$

where $\mathcal{L}\{\boldsymbol{\theta}_i | \mathcal{D}^*\}$ *is obtained by evaluating the likelihood function of Definition 6.5 at* ζ_l^*, $l \in \{1, ..., n_t\}$ *and with respect to the system dynamics* (6.40), *leads to parameters* $\hat{\boldsymbol{\theta}}_i$ *such that*

$$\mathbb{E}_{\mathrm{p}\left(\zeta | \hat{\boldsymbol{\theta}}_i\right)}\left\{J_j\left(\zeta | \boldsymbol{\theta}_j^*\right)\right\} = \mathbb{E}_{\mathrm{p}(\zeta | \boldsymbol{\theta}_i^*)}\left\{J_j\left(\zeta | \boldsymbol{\theta}_j^*\right)\right\} \tag{6.43}$$

holds for all $i, j \in \mathcal{P}$ *(cf. Theorem 6.1).*

Proof:
The cost functions J_i, $i \in \mathcal{P}$ can be rewritten using the modified features (6.41). Afterwards, the theorem can be proved analogously to Theorem 6.1. □

6.6 Practical Aspects

The results of the previous sections provide the theoretical foundation for the application of MaxEnt IRL for the solution of inverse dynamic game problems. The core of the method is the MLE based on the probability density functions $\mathrm{p}\left(\zeta | \boldsymbol{\theta}_i^*\right)$. The focus of this section is laid on the computation of the MLEs which yield cost function parameters $\hat{\boldsymbol{\theta}}_i$ explaining observed results of a dynamic game. This poses the practical challenge of evaluating the probability density function (6.24) and with that result, the likelihood function (6.25). This is the main objective approached in this section.

6.6.1 Approximation of the Probability Density Function

The integral in the denominator of (6.24) is computationally intractable and therefore, an approximation is necessary. This may be achieved by replacing the integral with a sum over several trajectory samples which have to be generated from a previously defined probability distribution [KPRS13, MHB16] or determined in each iteration from a forward optimal control or dynamic game solution with current cost function parameter candidates [AB11]. Which sampled trajectories are chosen has a great impact on the estimation of cost function parameters (cf. [AB11]). In order to avoid the problem of choosing adequate samples, in this thesis the integral and therewith, the probability density functions are approximated locally. The following procedure is inspired by the approach proposed in [LK12] for a single-player case. Nonetheless, some modifications are introduced and will be explained when suitable.

Consider any player $i \in \mathcal{P}$. Given an observed trajectory $\tilde{\zeta}_l$, $l \in \{1, 2, \dots, n_t\}$, and consequently, the control trajectories $\underline{\tilde{\boldsymbol{u}}}_{\neg i,l}$ of all other players, we can formulate the costs $J_i(\tilde{\zeta}_l, \boldsymbol{\theta}_i)$ of player i generated by $\tilde{\zeta}_l$ such that only variations of his own control trajectory $\underline{\tilde{\boldsymbol{u}}}_{i,l}$ are taken into account, i.e. the costs are formulated as $J_i\left(\underline{\boldsymbol{u}}_i, \underline{\tilde{\boldsymbol{u}}}_{\neg i,l}, \boldsymbol{\theta}_i\right)$. Local approximations of the observed trajectory $\tilde{\zeta}_l$ are considered which arise from the aforementioned variations of $\underline{\boldsymbol{u}}_{i,l}$ while the other players' controls $\underline{\tilde{\boldsymbol{u}}}_{\neg i,l}$ remain unchanged. Hence, we approximate the cost function $J_i\left(\underline{\boldsymbol{u}}_i, \underline{\tilde{\boldsymbol{u}}}_{\neg i,l}, \boldsymbol{\theta}_i\right)$ by means of a second-order Taylor series expansion around the observed controls $\underline{\tilde{\boldsymbol{u}}}_{i,l}$ corresponding to the trajectory $\tilde{\zeta}_l$. This results in

$$
\begin{aligned}
J_i\left(\underline{\boldsymbol{u}}_i, \underline{\tilde{\boldsymbol{u}}}_{\neg i,l}, \boldsymbol{\theta}_i\right) \approx J_i\left(\underline{\tilde{\boldsymbol{u}}}_{i,l}, \underline{\tilde{\boldsymbol{u}}}_{\neg i,l}, \boldsymbol{\theta}_i\right) &+ \left(\underline{\boldsymbol{u}}_i - \underline{\tilde{\boldsymbol{u}}}_{i,l}\right)^\top \tilde{\boldsymbol{g}}_{i,l}(\boldsymbol{\theta}_i) \\
&+ \frac{1}{2}\left(\underline{\boldsymbol{u}}_i - \underline{\tilde{\boldsymbol{u}}}_{i,l}\right)^\top \tilde{G}_{i,l}(\boldsymbol{\theta}_i)\left(\underline{\boldsymbol{u}}_i - \underline{\tilde{\boldsymbol{u}}}_{i,l}\right),
\end{aligned} \tag{6.44}
$$

where $\tilde{\boldsymbol{g}}_{i,l}(\boldsymbol{\theta}_i) \in \mathbb{R}^{m_i k_E}$ and $\tilde{G}_{i,l}(\boldsymbol{\theta}_i) \in \mathbb{R}^{m_i k_E \times m_i k_E}$ denote the first and second derivative of J_i with respect to $\underline{\boldsymbol{u}}_i$, respectively, i.e.

$$
\tilde{\boldsymbol{g}}_{i,l}(\boldsymbol{\theta}_i) := \left.\frac{\mathrm{d}J_i}{\mathrm{d}\underline{\boldsymbol{u}}_i}\right|_{\underline{\boldsymbol{u}}_i = \underline{\tilde{\boldsymbol{u}}}_{i,l}} \tag{6.45}
$$

$$
\tilde{G}_{i,l}(\boldsymbol{\theta}_i) := \left.\frac{\mathrm{d}^2 J_i}{\mathrm{d}\underline{\boldsymbol{u}}_i^2}\right|_{\underline{\boldsymbol{u}}_i = \underline{\tilde{\boldsymbol{u}}}_{i,l}} . \tag{6.46}
$$

In the following, $\tilde{\boldsymbol{g}}_{i,l}(\boldsymbol{\theta}_i)$ and $\tilde{G}_{i,l}(\boldsymbol{\theta}_i)$ are written as $\tilde{\boldsymbol{g}}_{i,l}$ and $\tilde{G}_{i,l}$, respectively, for brevity.

By reformulating (6.24) using the Taylor series based approximation (6.44) of the cost function and considering that the observed trajectory $\tilde{\zeta}_l$ is (with fixed $\boldsymbol{\theta}_i$) uniquely defined by the controls $\underline{\tilde{\boldsymbol{u}}}_{i,l}$ with given $\underline{\tilde{\boldsymbol{u}}}_{\neg i,l}$ and the initial state $\boldsymbol{x}^{(1)}$, the probability density function can be evaluated at $\tilde{\zeta}_l$ using the relation

$$
\begin{aligned}
\mathrm{p}\left(\underline{\tilde{\boldsymbol{u}}}_{i,l} \middle| \underline{\tilde{\boldsymbol{u}}}_{\neg i,l}, \boldsymbol{x}^{(1)}, \boldsymbol{\theta}_i\right) &= \frac{\mathrm{e}^{-J_i\left(\underline{\tilde{\boldsymbol{u}}}_{i,l} \middle| \underline{\tilde{\boldsymbol{u}}}_{\neg i,l}, \boldsymbol{x}^{(1)}, \boldsymbol{\theta}_i\right)}}{\displaystyle\int_{-\infty}^{\infty} \mathrm{e}^{-J_i\left(\underline{\boldsymbol{u}}_i \middle| \underline{\tilde{\boldsymbol{u}}}_{\neg i,l}, \boldsymbol{x}^{(1)}, \boldsymbol{\theta}_i\right)} \mathrm{d}\underline{\boldsymbol{u}}_i} \\
&\approx \mathrm{e}^{\left(-\frac{1}{2}\tilde{\boldsymbol{g}}_{i,l}^\top \tilde{G}_{i,l}^{-1} \tilde{\boldsymbol{g}}_{i,l}\right)} \det\left(\tilde{G}_{i,l}\right)^{\frac{1}{2}} (2\pi)^{-\frac{\dim(\underline{\tilde{\boldsymbol{u}}}_{i,l})}{2}} .
\end{aligned} \tag{6.47}
$$

This leads to the log-likelihood function

$$\ln\left(\mathcal{L}\{\boldsymbol{\theta}_i \mid \mathcal{D}\}\right) \approx \sum_{l=1}^{n_t}\left(-\frac{1}{2}\tilde{\boldsymbol{g}}_{i,l}^\top \tilde{\boldsymbol{G}}_{i,l}^{-1}\tilde{\boldsymbol{g}}_{i,l} + \frac{1}{2}\ln\left(\det(\tilde{\boldsymbol{G}}_{i,l})\right) - \frac{1}{2}\dim\left(\underline{\boldsymbol{u}}_{i,l}\right)\ln(2\pi)\right) \tag{6.48}$$

which can be used for the MLEs stated in Theorems 6.1 and 6.2. The detailed calculation steps are provided in Section B.5 of the Appendix.[40]

Therefore, in order to evaluate (6.24), the first derivative $\tilde{\boldsymbol{g}}_{i,l}$ and the second derivative $\tilde{\boldsymbol{G}}_{i,l}$ are needed. Their calculation is explained in the following.

6.6.2 Evaluation of the Log-Likelihood Function

By applying the chain rule, the first and second derivatives of the cost function are given by

$$\tilde{\boldsymbol{g}}_{i,l} = \nabla_{\underline{\boldsymbol{u}}_i} J_i + \left(\nabla_{\underline{\boldsymbol{u}}_i}\underline{\boldsymbol{x}}\right)^\top \nabla_{\underline{\boldsymbol{x}}} J_i \Big|_{\substack{\underline{\boldsymbol{u}}_i = \tilde{\underline{\boldsymbol{u}}}_{i,l} \\ \underline{\boldsymbol{x}} = \tilde{\underline{\boldsymbol{x}}}_l}} \tag{6.49}$$

$$\tilde{\boldsymbol{G}}_{i,l} = \nabla_{\underline{\boldsymbol{u}}_i\underline{\boldsymbol{u}}_i} J_i + \left(\nabla_{\underline{\boldsymbol{u}}_i}\underline{\boldsymbol{x}}\right)^\top \nabla_{\underline{\boldsymbol{x}}\underline{\boldsymbol{x}}} J_i \nabla_{\underline{\boldsymbol{u}}_i}\underline{\boldsymbol{x}} + \nabla_{\underline{\boldsymbol{u}}_i\underline{\boldsymbol{u}}_i}\underline{\boldsymbol{x}} \times_1 \nabla_{\underline{\boldsymbol{x}}} J_i + 2\nabla_{\underline{\boldsymbol{u}}_i\underline{\boldsymbol{x}}} J_i \nabla_{\underline{\boldsymbol{u}}_i}\underline{\boldsymbol{x}} \Big|_{\substack{\underline{\boldsymbol{u}}_i = \tilde{\underline{\boldsymbol{u}}}_{i,l} \\ \underline{\boldsymbol{x}} = \tilde{\underline{\boldsymbol{x}}}_l}} \tag{6.50}$$

where $\nabla_{\underline{\boldsymbol{u}}_i} J_i$ and $\nabla_{\underline{\boldsymbol{x}}} J_i$ denote the partial derivatives of J_i with respect to $\underline{\boldsymbol{u}}_i$ and $\underline{\boldsymbol{x}}$, respectively.[41] Likewise, $\nabla_{\underline{\boldsymbol{u}}_i\underline{\boldsymbol{u}}_i} J_i$, $\nabla_{\underline{\boldsymbol{x}}\underline{\boldsymbol{x}}} J_i$ and $\nabla_{\underline{\boldsymbol{u}}_i\underline{\boldsymbol{x}}} J_i$ represent second-order partial derivatives of J_i with respect to $\underline{\boldsymbol{u}}_i$ and $\underline{\boldsymbol{x}}$. The partial derivative $\nabla_{\underline{\boldsymbol{u}}_i}\underline{\boldsymbol{x}}$ is defined analogously. The term $\nabla_{\underline{\boldsymbol{u}}_i\underline{\boldsymbol{u}}_i}\underline{\boldsymbol{x}}$ is used with some abuse of notation to represent a third-order tensor such that $\times_1$ represents a 1-mode tensor multiplication [KB09, Section 2.5].[42]

In the following, we elaborate on the structure of the partial derivatives which form $\tilde{\boldsymbol{g}}_{i,l}$ and $\tilde{\boldsymbol{G}}_{i,l}$ as given in (6.49) and (6.50), with the partial derivatives with respect to $\underline{\boldsymbol{x}}$ as an example. With the assumed structure of the cost function (6.2), we obtain the first-order partial derivative

$$\nabla_{\underline{\boldsymbol{x}}} J_i = -\left[\left(\nabla_{\boldsymbol{x}}\boldsymbol{\phi}_i\right)\boldsymbol{\theta}_i\big|_{\boldsymbol{x}^{(k)}=\boldsymbol{x}^{(1)}} \quad \cdots \quad \left(\nabla_{\boldsymbol{x}}\boldsymbol{\phi}_i\right)\boldsymbol{\theta}_i\big|_{\boldsymbol{x}^{(k)}=\boldsymbol{x}^{(k_E)}}\right]^\top \in \mathbb{R}^{nk_E}, \tag{6.51}$$

where, unless otherwise specified, $\nabla_{\boldsymbol{x}}\boldsymbol{\phi}_i$ denotes the partial derivative of $\boldsymbol{\phi}_i$ with respect to $\boldsymbol{x}^{(k)}$. The second-order partial derivatives of the cost function $\nabla_{\underline{\boldsymbol{x}}\underline{\boldsymbol{x}}} J_i$, $\nabla_{\underline{\boldsymbol{u}}_i\underline{\boldsymbol{u}}_i} J_i$ and $\nabla_{\underline{\boldsymbol{u}}_i\underline{\boldsymbol{x}}} J_i$ are

---

40 Note that (6.44) and (6.48) yield equalities in the case of quadratic cost functions.

41 The last term in (6.50) was neglected in [LK12]. Nevertheless, it can only be neglected if there are no features which depend on both $\boldsymbol{x}$ and $\boldsymbol{u}_i$, i.e. $\phi_{i,(j)}(\boldsymbol{x}, \boldsymbol{u}_i)$ is equal to either $\phi_{i,(j)}(\boldsymbol{x})$ or $\phi_{i,(j)}(\boldsymbol{u}_i)$ for all $i \in \mathcal{P}$ and all $j \in \{1, \ldots, M_i\}$.

42 For the 1-mode tensor multiplication we obtain $\nabla_{\underline{\boldsymbol{u}}_i\underline{\boldsymbol{u}}_i}\underline{\boldsymbol{x}} \times_1 \nabla_{\underline{\boldsymbol{x}}} J_i = \left(\nabla_{\underline{\boldsymbol{x}}} J_i\right)^\top \nabla_{\underline{\boldsymbol{u}}_i\underline{\boldsymbol{u}}_i}\underline{\boldsymbol{x}} \in \mathbb{R}^{1\times m_i k_E \times m_i k_E}$, which can be represented as a matrix of dimensions $m_i k_E \times m_i k_E$.

block diagonal matrices since the costs at time step k only depend on the states $\boldsymbol{x}^{(k)}$ and controls $\boldsymbol{u}_i^{(k)}$ at time step k. Therefore, we obtain

$$\nabla_{\underline{\boldsymbol{x}}\underline{\boldsymbol{x}}} J_i = \operatorname{blkdiag}\left(-\sum_{l=1}^{M_i} \nabla_{\boldsymbol{x}\boldsymbol{x}} \phi_{i,(l)} \theta_{i,(l)} \Big|_{\boldsymbol{x}^{(k)}=\boldsymbol{x}^{(1)}}, \quad \dots \quad, -\sum_{l=1}^{M_i} \nabla_{\boldsymbol{x}\boldsymbol{x}} \phi_{i,(l)} \theta_{i,(l)} \Big|_{\boldsymbol{x}^{(k)}=\boldsymbol{x}^{(k_E)}} \right), \quad (6.52)$$

where blkdiag$(\cdot)$ denotes a block diagonal matrix. In this case, there are k_E blocks of dimension $n \times n$. The other partial derivatives can be computed analogously to (6.51) and (6.52).

The partial derivative $\nabla_{\underline{\boldsymbol{u}}_i} \underline{\boldsymbol{x}}$ describes the sensitivity of $\boldsymbol{x}$ with respect to $\boldsymbol{u}_i$ for all time steps as a consequence of the system dynamics. Since present actions are not influenced by future actions, the matrix

$$\boldsymbol{D}_i := \nabla_{\underline{\boldsymbol{u}}_i} \underline{\boldsymbol{x}} \Big|_{\substack{\underline{\boldsymbol{u}}_i = \underline{\tilde{\boldsymbol{u}}}_{i,l} \\ \underline{\boldsymbol{x}} = \underline{\tilde{\boldsymbol{x}}}_l}} \quad (6.53)$$

is defined, where $\boldsymbol{D}_i$ is a block upper triangular matrix. The blocks within $\boldsymbol{D}_i$ are given by

$$\boldsymbol{D}_i^{(k_2,k_1)} = \begin{cases} \nabla_{\boldsymbol{u}_i^{(k)}} \boldsymbol{x}^{(k+1)} \Big|_{k=k_1}, & \text{for } k_2 = k_1 + 1 \\ \left(\nabla_{\boldsymbol{x}^{(k)}} \boldsymbol{x}^{(k+1)} \right) \boldsymbol{D}_i^{(k_2-1,k_1)} \Big|_{k=k_2-1}, & \text{for } k_2 > k_1 + 1 \\ \boldsymbol{0}, & \text{else.} \end{cases} \quad (6.54)$$

The blocks $\boldsymbol{D}_i^{(k_2,k_1)}$, $k_1, k_2 \in \mathcal{K}$ have the dimension $n \times m_i$ and represent the influence of the player i's control at time step k_2 on the states at time step k_1. These partial derivatives can be interpreted as part of the numerical solution of the initial value problem which approximates the next state. The matrix $\boldsymbol{D}_i$ employs the partial derivatives with respect to $\boldsymbol{u}_i$ in each time step for the whole corresponding time interval between two time steps. Contrary to this approach, a modification of the matrix $\boldsymbol{D}_i$ is proposed here in order to improve the approximation. Inspired by the trapezoid method for solving initial value problems [Epp13, Section 6.5], the effect of $\boldsymbol{u}^{(k_2)}$ at k_2 on $\boldsymbol{x}^{(k_1)}$ is approximated by means of

$$\begin{aligned} \tilde{\boldsymbol{D}}_i^{(k_2,k_1)} &:= \frac{1}{2} \left(\nabla_{\boldsymbol{u}^{(k_1)}} \boldsymbol{x}^{(k_2)} + \nabla_{\boldsymbol{u}^{(k_1)}} \boldsymbol{x}^{(k_2+1)} \right) \\ &= \frac{1}{2} \left(\boldsymbol{D}_i^{(k_2,k_1)} + \boldsymbol{D}_i^{(k_2+1,k_1)} \right). \end{aligned} \quad (6.55)$$

The modified matrix $\tilde{\boldsymbol{D}}_i$, which is built with the blocks $\tilde{\boldsymbol{D}}_i^{(k_2,k_1)}$ analogously to $\boldsymbol{D}$ with (6.54), takes into account the effect of the control value $\boldsymbol{u}_i^{(k_1)}$ on the interval of $\boldsymbol{x}^{(k_2)}$ until $\boldsymbol{x}^{(k_2+1)}$ and yields a better approximation of the system dynamics.[43]

[43] This modification was applied in experimental work presented in [IEFH18].

Contrary to the aforementioned partial derivatives, the term $\nabla_{\underline{u}_i \underline{u}_i} \underline{x}$ is a third-order tensor and does not exhibit a convenient structure for its computation. Therefore, following the recommendations in [LK12], this term is neglected in favor of more efficient calculations.[44]

6.6.3 Algorithms

The results presented in the previous sections are condensed in two algorithms for the solution of inverse dynamic games by means of MaxEnt IRL. The algorithms summarize the procedure for cost function identification in a dynamic game when an open-loop information structure or a feedback information structure (and corresponding Nash equilibria) lie at hand. The following Algorithm 3 corresponds to the open-loop case.

Algorithm 3 IRL Method in Open-Loop Dynamic Games for Player i.

Input: Observed trajectory set $\mathcal{D}$, dynamics $\boldsymbol{f}$, basis functions $\boldsymbol{\phi}_i$.
Output: Computed player i cost function parameters $\boldsymbol{\theta}_i$.
1: Determine the derivatives of the features $\nabla_{\boldsymbol{x}} \boldsymbol{\phi}_i$, $\nabla_{\boldsymbol{u}_i} \boldsymbol{\phi}_i$, $\nabla_{\boldsymbol{xx}} \boldsymbol{\phi}_i$, and $\nabla_{\boldsymbol{u}_i \boldsymbol{u}_i} \boldsymbol{\phi}_i$.
2: Determine the matrix $\tilde{\boldsymbol{D}}_i$ with (6.55).
3: Determine the first and second derivatives $\tilde{\boldsymbol{g}}_{i,l}$ and $\tilde{\boldsymbol{G}}_{i,l}$ evaluated at the trajectories $\tilde{\zeta}_l$ by means of (6.49) and (6.50), respectively.
4: Calculate the MLE according to (6.29) using the log-likelihood function (6.48).
5: **return** $\boldsymbol{\theta}_i$.

The next Algorithm 4 gives the necessary steps for solving inverse dynamic games with a feedback information structure based on MaxEnt IRL.

Algorithm 4 IRL Method in Feedback Dynamic Games for Player i.

Input: Observed trajectory set $\mathcal{D}$, dynamics $\boldsymbol{f}$, basis functions $\boldsymbol{\phi}_i$.
Output: Computed player i cost function parameters $\boldsymbol{\theta}_i$.
1: Determine the system dynamics with respect to player i by means of (6.40) and the features according to (6.41).
2: Determine the derivatives of the features $\nabla_{\boldsymbol{x}} \boldsymbol{\phi}_i$, $\nabla_{\boldsymbol{u}_i} \boldsymbol{\phi}_i$, $\nabla_{\boldsymbol{xx}} \boldsymbol{\phi}_i$, and $\nabla_{\boldsymbol{u}_i \boldsymbol{u}_i} \boldsymbol{\phi}_i$.
3: Determine the matrix $\tilde{\boldsymbol{D}}_i$ with (6.55).
4: Determine the first and second derivatives $\tilde{\boldsymbol{g}}_{i,l}$ and $\tilde{\boldsymbol{G}}_{i,l}$ evaluated at the trajectories $\tilde{\zeta}_l$ by means of (6.49) and (6.50), respectively.
5: Calculate the MLE according to (6.42) using the log-likelihood function (6.48).
6: **return** $\boldsymbol{\theta}_i$.

[44] Neglecting this term does not have any effect for most problems. For example, this term is always zero for the broad class of nonlinear control-affine systems (3.11).

Remark 6.5:

Step 1 and Step 2 of Algorithms 3 and 4, respectively, can also be calculated prior to the identification procedure since they are independent of the observed data.

Remark 6.6:

The methods shown in this chapter are formulated for a finite-horizon problem, i.e. $k_E \in \mathbb{N}_{>0}$ in (6.2). However, all results can still be applied if the assumed underlying LQ dynamic game has an infinite horizon $k_E \to \infty$. The presented method solely requires the availability of observed state trajectories $\underline{x} \in \mathbb{R}^{nK_i}$ and $\underline{u}_i \in \mathbb{R}^{m_i K_i}$ where $K_i \ll \infty$ (cf. Definition 6.1). For adequate results, $[0; K_i]$ should be a sufficiently representative interval of the complete time span $[0, \infty)$.

6.7 Application to Inverse LQ Dynamic Games

This section presents an exemplary application of IRL for solving inverse LQ dynamic games in order to illustrate the procedures presented in Algorithms 3 and 4. In the following, both inverse open-loop dynamic games and inverse feedback dynamic games are examined.

6.7.1 Open-Loop

Consider N-player LQ dynamic games according to Definition A.7. Therefore, each player applies his controls to a system described by the difference equation

$$\boldsymbol{x}^{(k+1)} = \boldsymbol{A}_D^{(k)} \boldsymbol{x}^{(k)} + \sum_{j=1}^{N} \boldsymbol{B}_{D,j}^{(k)} \boldsymbol{u}_j^{(k)}. \tag{6.56}$$

Furthermore, each player $i \in \mathcal{P}$ selects an open-loop strategy $\boldsymbol{\gamma}_i^{(k)} = \boldsymbol{u}_i^{(k)}$ (cf. Definition A.4) based on a quadratic cost function of the form

$$J_i = -\frac{1}{2} \sum_{k=1}^{k_E} \left(\left(\boldsymbol{x}^{(k)}\right)^\top \boldsymbol{Q}_i \boldsymbol{x}^{(k)} + \left(\boldsymbol{u}_i^{(k)}\right)^\top \boldsymbol{R}_{ii} \boldsymbol{u}_i^{(k)} \right), \tag{6.57}$$

where $\boldsymbol{Q}_i$ and $\boldsymbol{R}_{ii} \prec \boldsymbol{0}$ are symmetric matrices.[45] The cost function (6.57) does not include the terms which penalize the controls $\boldsymbol{u}_j^{(k)}, j \neq i$ of all other players (cf. (A.16)). This is due to the fact that these controls do not have any influence on the solution of open-loop dynamic games and therefore can be neglected. This follows e.g. from the necessary conditions for Nash equilibria given in Theorem A.1.

[45] The negative sign is considered in this chapter according to (6.2) and thus $\boldsymbol{R}_{ii}$ is negative definite instead of positive definite to ensure a meaningful problem.

In order to apply the results of the previous sections to linear-quadratic dynamic games, it is necessary to reformulate quadratic objective functions such that the structure in (6.2) is obtained. Furthermore, the partial derivatives of the states with respect to the controls have a particular structure in the case of linear system dynamics. These aspects will be examined and presented in the following.

Features in LQ Open-Loop Dynamic Games

The features in the vector $\boldsymbol{\phi}_i$ which correspond to the $\frac{1}{2}(n^2+n)$ non-redundant elements of the matrix $\boldsymbol{Q}_i$ are given by

$$\phi_{i,rc}^{Q_i} = -\frac{1}{2} x_r^{(k)} x_c^{(k)}, \quad c = 1, \dots, n,\ r = 1, \dots, c. \tag{6.58}$$

Similarly, for the $\frac{1}{2}(m_i^2+m_i)$ parameters of the symmetric matrix $\boldsymbol{R}_{ii}$, we obtain the features

$$\phi_{i,rc}^{R_{ii}} = -\frac{1}{2} u_{i,r}^{(k)} u_{i,c}^{(k)}, \quad c = 1, \dots, m_i,\ r = 1, \dots, c. \tag{6.59}$$

For $r = c$, the parameters which are multiplied with $\phi_{i,rc}^{Q_i}$ and $\phi_{i,rc}^{R_{ii}}$ correspond to the r-th diagonal entry of the matrix $\boldsymbol{Q}_i$ and $\boldsymbol{R}_{ii}$, respectively. For the case where $c \neq r$, these parameters correspond to two times the off-diagonal (symmetric) entries $Q_{i,rc} = Q_{i,cr}$ and $R_{ii,rc} = R_{ii,cr}$, respectively.

System Dynamics

The linear system dynamics lead to the relations

$$\nabla_{\boldsymbol{x}^{(k)}} \boldsymbol{x}^{(k+1)} = \boldsymbol{A}_D^{(k)} \quad \text{and} \quad \nabla_{\boldsymbol{u}_i^{(k)}} \boldsymbol{x}^{(k+1)} = \boldsymbol{B}_{D,i}^{(k)}. \tag{6.60}$$

Then, $\tilde{\boldsymbol{D}}_i$ can be determined with (6.54) and (6.55).

The following example illustrates the solution of an inverse dynamic game with MaxEnt IRL to identify cost function parameters:

> **Example 6.1:**
>
> *Consider a two-player discrete-time dynamic game with system dynamics* (6.56) *defined by the matrices*
>
> $$\boldsymbol{A}_D^{(k)} = \begin{bmatrix} 1 & 0.02 \\ 0 & 1 \end{bmatrix}, \quad \boldsymbol{B}_{D,i}^{(k)} = \begin{bmatrix} 0.0002 \\ 0.02 \end{bmatrix}, \quad i \in \{1, 2\},\ k \in \mathcal{K} \tag{6.61}$$

and the initial value $\mathbf{x}^{(1)} = \begin{bmatrix} 1 & -1 \end{bmatrix}^\top$. *These matrices correspond to a continuous-time double-integrator system (cf. Example 5.2) sampled with* $\Delta T = 0.02$ s. *In addition, let the quadratic cost function of the players be given by* (6.57), *where*

$$\mathbf{Q}_1 = -\begin{bmatrix} 4 & 1 \\ 1 & 3 \end{bmatrix}, \quad \mathbf{R}_{11} = -1, \quad \mathbf{Q}_2 = -\begin{bmatrix} 10 & 1 \\ 1 & 2 \end{bmatrix}, \quad \mathbf{R}_{22} = -1. \tag{6.62}$$

Then, the features corresponding to the cost function of player i are given by

$$\begin{aligned} \phi_{i,11}^{\mathbf{Q}_i} &= -\frac{1}{2}\left(x_1^{(k)}\right)^2, & \phi_{i,12}^{\mathbf{Q}_i} &= -\frac{1}{2}x_1^{(k)}x_2^{(k)}, \\ \phi_{i,22}^{\mathbf{Q}_i} &= -\frac{1}{2}\left(x_2^{(k)}\right)^2, & \phi_{i,11}^{\mathbf{R}_{ii}} &= -\frac{1}{2}\left(u_i^{(k)}\right)^2. \end{aligned} \tag{6.63}$$

The cost functions J_i *of player i can be rewritten as*

$$J_i = -\sum_{k=1}^{k_E}\left[\theta_{i,(1)}\phi_{i,11}^{\mathbf{Q}_i} + \theta_{i,(2)}\phi_{i,12}^{\mathbf{Q}_i} + \theta_{i,(3)}\phi_{i,22}^{\mathbf{Q}_i} + \theta_{i,(4)}\phi_{i,11}^{\mathbf{R}_{ii}}\right], \quad i \in \{1,2\}. \tag{6.64}$$

with the cost function parameters

$$\begin{aligned} \boldsymbol{\theta}_1 &= \boldsymbol{\theta}_1^* = \begin{bmatrix} 4 & 2 & 3 & 1 \end{bmatrix}^\top, \\ \boldsymbol{\theta}_2 &= \boldsymbol{\theta}_2^* = \begin{bmatrix} 10 & 2 & 2 & 1 \end{bmatrix}^\top. \end{aligned}$$

Now we assume $k_E = 250$ *and use the coupled Riccati equations* (3.60) *to calculate the OLNE*[46] *and obtain the trajectory set* ζ^*. *The state and control trajectories belonging to this set are corrupted by Gaussian white noise such that the resulting trajectories have a signal-to-noise ratio (SNR) of 30 dB. A total number of 30 realizations are generated, leading to* $n_t = 30$ *trajectories* $\tilde{\zeta}_l$, $l \in \{1, ..., n_t\}$. *These are used to evaluate the log-likelihood function* (6.48), *for which we compute the necessary partial derivatives. The partial derivative* $\nabla_{\underline{\mathbf{x}}} J_i$ *is given by* (6.51), *where*

$$\left(\nabla_{\mathbf{x}}\boldsymbol{\phi}_i\right)\boldsymbol{\theta}_i = \begin{bmatrix} \theta_{i,(1)}x_1^{(k)} + \frac{1}{2}\theta_{i,(2)}x_2^{(k)} \\ \frac{1}{2}\theta_{i,(2)}x_1^{(k)} + \theta_{i,(3)}x_2^{(k)} \end{bmatrix}. \tag{6.65}$$

Similarly, $\nabla_{\underline{\mathbf{u}}_i} J_i \in \mathbb{R}^{k_E}$ *is determined by using the partial derivative*

$$\left(\nabla_{\mathbf{u}_i}\boldsymbol{\phi}_i\right)\boldsymbol{\theta}_i = \theta_{i,(4)}u_i^{(k)}. \tag{6.66}$$

For the second partial derivates we obtain

$$\nabla_{\underline{\mathbf{x}}\underline{\mathbf{x}}} J_i = \operatorname{blkdiag}\left(-\begin{bmatrix} \theta_{i,(1)} & \frac{1}{2}\theta_{i,(2)} \\ \frac{1}{2}\theta_{i,(2)} & \theta_{i,(3)} \end{bmatrix}, \dots, -\begin{bmatrix} \theta_{i,(1)} & \frac{1}{2}\theta_{i,(2)} \\ \frac{1}{2}\theta_{i,(2)} & \theta_{i,(3)} \end{bmatrix}\right) \tag{6.67}$$

$$\nabla_{\underline{\mathbf{u}}_i\underline{\mathbf{u}}_i} J_i = \operatorname{blkdiag}\left(-\theta_{i,(4)}, \dots, -\theta_{i,(4)}\right). \tag{6.68}$$

The MLE (6.29) is performed using a numerical optimization method, namely the Broyden-Fletcher-Goldfarb-Shannon (BFGS) method. We obtain, after normalizing with respect to $\theta_{i,(4)}$ for a better comparability, the estimated parameters

$$\begin{aligned}\hat{\boldsymbol{\theta}}_1 &= \begin{bmatrix}3.88 & -2.22 & 2.98 & 1.00\end{bmatrix}^\top \\ \hat{\boldsymbol{\theta}}_2 &= \begin{bmatrix}10.19 & -1.69 & 2.12 & 1.00\end{bmatrix}^\top .\end{aligned} \tag{6.69}$$

Consider now the feature count

$$\hat{\boldsymbol{\mu}} = \frac{1}{2}\sum_{k=1}^{k_E}\begin{bmatrix}(x_1^{(k)})^2 & x_1^{(k)}x_2^{(k)} & (x_2^{(k)})^2 & (u_1^{(k)})^2 & (u_2^{(k)})^2\end{bmatrix}^\top . \tag{6.70}$$

The feature count of the trajectory $\hat{\zeta}$ generated by solving an LQ dynamic game with the estimated parameters (6.69) is given by

$$\hat{\boldsymbol{\mu}} = \begin{bmatrix}9.88 & -12.75 & 17.04 & 32.46 & 12.66\end{bmatrix} . \tag{6.71}$$

The mean feature count of observed trajectories is

$$\tilde{\boldsymbol{\mu}} = \begin{bmatrix}9.88 & -12.76 & 17.05 & 32.90 & 12.87\end{bmatrix} , \tag{6.72}$$

suggesting, in consideration of (6.10), that the estimated parameters $\hat{\boldsymbol{\theta}}_i$ are different to the original parameters $\boldsymbol{\theta}_i^*$, but lead to very similar costs. The original trajectory ζ^* and the estimated $\hat{\zeta}$ are depicted in Figure 6.2, showing that the identified parameters are able to explain the observed behavior.

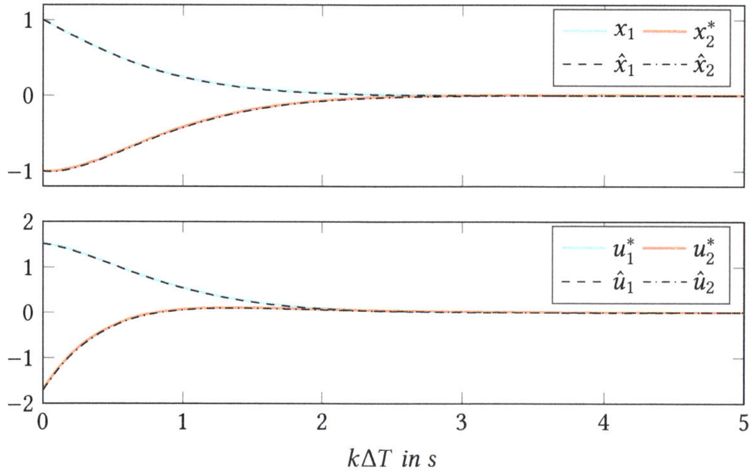

Figure 6.2: Observed trajectories and trajectories following from the estimated parameters of the LQ dynamic game in Example 6.1

6.7.2 Feedback Case

Consider now a LQ dynamic game where players choose their feedback strategies (cf. Definition A.5) based on a quadratic cost function. Since we consider a feedback (or MPS) information pattern, the general quadratic cost functions are given by

$$J_i = -\frac{1}{2}\sum_{k=1}^{k_E}\left(\boldsymbol{x}^{(k)\top}\boldsymbol{Q}_i\boldsymbol{x}^{(k)} + \sum_{j=1}^{N}\boldsymbol{u}_j^{(k)\top}\boldsymbol{R}_{ij}\boldsymbol{u}_j^{(k)}\right), \quad i \in \mathcal{P}, \tag{6.73}$$

and the resulting feedback strategies are given by (6.39). This relation can be used to obtain system dynamics from the point of view of player i given by

$$\begin{aligned}\boldsymbol{x}^{(k+1)} &= \boldsymbol{A}_D^{(k)}\boldsymbol{x}^{(k)} + \boldsymbol{B}_{D,i}^{(k)}\boldsymbol{u}_i^{(k)} - \sum_{\substack{j=1\\ j\neq i}}^{N}\boldsymbol{B}_{D,j}^{(k)}\boldsymbol{K}_j^{(k)}\boldsymbol{x}^{(k)}\\ &= \left(\boldsymbol{A}_D^{(k)} - \sum_{\substack{j=1\\ j\neq i}}^{N}\boldsymbol{B}_{D,j}\boldsymbol{K}_j^{(k)}\right)\boldsymbol{x}^{(k)} + \boldsymbol{B}_{D,i}^{(k)}\boldsymbol{u}_i^{(k)}\\ &=: \bar{\boldsymbol{A}}_{D,i}^{(k)}\boldsymbol{x}^{(k)} + \boldsymbol{B}_{D,i}^{(k)}\boldsymbol{u}_i^{(k)}.\end{aligned} \tag{6.74}$$

As described in Section 6.5, inverse feedback dynamic games can be solved by exploiting the knowledge of the strategies $\boldsymbol{\gamma}_i^{(k)}$. For the case of LQ dynamic games this means that the feedback matrices $\boldsymbol{K}_i^{(k)*}$, $i \in \mathcal{P}, k \in \mathcal{K}$ are given.

Remark 6.7:

In the typical case that $\boldsymbol{K}_i^{(k)*}$, $i \in \mathcal{P}$, $k \in \mathcal{K}$ are not known, it is possible to assume an infinite horizon, i.e. $k_E \to \infty$ and estimate a constant feedback law which approximates the relationship between the controls and the states (cf. Section 5.4.2).[47] In the case of an infinite-horizon inverse LQ dynamic game, then the estimation can be effectively done by means of (5.36).

---

[46] The continuous-time equations were used as the considered time step $\Delta T = 0.02$ s allows a quasi-continuous analysis instead of the use of discrete-time equations for determining Nash equilibria. The interested reader is referred to Section A.5 of the Appendix where references on discrete-time Riccati equations are given.

[47] If the limit of the Riccati matrix $\boldsymbol{P}_i^{(k)}$ for ($k = k_E \to \infty$) exists, then it corresponds to a FNE for the infinite-horizon dynamic game. In general, other FNE solutions may also exist which are not necessarily related to the aforementioned solution [BO99, P. 290].

Features in LQ Feedback Dynamic Games

By using the known feedback control matrices $K_i^{(k)*}$, the quadratic cost function (6.73) of player i can be rewritten as

$$J_i = \frac{1}{2}\sum_{k=1}^{k_E}\left(x^{(k)\top}Q_i x^{(k)} + \sum_{\substack{j=1\\ j\neq i}}^{N} x^{(k)\top}K_j^{(k)\top}R_{ij}K_j^{(k)}x^{(k)} + u_i^{(k)\top}R_{ii}u_i^{(k)}\right). \tag{6.75}$$

The features corresponding to the entries of Q_i und R_{ii} are identical to the open-loop case (cf. (6.58) und (6.59)). In the feedback case, we further have the features corresponding to the entries of R_{ij} which are given by

$$\phi_{i,rc}^{R_{ij}} = -\frac{1}{2}(K_j^{(k)*}x^{(k)})_r(K_j^{(k)*}x^{(k)})_c, \quad c = 1,\dots,m_j,\ r = 1,\dots,c, \tag{6.76}$$

where $(K_j^{(k)*}x^{(k)})_r$ denotes the r-th entry of the vector $K_j^{(k)*}x^{(k)}$. Similar to the matrices Q_{ii} and R_{ii}, the main diagonal elements of R_{ij} correspond to parameters which weight the features $\phi_{i,rr}^{R_{ij}}$, $r = 1,\dots,m_i$. For the case where $c \neq r$, these parameters correspond to two times the off-diagonal (symmetric) entries $R_{ij,rc} = R_{ij,cr}$, respectively.

System Dynamics

The linear system dynamics lead to the relations

$$\nabla_{x^{(k)}}x^{(k+1)} = \bar{A}_{D,i}^{(k)} \quad \text{and} \quad \nabla_{u_i^{(k)}}x^{(k+1)} = B_{D,i}^{(k)}. \tag{6.77}$$

Then, $\tilde{D}_i$ can be computed with (6.54) and (6.55).

Example 6.2:

Consider a two-player discrete-time dynamic game with the system dynamics (6.61)*, the initial value* $x^{(1)} = \begin{bmatrix}1 & -1\end{bmatrix}^\top$*, and cost functions of the form* (6.73) *with the cost function matrices*

$$\begin{aligned} Q_1 &= \begin{bmatrix}8 & 0\\ 0 & 2\end{bmatrix}, \quad R_{11} = 1, \quad R_{12} = 1,\\ Q_2 &= \begin{bmatrix}1 & 0\\ 0 & 4\end{bmatrix}, \quad R_{22} = 1, \quad R_{21} = 0.3. \end{aligned} \tag{6.78}$$

The LQ dynamic game leads to feedback strategies

$$u_i^{(k)*} = \gamma_i^{(k)*}(x) = \begin{bmatrix}k_{2,(1)}^{(k)*} & k_{2,(2)}^{(k)*}\end{bmatrix}\begin{bmatrix}x_1^{(k)}\\ x_2^{(k)}\end{bmatrix}. \tag{6.79}$$

We assume that $\boldsymbol{K}_i^* = \begin{bmatrix} k_{i,(1)}^{(k)*} & k_{i,(2)}^{(k)*} \end{bmatrix}$ *is not known and is approximated by a constant feedback law* $\tilde{\boldsymbol{K}}_i$ *to be identified, as mentioned in Remark 6.7. We obtain* $||\tilde{\boldsymbol{K}}_i \boldsymbol{x} - \boldsymbol{u}_i^*|| < 0.02$ *for all* $i = \{1, 2\}$*. The approximation of the time-variant control matrices* $\boldsymbol{K}_i$ *by means of the constant matrices* $\hat{\boldsymbol{K}}_i$ *is shown in Figure 6.3.*

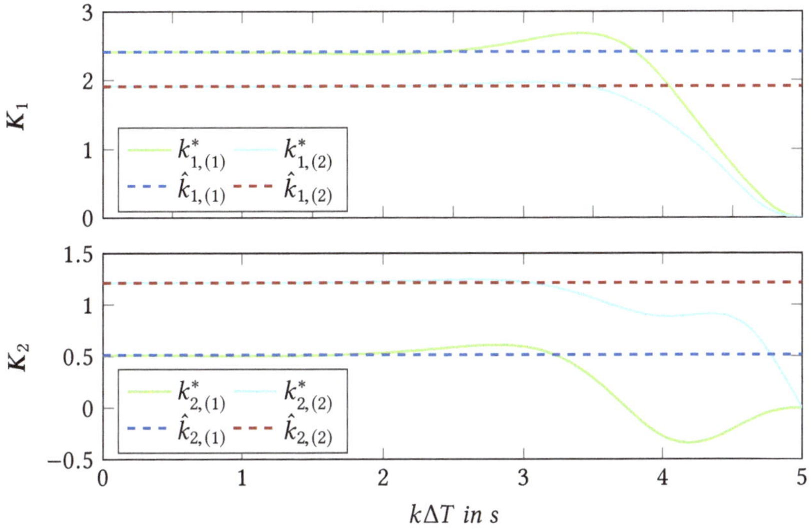

Figure 6.3: Nash equilibrium feedback matrices $\boldsymbol{K}_i^{(k)*}$ and their approximation by means of constant feedback matrices $\hat{\boldsymbol{K}}_i$ in Example 6.2

The features corresponding to the cost function of player i are given by:

$$\begin{aligned} \phi_{i,11}^{Q_i} &= -\frac{1}{2}\left(x_1^{(k)}\right)^2, \quad \phi_{i,22}^{Q_i} = -\frac{1}{2}\left(x_2^{(k)}\right)^2 \\ \phi_{i,11}^{R_{ii}} &= -\frac{1}{2}\left(u_i^{(k)}\right)^2, \quad \phi_{i,11}^{R_{ij}} = -\frac{1}{2}\left(k_{j,(1)}^* x_1^{(k)} + k_{j,(2)}^* x_2^{(k)}\right)^2 \end{aligned} \tag{6.80}$$

The cost functions J_i*,* $i \in \{1, 2\}$ *, can be rewritten as*

$$J_i = -\sum_{k=1}^{k_E} \left[\theta_{i,(1)}\phi_{i,11}^{Q_i} + \theta_{i,(2)}\phi_{i,22}^{Q_i} + \theta_{i,(3)}\phi_{i,11}^{R_{ii}} + \theta_{i,(4)}\phi_{i,11}^{R_{ij}}\right], \; i,j \in \{1,2\}, \; i \neq j, \tag{6.81}$$

where the cost function parameters are given by

$$\begin{aligned} \boldsymbol{\theta}_1 &= \begin{bmatrix} 8 & 2 & 1 & 1 \end{bmatrix}^\top, \\ \boldsymbol{\theta}_2 &= \begin{bmatrix} 1 & 4 & 1 & 0.3 \end{bmatrix}^\top. \end{aligned}$$

The calculated FNE trajectory ζ^* *is used to identify cost function parameters which explain it. However, this time the exact FNE trajectory* ζ^* *and one single demonstration, i.e.* $n_t = 1$*,* *are used. Using the MLE* (6.42) *which is determined again with the BFGS method, we obtain the cost function parameters*

$$\hat{\theta}_1 = \begin{bmatrix} 7.67 & 0.148 & 1.00 & 2.26 \end{bmatrix}^\top , \\ \hat{\theta}_2 = \begin{bmatrix} -1.44 & 2.47 & 1.00 & 0.72 \end{bmatrix}^\top . \tag{6.82}$$

Similar to last example, we consider the extended feature count

$$\hat{\tilde{\mu}} = \frac{1}{2} \sum_{k=1}^{k_E} \begin{bmatrix} \phi_{1,11}^{Q_1} & \phi_{1,22}^{Q_1} & \phi_{1,11}^{R_{11}} & \phi_{1,11}^{R_{12}} & \phi_{1,11}^{R_{22}} & \phi_{1,11}^{R_{21}} \end{bmatrix}^\top . \tag{6.83}$$

for both the observed trajectory ζ^* *and the trajectory* $\hat{\zeta}$ *corresponding to the parameters* (6.82), *obtaining*

$$\hat{\tilde{\mu}} = \begin{bmatrix} 10.44 & 15.92 & 1.34 & 10.37 & 10.41 & 1.34 \end{bmatrix}^\top \tag{6.84}$$

and

$$\tilde{\tilde{\mu}} = \begin{bmatrix} 10.44 & 15.93 & 1.34 & 10.37 & 10.37 & 1.34 \end{bmatrix}^\top , \tag{6.85}$$

and indicating that the identified parameters indeed approximate the observed trajectory adequately (cf. Example 6.1).

6.8 Method Limitations

Some potential limitations of the presented mehods shall be discussed before concluding this chapter. The introduced IRL-based inverse dynamic game methods can cope with truncated trajectories in $[0, K_i]$ with $K_i < K_E$ as long as these represent the complete trajectories adequately (cf. Remark 6.6). Small values of K_i compared to K_E may deteriorate the results, i.e. the results improve the closer K_i is to K_E.

Noise-corrupted trajectories can also represent an issue since the approach indirectly attempts to equalize the feature count values of observed trajectories with the ones which would arise from the probability density function with identified parameters. On the other hand, equalizing feature count values may lead to a greater robustness in case the features, i.e. the basis functions, are not specified correctly. The effects of these issues on the identification results will be examined in Chapter 7.

Finally, a further possible detriment can arise if the available trajectories do not constitute a Nash equilibrium. The method is based on the probability density function (5.1) which includes the implicit assumption that each player's decision was not directly affected by the choice of the other players' controls, a sufficient condition of which is given by the availability of trajectories representing a Nash equilibrium. In addition, the method for feedback information structures leverages the availability of feedback control laws. If the control laws describe the functional relationship between states and controls, then the modified system dynamics still reflect the actions of the other players. Therefore, the IRL methods have the potential of being robust to at least mild deviations from the Nash equilibrium. Indeed, the basis of the presented results is Assumption 6.2, which does not demand that the observed trajectories are exactly equal to a deterministic result of the dynamic game with cost function parameters $\boldsymbol{\theta}_i^*$. This allows for the estimation of cost function parameters $\hat{\boldsymbol{\theta}}_i$ from trajectories which represent and resemble Nash equilibrium trajectories, but may deviate from this optimality.

6.9 Conclusion

In this chapter, IRL was considered as a means to solve inverse problems in dynamic games. The principle of maximum entropy was applied to the dynamic game scenario and the obtained results were used to derive probability density functions to model the origin of observed dynamic game trajectories. Based on these, a maximum-likelihood estimation of the cost function parameters was proposed for the case when players apply Nash equilibrium strategies. Both open-loop and feedback strategies were regarded. In addition, the unbiasedness of this maximum-likelihood estimation was proved under typical IRL assumptions. The results of this chapter lay the theoretical foundation for the application of MaxEnt IRL for identifying cost function parameters of players in a dynamic game. Finally, solutions of inverse linear-quadratic dynamic games were shown to illustrate the presented methods and their applicability.

After this last chapter presenting theoretical results on inverse dynamic games and their solution, the following chapters present a comparison between different method classes in both simulations and a real application.

7 Simulations

In the previous chapters, inverse problems in dynamic game theory were introduced and two main classes of methods were proposed for their solution, namely the residual-based IOC method and an IRL-based approach. These classes of methods are different from a theoretical and conceptual point of view given their contrasting origins in automatic control and computer science. This chapter aims at presenting the capabilities of both classes of methods and comparing them by using different test scenarios in simulations. In this way, their strengths and weaknesses shall be examined. Moreover, the IOC and IRL methods are systematically compared with a Direct Bilevel (DB) approach which is based on the solution of a forward dynamic game in each iteration (see Section 2.1.1).

This chapter starts with a mathematical description of the DB approach used for comparison to the new inverse dynamic game methods. Afterwards, the considered scenarios are introduced before explaining the general evaluation procedure applied in this chapter, as well as the metrics used for comparison. Then, the simulation results are presented and discussed. These results include an evaluation of the methods' robustness to measurement noise and errors in the basis function vectors. After shortly analyzing the computation times of the methods, the chapter ends with conclusions based on the obtained insights.

7.1 Direct Bilevel Approach

The Direct Bilevel (DB) approach considered in this chapter is a direct extension of the method introduced in [MTL10] (see also Section 2.1.1), which was recently formulated in [MFP17a]. It aims to determine cost function parameters $\boldsymbol{\theta} = (\boldsymbol{\theta}_1, ..., \boldsymbol{\theta}_N)$ such that the corresponding Nash equilibrium trajectories approximate the observed state and control trajectories. For this objective, the following optimization problem can be formulated:

$$\min_{\boldsymbol{\theta}} J_{\mathrm{DB}} = \int_0^T ||\boldsymbol{x}_{\boldsymbol{\theta}}(t) - \tilde{\boldsymbol{x}}(t)||^2 + \sum_{j=1}^{N} ||\boldsymbol{u}_{\boldsymbol{\theta},j}(t) - \tilde{\boldsymbol{u}}_j(t)||^2 \, \mathrm{d}t, \tag{7.1}$$

where $\boldsymbol{x}_{\boldsymbol{\theta}}(t)$ and $\boldsymbol{u}_{\boldsymbol{\theta},i}(t)$ denote Nash equilibrium trajectories resulting from cost functions with parameters $\boldsymbol{\theta}$. The objective functional J_{DB} provides a natural squared-error metric between candidate state and control trajectories and the observed Nash equilibrium state $\tilde{\boldsymbol{x}}(t)$ and control trajectories $\tilde{\boldsymbol{u}}_i(t)$. Note that if the observed trajectories correspond to a Nash equilibrium with cost function parameters $\boldsymbol{\theta}^* \in \Theta$, then the optimization problem is solved for any $\boldsymbol{\theta}$ which also belongs to the solution set Θ according to the equivalence of cost functions (cf. Section B.2 in the Appendix) which imply identical Nash equilibrium trajectories. Some details need to be considered for practical implementation of this approach. These are given in Section B.6 in the Appendix.

7.2 Simulation Scenarios

In this chapter, two main simulation scenarios are considered:

1. a non-linear open-loop dynamic game with two players controlling a ball-on-beam system
2. a generic LQ feedback dynamic game with three players

In the first scenario, the ball-on-beam is chosen as a dynamic system. It is a well-known benchmark system in control engineering since it poses a challenging stabilization problem which is representative of the difficulties generated by growing nonlinearities [HSK92, BSLK97]. This scenario shall serve to show the solution of inverse dynamic games with open-loop strategies.

The second scenario consists of a LQ dynamic game with feedback strategies. Considering the class of LQ dynamic games allows for an analysis with the tools developed in Chapter 5. Furthermore, in order to increase the complexity of the LQ dynamic game, a generic dynamic game is considered where three players influence a system by means of two control variables each. This scenario is used for the examination of inverse feedback dynamic games.

For each scenario, one IOC-based method, one IRL method and a DB approach shall be compared. The performance comparison is first done with assumed perfect observations of the Nash trajectories. Nevertheless, an evaluation of the robustness of all methods to noise in the observations is also presented.

7.3 Evaluation Method

In the following, the evaluation method is presented. After describing the general steps constituting the whole evaluation process, the metrics used for the comparison are introduced.

7.3.1 General Steps

The evaluation procedure used in this chapter is summarized in Figure 7.1 and shall be explained in the following. For the simulation environment, a cost function structure defined by a linear combination of basis functions according to (4.2) is assumed. Therefore, it is first necessary to define a basis function vector $\boldsymbol{\phi}_i$ and a parameter vector $\boldsymbol{\theta}_i^*$ for each player $i \in \mathcal{P}$. These cost functions are used to **calculate the Nash equilibrium** trajectories of the states $\boldsymbol{x}^*(t)$ and the controls $\boldsymbol{u}_i^*(t)$.[48] For the case where perfect observations are assumed, the observations $\tilde{\boldsymbol{x}}(t)$ and $\tilde{\boldsymbol{u}}_i(t)$ correspond to the calculated Nash equilibrium trajectories $\boldsymbol{x}(t)$ and $\boldsymbol{u}_i(t)$. Otherwise, Gaussian white noise $\epsilon^{\boldsymbol{x}}$ and $\epsilon^{\boldsymbol{u}_i}$ is added to the Nash equilibrium state trajectories and control trajectories to form the observations, respectively. The generated observations $\tilde{\boldsymbol{x}}(t)$ and $\tilde{\boldsymbol{u}}(t)$ simulate dynamic game data which is measured and results from the interaction between the players. Based on these observations, in the **inverse dynamic game** step, one of the inverse dynamic game methods is applied to obtain estimations of the cost function parameters $\hat{\boldsymbol{\theta}}_i$ for all players $i \in \mathcal{P}$. At this point, the analysis of the identification results may be conducted based on the **parameter deviation**, i.e. the comparison of the estimated cost function parameters with the ground truth. Nevertheless, particularly for the robustness evaluation, it will be examined whether potentially inexact identification of the cost function parameters has a considerable impact on the capability to approximate the observations. For these cases, identified trajectories $\hat{\boldsymbol{x}}(t)$ and $\hat{\boldsymbol{u}}_i(t)$ are determined. This is done by **calculating the Nash equilibrium** again, yet this time based on the estimated parameters $\hat{\boldsymbol{\theta}}_i$ of all players. By comparing the identified trajectories with the ground truth trajectories, it is possible to evaluate if the estimated parameters can describe the observed outcome of the dynamic game despite a potential deviation from the real parameters. We **determine the trajectory deviation** by calculating the metrics $\delta^{\boldsymbol{x}}$, $\delta^{\boldsymbol{u}}$, and $\Delta^{\boldsymbol{\theta}}$ which are presented in the next section.

7.3.2 Evaluation Metrics

As previously mentioned, the results of the inverse dynamic game methods are evaluated with respect to the quality of the cost function parameter identification. Furthermore, the approximation of the observed trajectories by means of the trajectories of the identified model are also assessed. For these two objectives, two different metrics are used which are introduced in the following.

[48] All simulated Nash equilibrium trajectories $\boldsymbol{x}^*(t)$ and $\boldsymbol{u}_i^*(t)$ are calculated using a continuous-time formulation of the dynamic game using the different theorems from Section 3.6, depending on the information structure and strategy types. The IRL-based methods, which were developed considering a discrete-time formulation, shall be given equivalent system dynamics corresponding to the selected time step ΔT as shown in Examples 6.1 and 6.2. Furthermore, one single trajectory set will be used, i.e. $n_t = 1$ for the IRL methods.

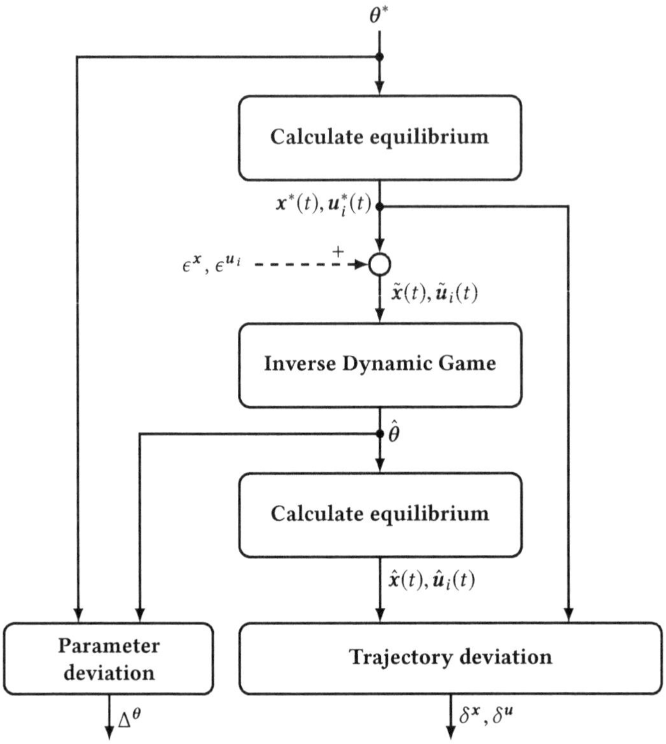

Figure 7.1: Evaluation procedure for simulation results

Cost Function Parameters

Since identification of the cost function parameters is only possible up to a scaling constant, the comparison is done after a normalization process. The ground truth parameters $\boldsymbol{\theta}_{i,\mathrm{GT}}$ and the identified parameters $\hat{\boldsymbol{\theta}}_i$ are normalized with respect to an arbitrary parameter. In this case and without loss of generality, the last entry of the vector $\boldsymbol{\theta}_i$ is chosen. This is done for all players $i \in \mathcal{P}$. Therefore, for the ground truth normalized parameter vectors $\boldsymbol{\theta}^*_{i,(\mathrm{norm})}$ and the normalized estimated parameter vectors $\hat{\boldsymbol{\theta}}_{i,(\mathrm{norm})}$ of player i, we have

$$\{\boldsymbol{\theta}^*_{i,(\mathrm{norm})}\}_p = \frac{\{\boldsymbol{\theta}^*_i\}_p}{\{\boldsymbol{\theta}^*_i\}_{M_i}} \quad \text{and} \quad \{\hat{\boldsymbol{\theta}}_{i,(\mathrm{norm})}\}_p = \frac{\{\hat{\boldsymbol{\theta}}_i\}_p}{\{\hat{\boldsymbol{\theta}}_i\}_{M_i}}, \qquad \forall p \in \{1, \ldots, M_i\}, \tag{7.2}$$

where $\{\boldsymbol{\theta}_i\}_p$ denotes the p-th entry of the parameter vector $\boldsymbol{\theta}_i$.[49] The parameter $\{\boldsymbol{\theta}_i\}_{M_i}$ is therefore the last entry of the vector $\boldsymbol{\theta}_i$. By using the normalized parameters, the relative parameter error is defined as

$$\delta_p^{\theta} = \frac{\{\hat{\boldsymbol{\theta}}_{i,(\mathrm{norm})}\}_p}{\{\boldsymbol{\theta}^*_{i,(\mathrm{norm})}\}_p}, \quad \forall p \in \{1,\ldots,M_i\}. \tag{7.3}$$

The comparison of the parameters is done by means of the absolute value of the relative error of the parameters

$$\Delta_p^{\theta} = \left|1 - \delta_p^{\theta}\right|, \quad \Delta_p^{\theta} \in [0,\infty)\,. \tag{7.4}$$

Therefore, the closer the absolute values of the relative error Δ_p^{θ} are to zero, the stronger the similarity is between identified and ground truth parameters. The mean and maximum value of Δ_p^{θ} will be considered. These are denoted with $\Delta_{p,\mathrm{mean}}^{\theta}$ and $\Delta_{p,\max}^{\theta}$, respectively.

Comparison of Trajectories

Before introducing the considered metrics for comparing trajectories, it is important to note that in the simulations, trajectories are available in the form of a series of K_i data points described by the set

$$\mathcal{T}_i = \{t_k \in [0,T] \mid 1 \le k \le K_i \wedge 0 \le t_k \le T\}. \tag{7.5}$$

In the following, $K_i = K$ is set for all $i \in \mathcal{P}$ to ease the comparison between ground truth and estimated trajectories. The estimated trajectories $\hat{\boldsymbol{x}}(t)$ and $\hat{\boldsymbol{u}}_i(t)$, $i \in \mathcal{P}$, are the ones which arise from the solution of the dynamic game with the estimated cost function parameters $\hat{\boldsymbol{\theta}}_i$. The different state and control trajectories may differ in maximal amplitude, which hinders a direct comparison between them. In order to be able to compare the error measures of all trajectories, it is reasonable to normalize each of them with respect to their respective maximum value. Therefore, we consider the **normalized sum of absolute trajectory errors (NSAE)**, which in case of the state error, is defined as

$$\delta^{x_j} = \frac{1}{\max_k \left|x_j^{*(k)}\right|} \sum_{k=1}^{K} \left|\tilde{x}_j^{(k)} - \hat{x}_j^{(k)}\right| \quad, j \in \{1,\ldots,n\}, \tag{7.6}$$

[49] The notation $\{\boldsymbol{\theta}_i\}_p$ is equivalent to the previously introduced $\theta_{i,(p)}$. These are used interchangeably in favor of better readability.

where $x_j^{(k)} = x_j^{t_k}$ denotes the k-th data point of the state x_j. For systems with more than one state, the sum of NSAEs of the state trajectories

$$\delta^x = \sum_{j=1}^{n} \delta^{x_j} \tag{7.7}$$

is considered. Similarly, the NSAE of the controls of player i is defined as

$$\delta^{u_i} = \sum_{j=1}^{m_i} \frac{1}{\max_k \left|u_{i,(j)}^{*(k)}\right|} \sum_{k=1}^{K} \left|\tilde{u}_{i,(j)}^{(k)} - \hat{u}_{1,(j)}^{(k)}\right| \quad , j \in \{1, \dots, m_i\}. \tag{7.8}$$

The overall sum of NSAEs of the control trajectories is given by

$$\delta^u = \sum_{i=1}^{N} \delta^{u_i}. \tag{7.9}$$

In the following, the error measures (7.7) and (7.9) will be used for trajectory comparison.

7.4 Inverse Open-Loop Dynamic Games

In this section, different classes of inverse dynamic game methods for identifying cost function parameters corresponding to an open-loop Nash equilibrium are evaluated and compared. The methods are

- the residual-based inverse differential game method of Section 4.3,
- the method of Section 6.4 based on IRL,
- the direct bilevel approach presented in Section 7.1 for the open-loop case, detailed in Section B.6.

These are abbreviated and referred to as IOC, IRL and DB methods, respectively.

7.4.1 Preliminaries

The considered system is a ball-on-beam system which was extended such the system is controlled by two players simultaneously instead of one. The task is to balance a ball in the middle of the beam.

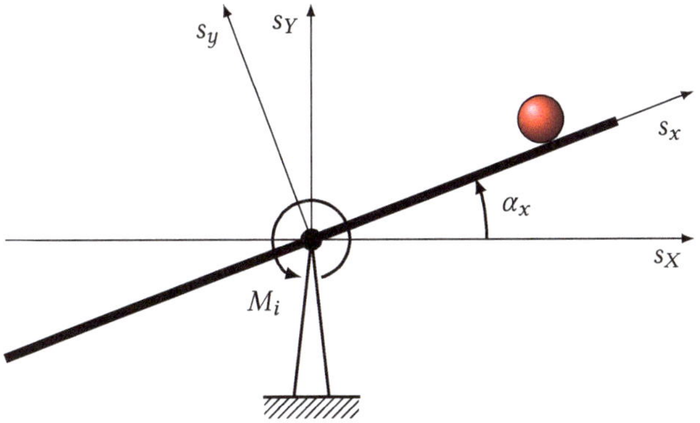

Figure 7.2: Ball-on-beam system

System Dynamics

The ball-on-beam system is shown schematically in Fig. 7.2. Here, α_x denotes the angle of the beam towards the horizontal. In addition, (s_X, s_Y) and (s_x, s_y) denote the positions of the ball in the earth-fixed and beam-fixed coordinate systems, respectively, both centered at the beam's center of rotation. Both players are allowed to interact with the system by applying a torque $u_i(t) = M_i(t)$, $i \in \{1, 2\}$, with respect to the beam's rotational axis. Let the system state be defined as

$$\boldsymbol{x}(t) = \begin{bmatrix} s_x(t) & \dot{s}_x(t) & \alpha_x(t) & \dot{\alpha}_x(t) \end{bmatrix}^\top . \tag{7.10}$$

Then, the system dynamics are described by the nonlinear differential equation (cf. [BVBB14])

$$\dot{\boldsymbol{x}} = \begin{bmatrix} x_2 \\ \frac{m_b r_b^2 (x_1 x_4^2 - g_e \sin(x_3))}{\Theta_b + m_b r_b^2} \\ x_4 \\ \frac{-2 m_b x_1 x_2 x_4 - m_b g_e x_1 \cos(x_3) + u_1 + u_2}{m_b x_1^2 + \Theta_w} \end{bmatrix} \tag{7.11}$$

where the time dependence of the states and controls was dropped for a better readability. The variable g_e is the gravitational constant, Θ_w is the inertia of the beam and r_b, m_b and Θ_b are the radius, mass and inertia of the ball, respectively. The parameter values are given in Table 7.1.

Table 7.1: Parameters of the ball-on-beam system used for simulation

g_e	m_b	r_b	Θ_b	Θ_w
$9.81\,\mathrm{m/s^2}$	$0.02\,\mathrm{kg}$	$25\,\mathrm{mm}$	$5 \cdot 10^{-6}\,\mathrm{kg\,m^2}$	$0.667\,\mathrm{kg\,m^2}$

The inertia of the beam was calculated assuming an equally distributed mass $m_w = 1.3\,\text{kg}$, a width $d_w = 0.01\,\text{m}$ and a length $l_w = 2\,\text{m}$.

Cost Functions and Data Generation

Each player acts based on an individual cost function of the form (4.2), where the basis function vector is given by

$$\boldsymbol{\phi}_i = \begin{bmatrix} x_1^2 & x_2^2 & x_3^2 & x_4^2 & u_i^2 \end{bmatrix}^\top, \quad \forall i \in \{1, 2\}. \tag{7.12}$$

This feature vector describes both players' individual preferences to zero the ball's displacement from the center of the beam, its velocity, the beam's angle and angular velocity, respectively. Furthermore, it represents the desire to keep their individual torques small. In the following, units are neglected as all quantities are given in SI units. To model the players' behavior by means of cost functions, let the ground truth parameters be given by

$$\boldsymbol{\theta}_1^* = \begin{bmatrix} 20 & 1 & 1 & 1 & 2 \end{bmatrix} \quad \text{and} \quad \boldsymbol{\theta}_2^* = \begin{bmatrix} 1 & 1 & 10 & 1 & 1 \end{bmatrix}. \tag{7.13}$$

In this way, the first player focuses on bringing the ball to the center of the beam whereas the second player mainly focuses on bringing the beam to a horizontal position (see state definition in (7.10)).

For the **calculate equilibrium** step, the system dynamics and cost functions with ground truth parameters are used to solve for open-loop Nash equilibrium trajectories by applying Pontryagin's minimum principle and then solving the resulting two-point boundary value problem, where the initial state

$$\boldsymbol{x}(0) = \begin{bmatrix} 0.5 & 0 & 0 & 0 \end{bmatrix}^\top, \tag{7.14}$$

was used. The solution leads to trajectories $\boldsymbol{x}^*(t)$ and $\boldsymbol{u}_i^*(t)$ corresponding to the open-loop Nash equilibrium (OLNE). Further details on the calculation are given in Section B.4 of the Appendix. The equilibrium state is illustrated in Figure 7.3, where the trajectories of the ball position and beam angle, i.e. of the states $x_1(t)$ and $x_3(t)$ are depicted. The applied torques of each player, i.e. the controls $u_1(t)$ and $u_2(t)$ are also shown. The different preferences of the players modeled by the cost function parameters in (7.13) can be recognized. Player 1 applies a positive torque such that the ball is moved towards the zero position, whereas player 2 counteracts this action since his focus is to regulate the beam angle towards zero.

7.4.2 Noisefree Case

The inverse methods are first tested under the assumption that the observed trajectories correspond exactly to the OLNE trajectories generated by the ground truth cost function param-

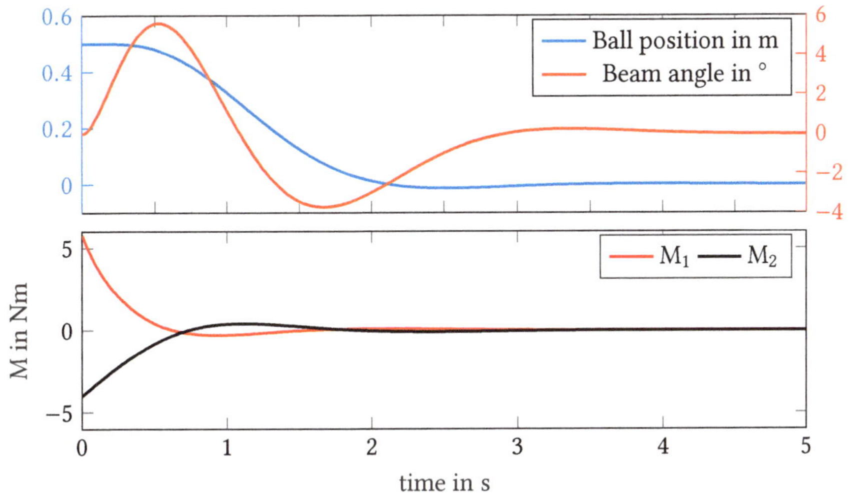

Figure 7.3: Open-loop Nash equilibrium trajectories of the ball-on-beam system

eters $\boldsymbol{\theta}_i^*$. This represents an ideal condition to analyze the extent up to which the real parameters $\boldsymbol{\theta}_i$ can be obtained. The cost function parameter values are given with a precision of 2 decimal values. More precision is not needed since, as it will be shown later, differences of less order of magnitude barely have an effect on the corresponding trajectories. Nevertheless, the parameter errors $\Delta_p^{\boldsymbol{\theta}}$ are calculated with the highest possible precision.

Inverse Optimal Control Based Method

The trajectories of the open-loop Nash equilibrium are used to determine the parameters $\boldsymbol{\theta}_i$ of each player by means of Algorithm 1. The solution of the RDE appearing in the method was calculated by means of a numerical solver of MATLAB (`ode45`).

The estimated parameters are[50]

$$\begin{aligned} \hat{\boldsymbol{\theta}}_1 &= \begin{bmatrix} 19.99 & 1.00 & 1.00 & 1.00 & 2.00 \end{bmatrix} \\ \hat{\boldsymbol{\theta}}_2 &= \begin{bmatrix} 1.01 & 1.00 & 10.00 & 1.00 & 1.00 \end{bmatrix}. \end{aligned} \tag{7.15}$$

which lead to a mean parameter error $\Delta_{p,\text{mean}}^{\boldsymbol{\theta}} = 0.16\%$ and a maximum parameter error $\Delta_{p,\text{mean}}^{\boldsymbol{\theta}} = 0.76\%$. The NSAE of the states is $\delta^{\boldsymbol{x}} = 0.0271$. The NSAE of the controls is $\delta^{\boldsymbol{u}} = 0.025$.

50 For the presented inverse open-loop dynamic game results, the parameter vectors $\boldsymbol{\theta}_i$, $\forall i \in \mathcal{P}$ were multiplied with a constant factor $c \in \mathbb{R}^+$ such that the last entry corresponds to the ground truth, i.e. $\hat{\theta}_{i,(5)} = \theta^*_{i,(5)}$, $\forall i \in \mathcal{P}$. This was done in favor of higher clearness in the comparison.

Inverse Reinforcement Learning Based Method

In order to solve the inverse dynamic game problem, Algorithm 3 was applied. The optimization problem corresponding to the MLE (6.29) was solved with the MATLAB solver `fminunc`, using a BFGS Quasi-Newton method. The estimated parameters are

$$\begin{aligned}\hat{\boldsymbol{\theta}}_1 &= \begin{bmatrix}19.51 & 0.95 & 0.73 & 0.77 & 2.00\end{bmatrix}\\ \hat{\boldsymbol{\theta}}_2 &= \begin{bmatrix}1.04 & 1.01 & 9.99 & 1.02 & 1.00\end{bmatrix}.\end{aligned} \tag{7.16}$$

We obtain a mean parameter error $\Delta^{\boldsymbol{\theta}}_{p,\text{mean}} = 8.1\%$ and a maximum parameter error $\Delta^{\boldsymbol{\theta}}_{p,\max} = 27.0\%$. The NSAE of the states is $\delta^{\boldsymbol{x}} = 0.664$. The NSAE of the controls is $\delta^{\boldsymbol{u}} = 0.554$. The parameter error is bigger than the one generated by the IOC approach. The NSAE values are also higher than the ones corresponding to the IOC based identification.

Direct Bilevel Approach

For this method, the optimization problem (7.1) was solved using the procedure in Section B.6 with an interior-point method of MATLAB's `fmincon` solver.

The estimated parameters are

$$\begin{aligned}\hat{\boldsymbol{\theta}}_1 &= \begin{bmatrix}20.11 & 0.89 & 3.91 & 0.85 & 2.00\end{bmatrix}\\ \hat{\boldsymbol{\theta}}_2 &= \begin{bmatrix}1.14 & 1.01 & 10.13 & 1.09 & 1.00\end{bmatrix}.\end{aligned} \tag{7.17}$$

The mean parameter error $\Delta^{\boldsymbol{\theta}}_{p,\text{mean}} = 42.9\%$ and a maximum parameter error $\Delta^{\boldsymbol{\theta}}_{p,\max} = 290.9\%$. The NSAE of the states is $\delta^{\boldsymbol{x}} = 1.4322$. The NSAE of the controls is $\delta^{\boldsymbol{u}} = 0.122$. The parameter error is bigger than the one generated by both the IOC and IRL approaches.

Comparison

The following Table 7.2 summarizes the results of the parameter identification with all methods. In addition, the identified parameters $\hat{\boldsymbol{\theta}}_i$ of all methods are used to generate OLNE trajectories $\hat{\boldsymbol{x}}(t)$ and $\hat{\boldsymbol{u}}_i(t)$. Both the original and identified trajectories of the controls as well as the ball position and beam angle (states x_1 and x_3, respectively) are depicted in Figure 7.4. While the parameter errors of the identification with IRL and the DB approach are higher than the ones corresponding to the IOC method, they do not have a big impact on the trajectory approximation in this setup. The OLNE of all identified cost functions is practically identical to the original OLNE trajectories. The differences are imperceptible even though there is a slight difference in the estimation accuracy by all methods. This also confirms that the presented parameter precision of two decimal values is sufficient for an adequate comparison.

Table 7.2: Ground truth and cost function parameters of the nonlinear OL differential game identified with all methods using noiseless trajectories

	θ_1	θ_2
GT	[20.00 1.00 1.00 1.00 2.00]	[1.00 1.00 10.00 1.00 1.00]
IOC	[19.99 0.99 1.00 0.99 2.00]	[1.01 1.00 9.99 1.00 1.00]
IRL	[19.51 0.95 0.73 0.77 2.00]	[1.04 1.01 9.99 1.02 1.00]
DB	[20.11 0.89 3.91 0.85 2.00]	[1.14 1.01 10.13 1.09 1.00]

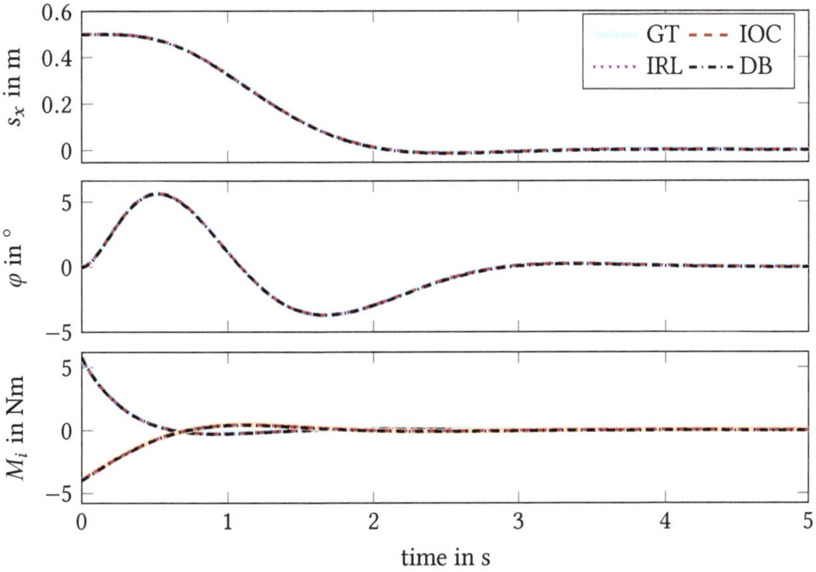

Figure 7.4: Trajectories resulting from the nonlinear inverse dynamic game solutions with IOC, IRL and DB methods

7.4.3 Robustness to Measurement Noise

In practice, measurements of the states and controls corresponding to a dynamic game may not be ideal. For example, the measurements may be affected by noise, which can be detrimental for the identification of cost function parameters. Therefore, the results of the inverse dynamic game methods should ideally be robust to measurement noise. In order to evaluate this property for the considered open-loop methods, Gaussian white noise is artificially added to the state and control trajectories. Hence, the new measurements which are used for identification of cost function parameters are given by

$$\tilde{x}_z(t) = x_z^*(t) + \epsilon_z^{\boldsymbol{x}}, \qquad \forall z \in \{1, ..., n\}, \tag{7.18}$$

$$\tilde{u}_{i,z}(t) = u_{i,z}^*(t) + \epsilon_{i,z}^{\boldsymbol{u}}, \qquad \forall z \in \{1, ..., m_i\}, \forall i \in \mathcal{P}. \tag{7.19}$$

The noise $\epsilon_z^{\boldsymbol{x}}$ and $\epsilon_{i,z}^{\boldsymbol{u}}$ was chosen in such a way that all signals have a particular signal-to-noise ratio (SNR). Different SNR levels from 20 dB to 40 dB were considered for trajectory generation. In order to examine the consistency of the results, 100 samples of Gaussian white noise are generated for each of the considered SNR levels such that we obtain the trajectories $\tilde{\zeta}_s$, $s \in \{1, ..., 100\}$ (cf. Definition 6.2). Figure 7.5 shows examples of noise-corrupted Nash equilibrium trajectories with different SNR values. The generated noisy trajectories are used to identify cost function parameters with all methods. Therefore, for each of the methods, we obtain 100 sets of identified parameters $\underline{\hat{\boldsymbol{\theta}}}_s$, $s \in \{1, ..., 100\}$. In turn, each of these is used to compute corresponding OLNE trajectories denoted by $\hat{\zeta}_s$, $s \in \{1, ..., 100\}$. The mean over all 100 values of the identified parameters of each player, denoted by $\hat{\boldsymbol{\theta}}_{i,\text{mean}}$ is computed for the following analysis. Moreover, the comparison of the estimated parameters and trajectories with the original ones is assessed with the mean of the NSAE (defined in (7.7), (7.8) and (7.9)) over all 100 trajectories. These are denoted by $\delta_{\text{mean}}^{\boldsymbol{x}}$, $\delta_{\text{mean}}^{\boldsymbol{u}_i}$ and $\delta_{\text{mean}}^{\boldsymbol{u}}$, respectively. Similarly, the maximum and mean parameter errors over all 100 results, denoted by $\Delta_{\text{max}}^{\boldsymbol{\theta}}$ and $\Delta_{\text{mean}}^{\boldsymbol{\theta}}$, are considered (cf. (7.4)).

Inverse Optimal Control

The mean values of the identified cost function parameters are given in Table 7.3, where the noisefree case is listed for comparison and is denoted by an infinite SNR. The parameter error increases considerably with the presence of noise. Even with a SNR value of 30 dB which implies a rather low magnitude of the noise, the parameters deviate significantly from the ground truth. In particular, from this SNR value on, the parameter $\hat{\theta}_{i,(3)}$ becomes negative which implies a reward of the deviations from zero, instead of a penalty as originally stated. This trend is confirmed by the mean values of the parameter and trajectory errors which are summarized in Table 7.4. The table shows very high errors for an SNR value equal to 30 dB or below.

Table 7.3: Mean values of the cost function parameters of the inverse nonlinear OL dynamic game which were identified with the IOC method

SNR in dB	$\hat{\boldsymbol{\theta}}_{1,\text{mean}}$					$\hat{\boldsymbol{\theta}}_{2,\text{mean}}$				
20	32.83	3.71	-29.72	10.29	2.00	52.57	12.11	-115.97	38.37	1.00
25	24.19	1.90	9.47	4.08	2.00	16.51	4.42	-30.00	12.44	1.00
30	21.33	1.31	-2.78	2.01	2.00	6.02	2.12	-3.12	4.71	1.00
35	20.40	1.10	-0.13	1.30	2.00	2.58	1.35	5.87	2.16	1.00
40	20.14	1.03	0.64	1.10	2.00	1.50	1.11	8.77	1.36	1.00
∞	19.99	0.99	1.00	0.99	2.00	1.01	1.00	9.99	1.00	1.00

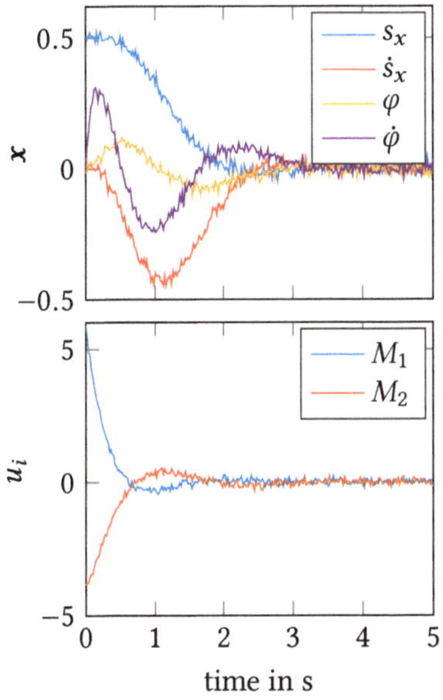

(a) SNR = 20 dB

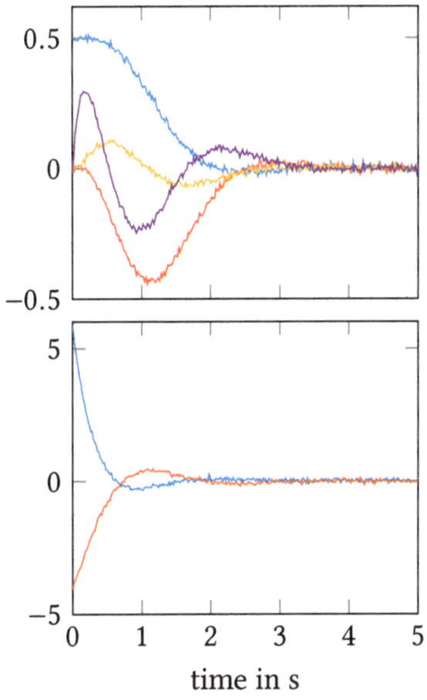

(b) SNR = 25 dB

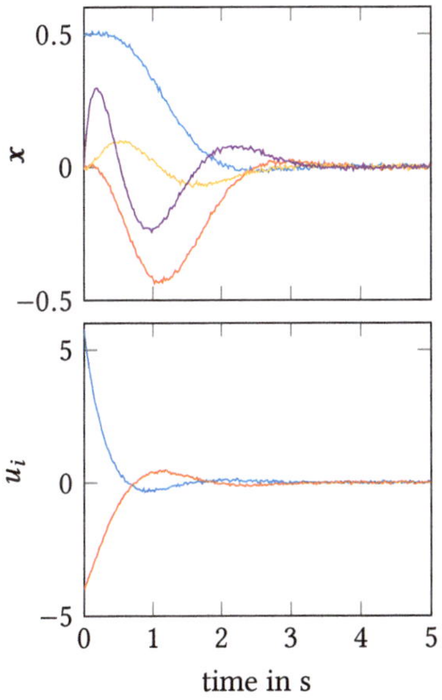

(c) SNR = 30 dB

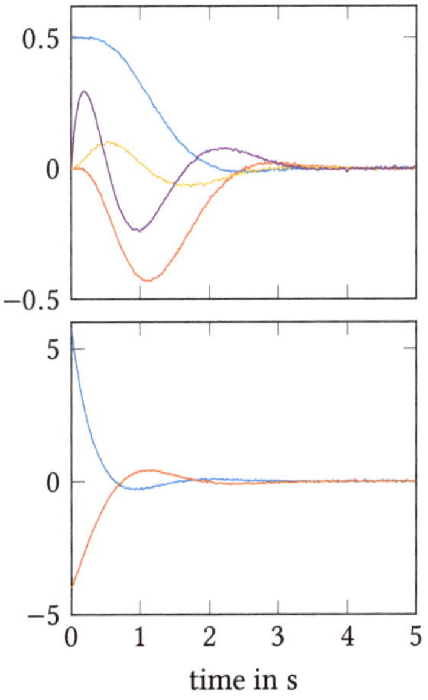

(d) SNR = 35 dB

Figure 7.5: Noise-corrupted open-loop Nash equilibrium trajectories of the two-player dynamic game with the nonlinear ball-on-beam system

Table 7.4: Mean parameter errors and NSAE of trajectories obtained with the IOC method

SNR in dB	δ^{x}_{mean}	$\delta^{u_1}_{\text{mean}}$	$\delta^{u_2}_{\text{mean}}$	$\Delta^{\theta}_{\text{max}}$	$\Delta^{\theta}_{\text{mean}}$
20	69.522	129.948	221.314	9206.2	49.28
25	20.706	69.119	100.339	47.59	6.56
30	10.449	38.274	55.584	10.05	2.10
35	4.123	15.534	22.563	4.45	0.68
40	1.469	5.177	7.519	2.57	0.25
∞	0.027	0.010	0.0147	0.01	0.002

Inverse Reinforcement Learning

The mean values of the identified cost function parameters are given in Table 7.5. The order of magnitude of the parameters is similar for all SNR values, but the results are also negatively affected by lower SNR values. For an SNR value of 20 dB, the parameter $\hat{\theta}_{1,(3)}$ of player 1 becomes slightly negative, leading to a reward of the deviation of x_3 from zero. The mean values of the errors listed in Table 7.6 are moderately low compared to the IOC results, especially the mean parameter error and the mean NSAE of the states.

Table 7.5: Mean values of the cost function parameters identified with the IRL method

SNR in dB	$\hat{\theta}_{1,\text{mean}}$					$\hat{\theta}_{2,\text{mean}}$				
20	20.62	1.19	-2.58	1.79	2.00	1.53	1.16	7.03	1.58	1.00
25	19.85	0.97	0.79	0.99	2.00	1.23	1.06	9.13	1.20	1.00
30	19.60	0.94	1.16	0.81	2.00	1.09	1.02	9.63	1.08	1.00
35	19.53	0.93	1.17	0.76	2.00	1.05	1.01	9.88	1.04	1.00
40	19.51	0.92	1.41	1.72	2.00	1.05	1.01	9.98	1.03	1.00
∞	19.51	0.95	0.73	0.77	2.00	1.04	1.01	9.99	1.02	1.00

Table 7.6: Parameter errors and NSAE obtained with the IRL method

SNR in dB	δ^{x}_{mean}	$\delta^{u_1}_{\text{mean}}$	$\delta^{u_2}_{\text{mean}}$	$\Delta^{\theta}_{\text{max}}$	$\Delta^{\theta}_{\text{mean}}$
20	4.808	10.915	15.854	23.05	1.07
25	2.522	4.256	6.182	10.24	0.39
30	1.345	2.088	3.033	6.83	0.20
35	0.920	1.109	1.610	5.42	0.14
40	0.724	0.591	0.855	2.97	0.15
∞	0.664	0.227	0.327	0.27	0.08

Direct Bilevel Approach

The mean values of the identified cost function parameters are given in Table 7.7. The identified parameters are very similar for all SNR values and no clear SNR-dependent trend can be recognized. Almost all parameters are very similar to the ground truth. Only the parameter $\hat{\theta}_{1,(3)}$ of the first player could not be recovered exactly. The mean values of the errors listed in Table 7.8 show that the parameter and trajectory error overall do increase with smaller SNR values. However, even for the lowest SNR value of 20 dB, the errors, especially the NSAE of the controls, are considerably low.

Table 7.7: Mean values of the identified cost function parameters obtained from noisy trajectories using the DB method

SNR in dB	$\hat{\theta}_{1,\text{mean}}$					$\hat{\theta}_{2,\text{mean}}$				
20	20.12	0.86	3.16	0.91	2.00	1.06	1.00	9.97	1.04	1.00
25	20.13	0.90	3.03	0.93	2.00	1.09	1.00	10.06	1.06	1.00
30	20.05	0.94	2.34	0.94	2.00	1.05	1.00	10.03	1.04	1.00
35	20.02	0.92	2.19	0.92	2.00	1.01	1.00	9.96	1.01	1.00
40	20.03	0.96	1.99	0.95	2.00	1.04	1.00	10.04	1.03	1.00
∞	20.11	0.89	3.91	0.85	2.00	1.14	1.01	10.13	1.09	1.00

Table 7.8: Mean parameter errors and NSAE obtained from noisy trajectories using the DB method

SNR in dB	δ^{x}_{mean}	$\delta^{u_1}_{\text{mean}}$	$\delta^{u_2}_{\text{mean}}$	$\Delta^{\theta}_{\text{max}}$	$\Delta^{\theta}_{\text{mean}}$
20	4.424	0.239	0.355	12.203	0.547
25	2.887	0.144	0.214	12.871	0.451
30	1.872	0.087	0.128	8.743	0.312
35	1.329	0.056	0.081	5.420	0.259
40	0.881	0.037	0.053	7.034	0.192
∞	1.432	0.050	0.072	2.909	0.429

Comparison

The results of cost function identification with noisy measurements are now compared. The mean values of the parameter error corresponding to the SNR values of 20 dB to 40 dB are illustrated in Figure 7.6. In a similar way, Figure 7.7 contrasts the mean values of the NSAE of the states and controls.

Figure 7.6 shows that the IOC approach outperforms the IRL and DB methods in the case of perfect observations of the Nash equilibrium trajectories, but its parameter estimation becomes notoriously worse as the SNR values become smaller. In contrast, both the IRL and DB method yield similar results across all SNR values and demonstrate being less affected

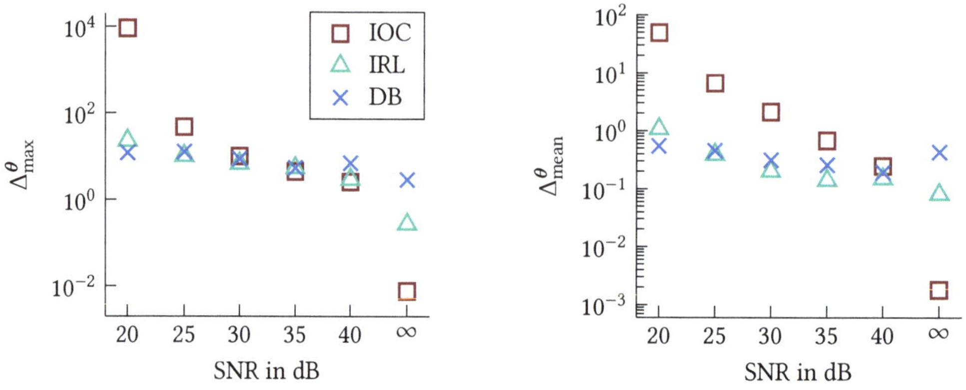

(a) Sum of absolute state errors **(b)** Maximum parameter error

Figure 7.6: Comparison of parameter errors of identification for all SNR values and all methods

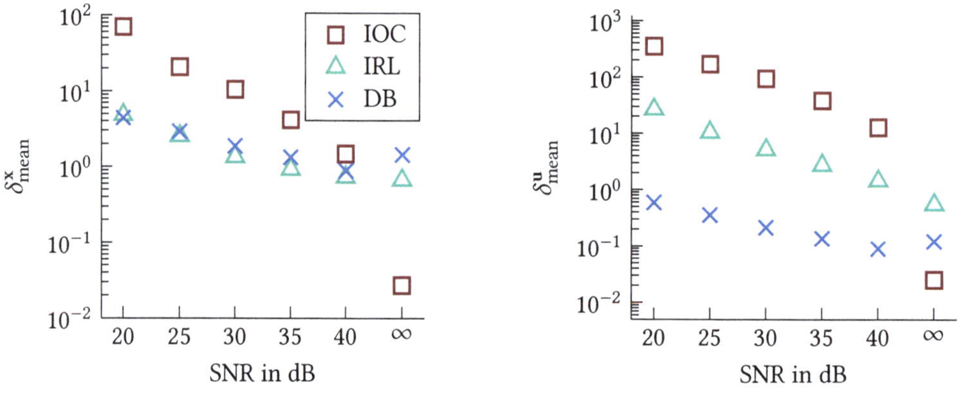

(a) Sum of normalized absolute state errors **(b)** Sum of normalized absolute control errors

Figure 7.7: Comparison of trajectory errors of identification for all SNR values and all methods

by measurement noise. The DB method is slightly better than the IRL approach only for the 20 dB case. A similar trend is observed in Figure 7.7. Nevertheless, it is noticeable that the differences in the parameter estimation can lead to big dissimilarities in the mean NSAE. The superiority of IRL and the DB method in terms of robustness to measurement noise is confirmed. However, it can be observed that the DB approach yields the lowest NSAE of the controls.

In order to obtain a better insight into the quality of the trajectory approximation, the mean values of the identified parameters with each method, i.e. the parameters in Tables 7.3, 7.5 and 7.7, are used to generate corresponding model state and control trajectories. Figure 7.8 shows an example for an SNR value of 30 dB. The IRL and DB methods yield very similar results.

The IOC approach is able to explain the state trajectories adequately, but fails to reproduce the course of the control trajectories. For SNR values lower than 30 dB, the control trajectory approximation by the IRL method starts to deteriorate while the DB approach maintains its robustness. Plots of this comparison for all SNR values can be found in Section E.1.1 of the Appendix.

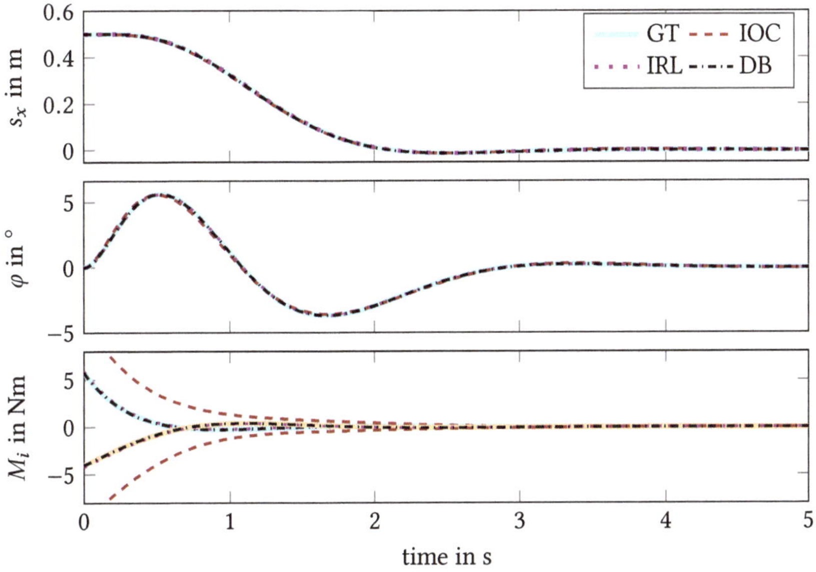

Figure 7.8: Observed trajectories and estimations based on mean identification results of all methods, SNR = 30 dB

7.4.4 Robustness to a Basis Function Mismatch

Especially in practical applications, it cannot be assured that the observed trajectories constitute Nash equilibrium trajectories generated by the considered basis functions $\boldsymbol{g}_i$. In order to give a first evaluation of the limits of the presented methods, a mismatch of the original ground truth (GT) basis functions and the ones used in the inverse dynamic game methods is regarded in this section. The following analysis utilizes the noisefree trajectories generated by the parameters $\boldsymbol{\theta}_1^*$ and $\boldsymbol{\theta}_2^*$, as given in Section 7.4.1, for identification. However, for both the inverse dynamic game step and the subsequent forward solution to obtain estimated trajectories $\hat{\boldsymbol{x}}(t)$, $\hat{\boldsymbol{u}}_1(t)$, and $\hat{\boldsymbol{u}}_2(t)$ (cf. Figure 7.1), four different basis function vectors shall be considered which differ from the original ones. These are given in Table 7.9.

The choice is motivated by the control task and the ground truth parametrization (cf. (7.13)). The basis functions x_2^2 and x_4^2 corresponding to the ball velocity and the beam angle velocity are both weakly weighted by $\theta_{i,(2)}$ and $\theta_{i,(4)}$, respectively. Therefore, case I neglects these

Table 7.9: Considered cases in the basis function mismatch analysis of inverse open-loop dynamic games

Case	g_i				
GT	$[x_1^2$	x_2^2	x_3^2	x_4^2	$u_i^2]$
I	$[x_1^2$		x_3^2		$u_i^2]$
II	$[x_3^2$				$u_i^2]$
III	$[x_1^2$				$u_i^2]$
IV	$[x_1^2$	x_2	x_3^2	x_4^2	$u_i^2]$

basis functions to evaluate their significance for identification. Cases II and III disregard one additional basis function, either x_1^2 or x_3^2, corresponding to the ball position and beam angle, respectively. Finally, case IV represents a situation where one of the basis functions is incorrectly specified.

The basis functions are assumed as different from the ground truth and hence, the parameters are not comparable. Therefore, only the NSAE of the trajectories shall be considered for the evaluation. The NSAE arising from identification with each method is given in Table 7.10 for each case. For case I we observe a low NSAE of the states and a higher NSAE of the controls. Cases II and III lead to worse results in terms of the state trajectory approximation. Lastly, for case IV only the IRL method yields low NSAE values for the states, whereas the DB method can only approximate the control trajectories adequately. The observed trajectories and the estimated trajectories are exemplarily shown for cases I and IV in Figures 7.9 and 7.10. Additional plots describing the results of the other cases can be found in Section E.1.2 of the Appendix.

Table 7.10: NSAE in case of basis function mismatch

Case	Method	δ^x	δ^{u_1}	δ^{u_2}	δ^u
	IOC	17.395	35.338	50.936	86.274
I	IRL	14.993	41.280	59.513	100.793
	DB	129.579	10.909	15.040	25.949
	IOC	339.288	17.472	23.152	40.624
II	IRL	315.659	13.063	21.286	34.348
	DB	338.988	12.635	20.357	32.992
	IOC	117.505	15.356	22.504	37.859
III	IRL	119.463	14.479	19.814	34.292
	DB	128.401	15.429	20.357	35.786
	IOC	521.423	11036.171	15928.605	26964.776
IV	IRL	15.394	47.481	68.439	115.920
	DB	124.605	4.994	4.595	9.589

7.4.5 Discussion of Inverse Open-Loop Dynamic Game Results

By comparing the results of both methods based on noisefree trajectories, it is recognizable that the method based on IOC offers the best results in terms of parameter accuracy. This also leads to a better performance considering the approximation of the ground truth trajectories. Nevertheless, even though the IRL method and the DB approach exhibit a lower parameter approximation accuracy, both are still able to explain the Nash equilibrium trajectories. While there is computationally a minor difference between their trajectory approximation errors, it is so low that it is imperceptible, as shown by Figure 7.4.

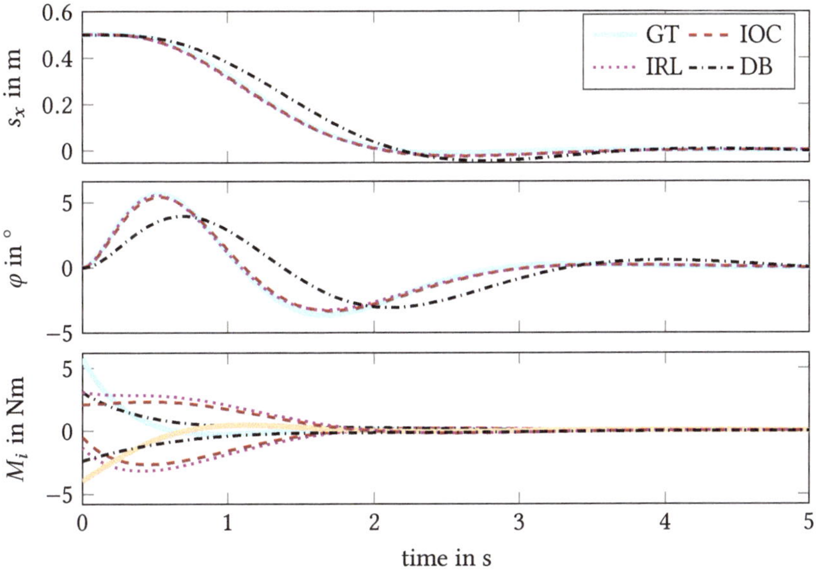

Figure 7.9: Inverse open-loop dynamic game results for all methods, basis function mismatch case I.

The differences in the parameter identification results can be explained by the different characteristics of each of the methods. All methods are based on the solution of an optimization problem. In the case of the IOC approach, the parameters which exactly fulfill the conditions for Nash equilibria are sought. Since the observations are perfect, i.e. they correspond to an exact Nash equilibrium, the corresponding cost function parameters can be found with great precision. The IRL approach is based on the maximization of a likelihood function which indirectly considers the requirement of matching the cost function values of the Nash equilibrium trajectories. The slight deviation to the true parameters arise given the fact that a sufficient match of trajectories, which correlates to a peak in the likelihood function, may not require a precise estimation of parameters. Finally, the DB approach similarly searches for parameters such that the deviation between the costs of observed and estimated trajectories is minimal. This also potentially does not require an exact estimation of parameters.

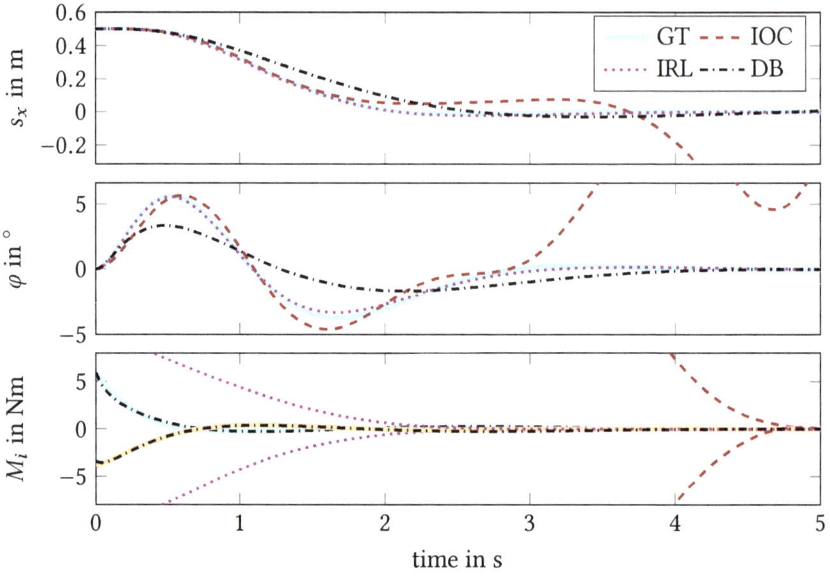

Figure 7.10: Inverse open-loop dynamic game results for all methods, basis function mismatch case IV.

Having discussed these differences in the noisefree case, it is possible to find similar explanations for the results of identification in the presence of measurement noise in the observed Nash equilibrium trajectories. In this case, we observe that the IRL method and the DB approach are more robust towards measurement noise. Even up to SNR values of 20 dB and 25 dB, cost function parameters can be found which explain the observed trajectories. This can also be explained by the different principles each method is based on. The probabilistic formulation of the inverse dynamic game problem in the IRL-based method with the indirect requirement of matching trajectory costs leads to a higher robustness to noise. On the contrary, the IOC approach is strongly affected by measurement noise. The parameter deviations of the IOC approach especially lead to a poor approximation of the control trajectories. The approximation of the state trajectories is not strongly affected by the parameters deviations due to higher trajectory noise.[51]

Finally, the analysis of basis function mismatch indicates that all methods are mildly robust towards a small mismatch of the basis function vectors, especially regarding the state trajectory approximation. All methods yield greater errors if an originally relevant basis function (e.g. x_1^2 and x_3^2 in the example) is neglected. The results suggest that the task can, to some extend, still be described by the other basis functions with a corresponding adequate parameterization which compensates the missing basis functions. However, this possibility may

[51] Similar results were reported in [MTFP16], where a one-player inverse optimal control problem was similarly solved by leveraging the minimum principle and where only the state were corrupted with noise in the evaluation.

depend on the real parametrization of the basis functions. This means that a missing basis function which was weighted by a high value of the corresponding parameter cannot be compensated with other basis functions. In addition, a misspecified basis function as in case IV can affect the results of all methods considerably, especially for the IOC method. This is due to the fact that the basis function x_2 is not appropriate for the task at hand which consists of regulating all states to zero. The other methods, IRL and DB, are less affected since they, either directly or indirectly, take the deviation between trajectories into consideration. This is further illustrated by Table 7.11 where the parameters identified by each method in case IV are listed. The table indicates that the IRL and DB methods correctly estimate the parameter $\theta_{i,(2)}$—the one which corresponds to x_2—as a value which has to be at least close to zero such that trajectories similar to the observed ones can be obtained.

Table 7.11: Identified cost function parameters for basis function mismatch case IV

	θ_1					θ_2				
GT	20.00	1.00	1.00	1.00	2.00	1.00	1.00	10.00	1.00	1.00
IOC	18.22	23.90	19.40	-1.30	2.00	-5.56	-7.37	16.73	-3.07	1.00
IRL	18.78	0.00	16.10	-0.62	2.00	1.55	0.02	33.17	-0.20	1.00
DB	29.42	0.17	199.12	43.19	2.00	8.11	0.01	95.59	18.82	1.00

7.5 Inverse Feedback Dynamic Games

After comparing inverse dynamic game methods for identification in open-loop dynamic games, this section is devoted to an evaluation of inverse feedback dynamic games in a Nash equilibrium, i.e. the players applied linear feedback strategies which led to a FNE. Analogously to last section, one method of each class is analyzed and compared in the following. In particular,

- the inverse LQ differential game method of Section 5.4,
- the method of Section 6.5 based on IRL,
- the direct bilevel approach presented in Section 7.1 for the feedback case, detailed in Section B.6.

Again, these are be abbreviated and referred to as IOC, IRL and DB methods, respectively.

7.5.1 Preliminaries

The following analysis is conducted by means of an infinite-horizon linear-quadratic dynamic game with the following system dynamics and cost functions.

System Dynamics

The system is described by the differential equation

$$\dot{\boldsymbol{x}}(t) = \boldsymbol{A}\boldsymbol{x}(t) + \sum_{i=1}^{3} \boldsymbol{B}_i \boldsymbol{u}_i(t) \tag{7.20}$$

with

$$\boldsymbol{A} = \begin{bmatrix} -8 & -6 & 1 & 0 \\ 1 & 0 & 2 & 1 \\ 0 & -2 & 0 & 1 \\ 0 & 1 & 0 & -1 \end{bmatrix}, \quad \boldsymbol{B}_1 = \begin{bmatrix} 0 & 1 \\ 0 & 0 \\ 0 & 0 \\ 1 & 0 \end{bmatrix}, \quad \boldsymbol{B}_2 = \begin{bmatrix} 0 & 0 \\ 0 & 0 \\ 1 & 0 \\ 0 & 1 \end{bmatrix}, \quad \boldsymbol{B}_3 = \begin{bmatrix} 0 & 0 \\ 1 & 0 \\ 0 & 1 \\ 0 & 1 \end{bmatrix}.$$

Therefore, in this case, each player i has a control vector $\boldsymbol{u}_i \in \mathbb{R}^{m_i}$ with $m_i = 2$ to apply at each time t. The system $(\boldsymbol{A}, \begin{bmatrix} \boldsymbol{B}_1 & \cdots & \boldsymbol{B}_N \end{bmatrix})$ is stabilizable and therefore, the existence of stabilizing linear feedback strategies of the form

$$\boldsymbol{u}_i(t) = -\boldsymbol{K}_i \boldsymbol{x}(t), \quad \forall\, i \in \mathcal{P}$$

is guaranteed [EBS00].

Cost Functions

Each player $i \in \mathcal{P}$ aims to minimize an individual quadratic performance index

$$J_i = \frac{1}{2} \int_0^{\infty} \boldsymbol{x}^\top(t) \boldsymbol{Q}_i \boldsymbol{x}(t) + \boldsymbol{u}_i^\top(t) \boldsymbol{R}_{ii} \boldsymbol{u}_i(t)\, \mathrm{d}t. \tag{7.21}$$

The ground truth parameters of the cost functions were set to

$$\begin{aligned} \boldsymbol{Q}_1^* &= \mathrm{diag}(1, 0.4, 2, 1), & \boldsymbol{R}_{11}^* &= \mathrm{diag}(1, 1), \\ \boldsymbol{Q}_2^* &= \mathrm{diag}(1, 0.6, 1, 2), & \boldsymbol{R}_{22}^* &= \mathrm{diag}(1, 1), \\ \boldsymbol{Q}_3^* &= \mathrm{diag}(1, 1, 0.5, 1), & \boldsymbol{R}_{33}^* &= \mathrm{diag}(1, 2). \end{aligned} \tag{7.22}$$

Using the ground truth cost function parameters, the feedback Nash equilibrium trajectories $\boldsymbol{x}^*(t)$ and $\boldsymbol{u}^*(t)$ were calculated by means of the coupled matrix Riccati equations [Eng05, Theorem 8.5]. The theorem allows to confirm the Nash character of the trajectories given the stability of the controlled system. The resulting Nash equilibruim feedback matrices are given by

$$
\begin{aligned}
K_1^* &= \begin{bmatrix} 0.012 & 0.123 & 0.114 & 0.318 \\ 0.066 & 0.028 & -0.006 & 0.012 \end{bmatrix}, \\
K_2^* &= \begin{bmatrix} 0.004 & -0.041 & 0.541 & 0.130 \\ 0.018 & 0.197 & 0.130 & 0.644 \end{bmatrix}, \\
K_3^* &= \begin{bmatrix} 0.025 & 0.650 & 0.115 & 0.149 \\ 0.020 & 0.132 & 0.384 & 0.301 \end{bmatrix}.
\end{aligned}
\tag{7.23}
$$

Properties of the Inverse LQ Dynamic Game

Before solving the inverse LQ dynamic game, the LQ character of the problem allows for its analysis by means of the results of Chapter 5. We first use the results of Lemma 5.2 to determine the matrices $\boldsymbol{M}_i \in \mathbb{R}^{8\times 6}$ with (5.14) using the control matrices $\boldsymbol{K}_i^*$. Now consider the rank of $\boldsymbol{M}_i$ and obtain $\text{rank}(\boldsymbol{M}_i) = 6$ for all $i \in \mathcal{P}$. By the results of Theorem 5.3, the necessary and sufficient conditions for a unique solution of the inverse LQ dynamic game up to a multiplying constant parameter are fulfilled.

7.5.2 Noisefree Case

The inverse dynamic game methods are first tested under ideal conditions, i.e. the observed trajectories are free of measurement noise and therefore correspond exactly to the FNE which arise out of the dynamic game consisting of the system dynamics (7.20) and cost functions (7.21) with ground truth parameters (7.22). Since both the IOC and IRL methods rely on the estimation of the Nash equilibrium feedback matrices, this is carried out for both players using a least-squares approach presented in Section 5.4.2 and the given trajectories $\boldsymbol{x}^*(t)$, $\boldsymbol{u}_i^*(t)$. The estimation yields very good results for $\hat{\boldsymbol{K}}_i$ as we obtain deviations where $||\hat{\boldsymbol{K}}_i - \boldsymbol{K}_i^*|| < 10^{-4}$, $i = \{1, 2, 3\}$, from the original Nash feedback matrices.

Inverse Optimal Control

The inverse dynamic game is solved by determining the solution of the quadratic static optimization problem (5.33) using the estimated feedback matrices $\hat{\boldsymbol{K}}_i$. The parameters in (7.22) are exactly identified exactly up to two decimal values and are therefore not explicitely given. The mean parameter error $\Delta^{\boldsymbol{\theta}}_{p,\text{mean}}$ is 0.05% and the maximum parameter error $\Delta^{\boldsymbol{\theta}}_{p,\text{max}}$ is 0.26%. The NSAE of the states is $\delta^{\boldsymbol{x}} = 0.002$ while the NSAE of the controls is $\delta^{\boldsymbol{u}} = 0.010$.

Inverse Reinforcement Learning

The IRL approach leads to identified cost function parameters which approximate the original ground truth paramters up to two decimal values. The mean parameter error is $\Delta^{\theta}_{p,\mathrm{mean}} = 0.1\%$ and the maximum parameter error is $\Delta^{\theta}_{p,\mathrm{max}} = 0.6\%$. The NSAE of the states is $\delta^{x} = 0.019$. The NSAE of the controls is $\delta^{u} = 0.073$. All errors are slightly bigger than the errors obtained with the IOC method.

Direct Bilevel Approach

The DB approach leads to a mean parameter error of $\Delta^{\theta}_{p,\mathrm{mean}} = 0.43\%$ and a maximum parameter error of $\Delta^{\theta}_{p,\mathrm{max}} = 3.85\%$. The NSAE of the states is $\delta^{x} = 0.028$ and the NSAE of the controls is $\delta^{u} = 0.151$. The DB approach yields greater errors than both the IOC and IRL methods.

Comparison

The following Tables 7.12 and 7.13 summarize the results of the parameter identification with all methods.[52]

Table 7.12: Ground truth and cost function matrices $\boldsymbol{Q}_i$ identified from noiseless trajectories with all methods

Case	Q_1	Q_2	Q_3
GT	(1.00, 0.40, 2.00, 1.00)	(1.00, 0.60, 1.00, 2.00)	(1.00, 1.00, 0.50, 1.00)
IOC	(1.00, 0.40, 2.00, 1.00)	(1.00, 0.60, 1.00, 2.00)	(1.00, 1.00, 0.50, 1.00)
IRL	(1.00, 0.40, 2.00, 1.00)	(1.00, 0.60, 1.00, 2.00)	(0.99, 1.00, 0.50, 1.00)
DB	(1.00, 0.40, 2.00, 1.00)	(1.04, 0.58, 1.00, 2.00)	(0.99, 1.00, 0.50, 1.00)

Table 7.13: Ground truth and cost function matrices $\boldsymbol{R}_{ii}$ identified from noiseless trajectories with all methods

Case	$R_{1,(22)}$	$R_{2,(22)}$	$R_{3,(22)}$
GT	1.00	1.00	2.00
IOC	1.00	1.00	2.00
IRL	1.00	1.00	2.00
DB	1.01	1.00	2.00

Even though the metrics show that the DB leads to the highest mean and maximum parameter errors as well as the highest NSAE, thus suggesting a superiority of IOC and IRL in the quality

[52] All results were normalized with respect to the parameter $\boldsymbol{R}_{i,(11)}$ for a better comparison. Therefore, this parameter is not explicitely given in Table 7.13.

of the estimation, all errors are relatively small. The values in Tables 7.12 and 7.13 confirm that all methods lead to an excellent estimation of the cost function parameters. For the sake of completeness and in order to see potential differences in the approximation of the observed trajectories, we solve the LQ differential game with the estimated parameters and determine the corresponding FNE trajectories for all methods. The ground truth and model state trajectories are depicted in Figure 7.11. Likewise, the control trajectories are shown in Figure 7.12. All methods are able to perfectly approximate the observed trajectories.

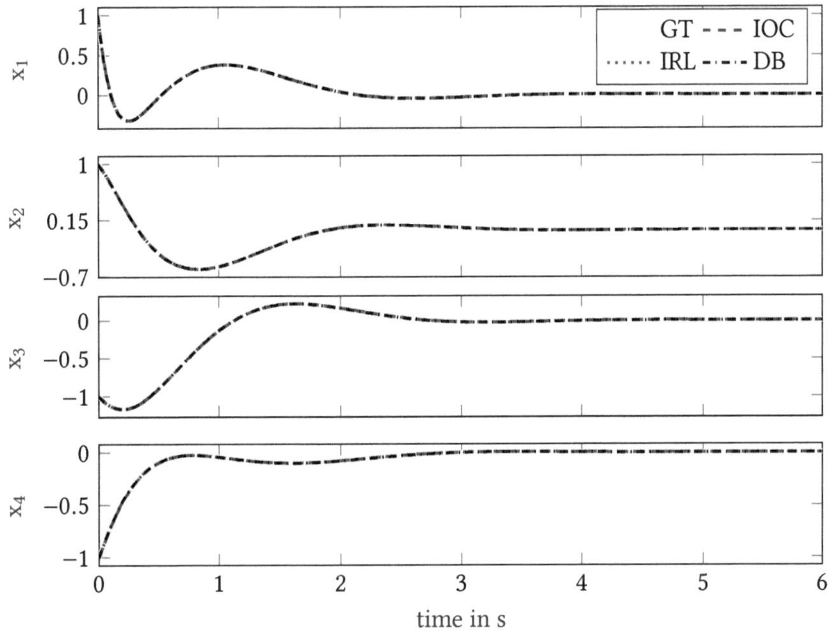

Figure 7.11: Ground truth and estimated state trajectories of the inverse LQ feedback dynamic game with each method

7.5.3 Robustness to Measurement Noise

This section presents simulation results on the influence of the presence of noise in the observed trajectories on the results of the inverse dynamic game methods. Similar to the evaluation in Section 7.4.3 for the open-loop case, Gaussian white noise is added to the state trajectories and the control trajectories according to (7.18) and (7.19), respectively. Once more, the added noise is generated such that the corrupted trajectories have a particular SNR value. The considered SNR values range from 20 dB to 40 dB. 100 samples of Gaussian white noise were generated and therefore, the noisy trajectories $\tilde{\zeta}_s$, $s \in \{1, \dots, 100\}$ (cf. Definition 6.2), are obtained for each of the different SNR values. Each one of these trajectories was used

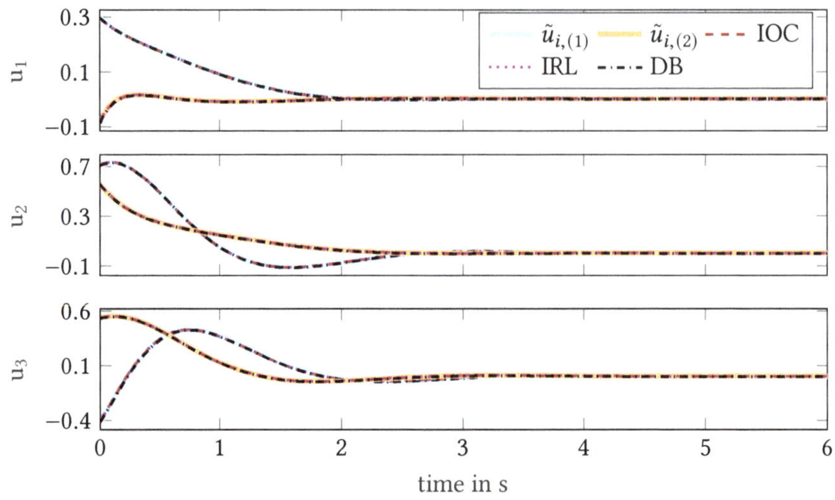

Figure 7.12: Ground truth and estimated control trajectories of the inverse LQ feedback dynamic game with each method

to identify cost function parameters. Therefore, we obtain for each method the parameter sets $\hat{\boldsymbol{\theta}}_s$, $s \in \{1, ..., 100\}$. Each of the parameter sets can be used to determine corresponding FNE trajectories which are denoted by $\hat{\zeta}_s$, $s \in \{1, ..., 100\}$. Analogously to Section 7.4.3, the metrics $\delta^{\boldsymbol{x}}_{\text{mean}}$, $\delta^{\boldsymbol{u}_i}_{\text{mean}}$ and $\delta^{\boldsymbol{u}}_{\text{mean}}$ for the trajectory comparison as well as $\Delta^{\boldsymbol{\theta}}_{\text{max}}$ and $\Delta^{\boldsymbol{\theta}}_{\text{mean}}$ for parameter comparison are considered.

Inverse Optimal Control

The parameter and trajectory errors are given in Table 7.14. The errors increase moderately with lower values of the SNR. The worst case mean parameter error is 18.2%. It is noticeable that the NSAE of the control $\boldsymbol{u}_1(t)$ is always bigger than the NSAE of the other players' controls.

Table 7.14: Parameter errors and NSAE between ground truth trajectories and trajectories obtained with IOC from noisy trajectories

SNR in dB	$\delta^{\boldsymbol{x}}_{\text{mean}}$	$\delta^{\boldsymbol{u}_1}_{\text{mean}}$	$\delta^{\boldsymbol{u}_2}_{\text{mean}}$	$\delta^{\boldsymbol{u}_3}_{\text{mean}}$	$\Delta^{\boldsymbol{\theta}}_{\text{max}}$	$\Delta^{\boldsymbol{\theta}}_{\text{mean}}$
20	4.380	8.082	4.454	4.837	0.977	0.182
25	1.668	3.437	1.614	1.878	0.843	0.162
30	0.714	1.168	0.687	0.842	0.379	0.080
35	0.330	0.455	0.337	0.408	0.342	0.045
40	0.173	0.257	0.182	0.233	0.299	0.017
∞	0.002	0.003	0.003	0.004	0.003	$4.67 \cdot 10^{-4}$

Inverse Reinforcement Learning

The error measures for each SNR value are given in Table 7.15. In this case, it can be observed again that the NSAE of the control $\boldsymbol{u}_1(t)$ is always bigger than the NSAE of the other players' controls. The worst case mean parameter error is 11.4%.

Table 7.15: Parameter errors and NSAE between ground truth trajectories and trajectories obtained with IRL from noisy trajectories

SNR in dB	δ^{x}_{mean}	$\delta^{u_1}_{\text{mean}}$	$\delta^{u_2}_{\text{mean}}$	$\delta^{u_3}_{\text{mean}}$	$\Delta^{\theta}_{\text{max}}$	$\Delta^{\theta}_{\text{mean}}$
20	1.556	4.947	1.449	1.144	1.859	0.114
25	0.693	2.136	0.717	0.579	1.327	0.061
30	0.291	0.839	0.361	0.309	0.631	0.030
35	0.158	0.475	0.183	0.181	0.467	0.018
40	0.085	0.250	0.098	0.096	0.175	0.008
∞	0.019	0.045	0.011	0.017	0.006	0.001

Direct Bilevel Approach

Table 7.16 gives the resulting NSAE and the parameter errors. The trend of less accurate estimations of the control $\boldsymbol{u}_1(t)$ is visible in this case as well. The worst case mean parameter error is 19.4%.

Table 7.16: Parameter errors and NSAE between ground truth trajectories and trajectories obtained with the DB method from noisy trajectories

SNR in dB	δ^{x}_{mean}	$\delta^{u_1}_{\text{mean}}$	$\delta^{u_2}_{\text{mean}}$	$\delta^{u_3}_{\text{mean}}$	$\Delta^{\theta}_{\text{max}}$	$\Delta^{\theta}_{\text{mean}}$
20	1.255	3.640	0.846	3.843	1.866	0.194
25	0.699	1.829	0.504	2.094	0.647	0.093
30	0.438	1.049	0.308	1.328	0.579	0.061
35	0.266	0.678	0.192	0.746	9.615	0.054
40	0.148	0.334	0.112	0.430	0.253	0.022
∞	0.028	0.062	0.026	0.063	0.039	0.004

Comparison

Figure 7.13 shows a comparison of the mean NSAE obtained with each method and for all SNR values. It is noticeable that the IRL approach leads to the least NSAE of the states for the SNR values 30 dB to 40 dB. For an SNR of 25 dB, the IRL method and the DB approach obtain almost the same results. Finally, for highly corrupted trajectories with an SNR of 20 dB, the DB approach offers the best results, closely followed by the IRL method. The IOC method leads for all SNR values to a higher state error than the other approaches. Similar results can be observed in the mean NSAE control errors $\delta^{\boldsymbol{u}_i}_{\text{mean}}$, $i \in \{1,2,3\}$. In this case, the IRL approach offers better results consistently across all SNR values. For little noise, i.e. for SNR values of 30 dB to 40 dB, the IOC method leads to better results than the DB approach.

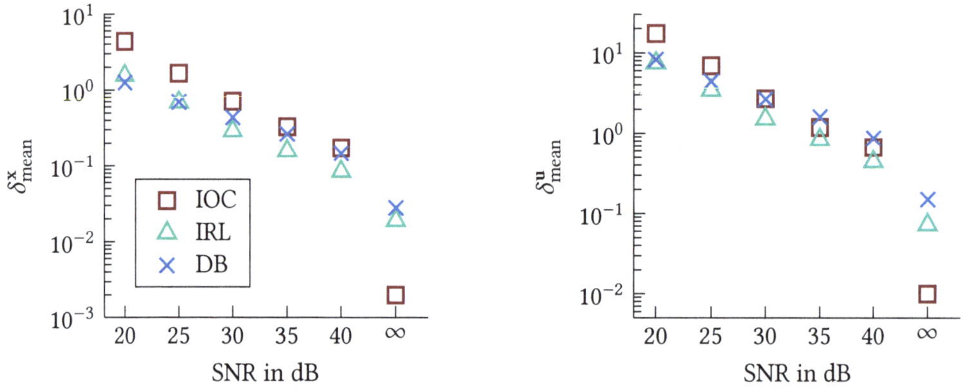

(a) Sum of absolute state errors

(b) Sum of absolute control errors

Figure 7.13: Mean NSAE obtained with each method for all trajectory SNR values

Regarding the parameter errors, Figure 7.14 shows that the least mean parameter error is obtained by the IRL method for all SNR values. However, the maximum parameter error does not show a clear trend, but suggests that the IOC method yields more consistent results, as the other methods have greater maximum parameter errors. The IOC method and the DB approach have similar mean parameter errors. However, by inspecting the maximum parameter error, it can be discerned that the IOC approach does not lead to great differences as the SNR value changes. On the contrary, the maximum parameter error of the IRL is always higher and varies considerably more with the exception of the case of an SNR value of 40 dB. The DB method results do not allow a particular interpretation as no clear trend can observed, except for the bigger error with less SNR which is common for all methods. Nevertheless, an outlier value can be observed for an SNR of 35 dB caused by an anomalously poor identification result.

Once more, for a better understanding of these results, the mean values of the identified parameters were used to determine mean estimated Nash equilibrium trajectories. These parameters are listed in the Appendix: Tables E.1 and E.2 correspond to the IOC method, Tables E.3 and E.4 to the IRL method and Tables E.5, E.6 show the results of the DB approach. The resulting estimated FNE trajectories are compared with the original noiseless trajectories in ζ^*. Figures 7.15 and 7.16 show this comparison for the FNE state and control trajectories, respectively, which were estimated from noisy observations with 20 dB. It is noticeable that, despite the low SNR, all methods lead to good approximations of the states and control tra-

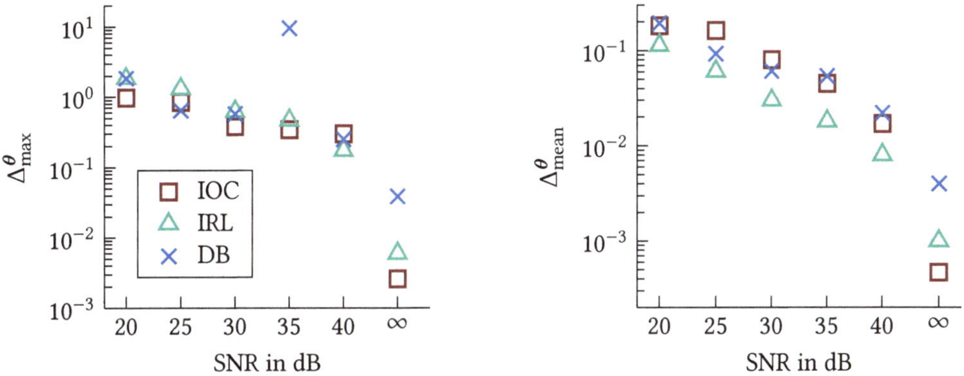

(a) Maximum parameter error

(b) Mean parameter error

Figure 7.14: Parameter error of identification for all SNR values and all methods

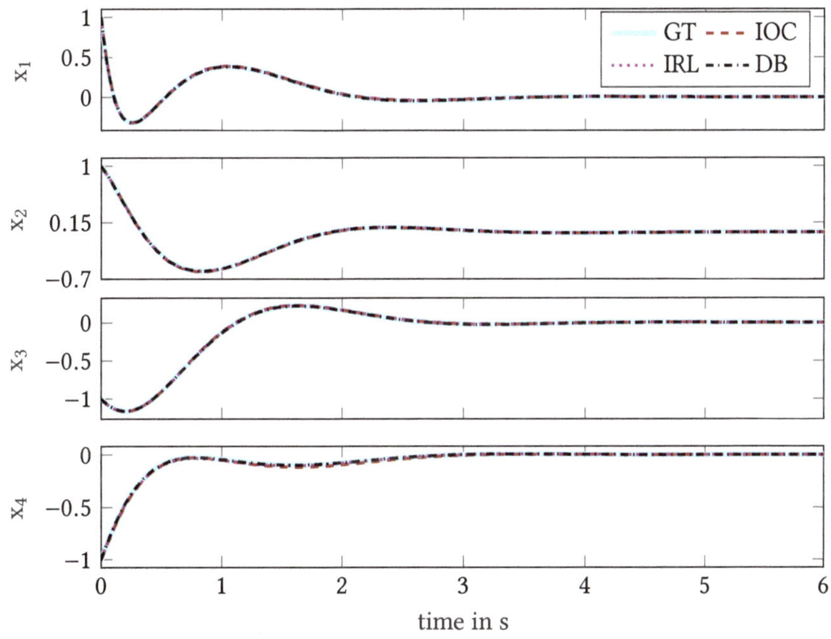

Figure 7.15: Ground truth and estimated state trajectories of the inverse LQ feedback dynamic game with each method. The identification was conducted using noise-corrupted trajectories with SNR = 20 dB.

jectories. In a detailed view of the results, there is a better agreement between the original trajectories and the estimated ones in the case of the state variables. Furthermore, we can observe that the DB method performs slightly better than the IRL and IOC methods. While this minor difference are visible in this case, these are even tinier for greater SNR values. The corresponding figures are given in Section E.2.1 of the Appendix.

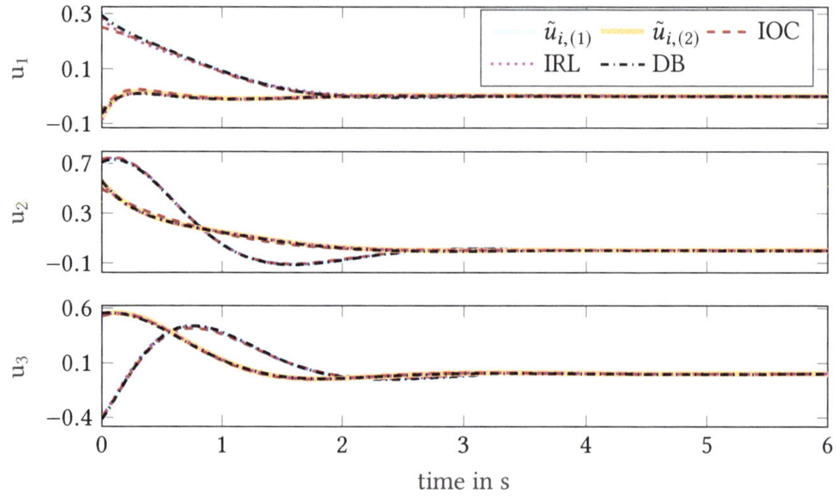

Figure 7.16: Ground truth and estimated state trajectories of the inverse LQ feedback dynamic game with each method. The identification was conducted using noise-corrupted trajectories with SNR = 20 dB.

7.5.4 Robustness to a Basis Function Mismatch

This section presents an evaluation of the robustness of the inverse LQ dynamic game methods to a mismatch in the basis functions, similar to the analysis conducted in Section 7.4.4 for the open-loop case. The noisefree trajectories generated by the cost function matrices $\boldsymbol{Q}_i^*$ and $\boldsymbol{R}_{ij}^*$, $i, j \in \mathcal{P}$, as given in Section 7.5.1 are used for identification. For both the inverse dynamic game step and the subsequent forward solution to obtain estimated trajectories $\hat{\boldsymbol{x}}(t)$ and $\hat{\boldsymbol{u}}_i(t)$, $i \in \mathcal{P}$, it shall be assumed that certain elements of the matrix $\boldsymbol{Q}_i$ are neglected and therefore not identified. The considered cases are described in Table 7.17. These describe an increasing number of parameters of the diagonal matrix $\boldsymbol{Q}_i$ which are neglected. Analogously to the open-loop case, only the NSAE of the trajectories shall be considered for the evaluation. The NSAE arising from identification with each method are given in Table 7.18. Similar error values can be observed for the cases I to III for all methods, with the DB method presenting slightly lower values. In turn, case IV shows a very high error for all methods. The observed trajectories and the estimated trajectories are exemplarily shown for case I in Figures 7.17 and 7.18. Additional plots describing the results of the other cases can be found in Section E.2.2 of the Appendix.

Table 7.17: Considered cases in the basis function mismatch analysis of inverse LQ feedback dynamic games

Case	θ_i					
GT	$[Q_{i,(1,1)}$	$Q_{i,(2,2)}$	$Q_{i,(3,3)}$	$Q_{i,(4,4)}$	$R_{ii,(1,1)}$	$R_{ii,(2,2)}]$
I	$[Q_{i,(1,1)}$	$Q_{i,(2,2)}$	$Q_{i,(3,3)}$	0	$R_{ii,(1,1)}$	$R_{ii,(2,2)}]$
II	$[Q_{i,(1,1)}$	$Q_{i,(2,2)}$	0	0	$R_{ii,(1,1)}$	$R_{ii,(2,2)}]$
III	$[Q_{i,(1,1)}$	0	0	0	$R_{ii,(1,1)}$	$R_{ii,(2,2)}]$
IV	[0	0	0	0	$R_{ii,(1,1)}$	$R_{ii,(2,2)}]$

Table 7.18: NSAE in case of basis function mismatch in inverse LQ dynamic games

Case	Method	δ^x	δ^{u_1}	δ^{u_2}	δ^{u_3}	δ^u
	IOC	37.622	19.944	76.219	30.861	127.024
I	IRL	11.978	22.173	17.352	8.489	48.013
	DB	11.224	16.388	14.781	8.625	39.795
	IOC	35.289	39.714	94.681	29.867	164.262
II	IRL	15.659	21.337	28.352	17.377	67.065
	DB	13.497	20.699	27.027	21.121	68.848
	IOC	29.423	61.954	88.718	22.088	172.759
III	IRL	47.402	30.572	50.260	59.949	140.782
	DB	13.870	24.111	26.942	21.399	72.451
	IOC	438.410	49.741	72.837	100.270	222.848
IV	IRL	438.410	49.741	72.837	100.270	222.848
	DB	201.243	49.741	158.091	100.270	308.102

7.5.5 Discussion of Inverse LQ Dynamic Game Results

The inverse LQ differential game was solved by means of an IOC based method, an IRL based method and the DB approach. All methods were shown to lead to good identification results both in terms of trajectory approximation and parameter estimation. The IOC method presented the highest parameter estimation precision in the case of noiseless trajectories.

The analysis with noise-corrupted trajectories demonstrated that the IRL based method offers the best results across all SNR values. Only for the mean NSAE of the states, the DB method is slightly better than IRL. The results indicate that the DB and IRL methods are more robust towards measurement noise than IOC. As for the parameter error, we observe that the mean parameter error reflects the fact that the IRL method performed the best with all SNR values. The higher robustness of the DB approach in low SNR regions compared to IOC can also be noticed. However, an interesting result of IOC is the lower variability in the maximum

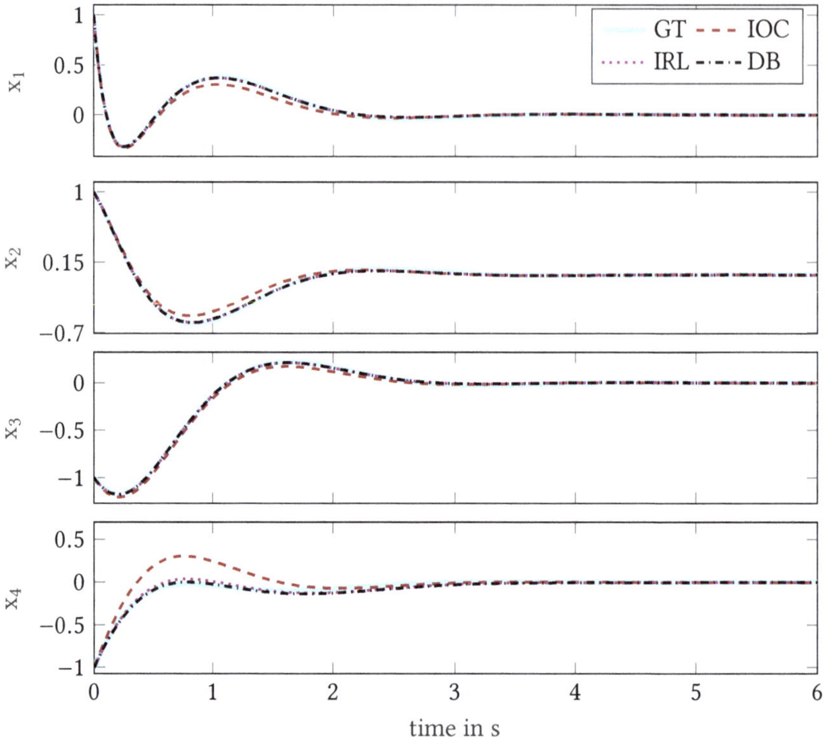

Figure 7.17: Ground truth and estimated state trajectories of the inverse LQ feedback dynamic game with each method. The identification was conducted with the wrong assumption that $Q_{i,(4,4)} = 0$ for $i \in \{1, 2\}$ (case I).

parameter error. This suggests that even though the DB approach and IRL performed better in the mean, they are not guaranteed to always lead to better results.

Regarding the robustness to a basis function mismatch, the resuls of Table 7.18 show that the methods are fairly robust to a mismatch caused by the neglection of features. However, not including any basis function which penalizes the states (as in case IV) leads to major deviations of both states and controls with respect to the original trajectories. The original parameters describe a behavior which aims at regulating all states to zero and has to be considered in the choice of the basis functions. Similarly to the analysis of the effects of measurement noise on the results, it can be discerned that the IRL and DB method are slightly more robust than the IOC method in case of a basis function mismatch. Finally, it can also be noted that the control trajectory approximation is corrupted more than the state approximation, especially for the IOC and IRL methods. In general, the approximations of the controls are affected more, independent of whether the perturbation lies in the basis functions or the trajectories.

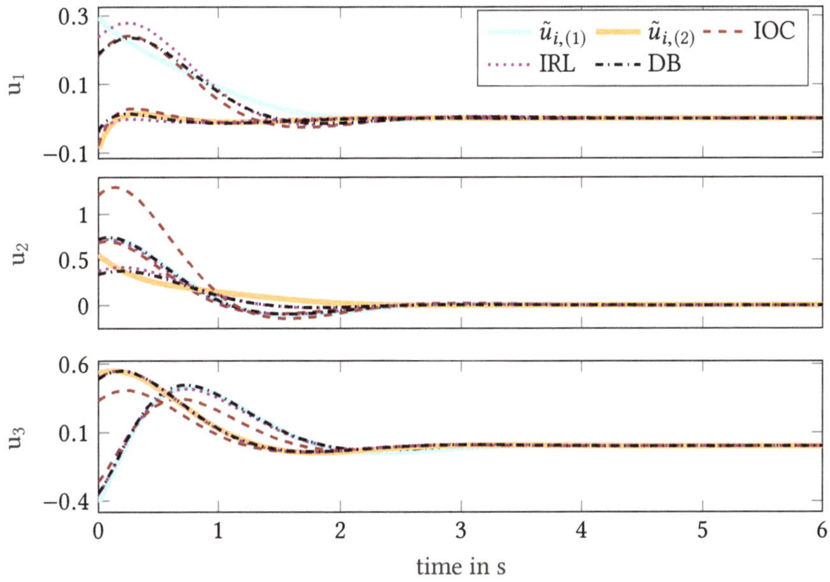

Figure 7.18: Ground truth and estimated state trajectories of the inverse LQ feedback dynamic game with each method. The identification was conducted with the wrong assumption that $Q_{i,(4,4)} = 0$ for $i \in \{1, 2\}$ (case I).

Analogously to the open-loop case, the results of this section can be explained by the different concepts behind each of the methods. IOC depends on the fact that the trajectories correspond to a feedback Nash equilbrium. The IRL method is based on the maximization of a likelihood function which indirectly includes the requirement of matching costs of the observed trajectories and therefore is more robust towards mild violations of the Nash equilibrium assumption generated by the measurement noise or by basis function errors. Finally, the objective function of the DB approach which explicitely considers the deviation between trajectories is responsible for its good results.

7.6 Computation Time

Before concluding on the observed results, the computation time of all approaches is briefly examined, as computational efficiency is an important issue towards the application of these methods for an online estimation of cost function parameters. The computational effort is exemplarily shown for the case with noisy trajectories with SNR = 25 dB to give an impression of the computational demands of each of the methods. Table 7.19 presents the computation times of the different methods in the case of an identification in an open-loop and feedback scenario.[53] The DB method yields the highest computation time, followed by the IRL and IOC methods. The DB method's computation time in the open-loop nonlinear case is approximately 26% higher than the one corresponding to the linear-quadratic feedback dynamic game. This can be explained by the fact that the first demands the repeated solution of a nonlinear dynamic game which is generally harder to solve than a linear-quadratic dynamic game. The IOC method is the fastest since it relies on the solution of a conventional RDE or a quadratic program, which can usually be efficiently solved with numerical techniques. Finally, the IRL method stands inbetween. The conceptually abstract likelihood function and its convergence properties are hard to analyze. However, the fact that it consists of one single static optimization problem yields a great chance of being faster than the DB method.

Table 7.19: Computation times for inverse dynamic games

Method	T_{CPU} **in** s	
	OL	**FB**
IOC	4.2	0.087
IRL	161.3	1060.2
DB	2435.8	1805.1

7.7 Conclusion

In this chapter, a systematic comparison between IOC, IRL and DB methods for solving inverse dynamic games was conducted. Both open-loop and feedback structures were considered. Moreover, the robustness of the approaches with respect to the presence of noise in the observed trajectories was examined. In addition to the quality of cost function parameter identification, the capability of the identified cost functions to describe observed data was also assessed.

In the **open-loop** case, the IOC method was shown to lead to the most accurate results in the parameter estimation if the observed trajectories correspond to a Nash equilibrium. Nevertheless, if the observations are noise-corrupted, the IOC method's results deteriorate. The state trajectory approximation is still adequate, but the control trajectories deviate considerably from the ground truth. The IRL and DB methods showed a higher robustness to measurement noise and yield to similar results. Only in the lowest considered SNR value case, the DB method led to slightly better approximations. In addition, all methods show a slight robustness to missing relevant basis functions as long as the other ones are meaningful and related to the control task at hand. In case a non-adequate basis function is provided, only the IRL and DB methods are able to neglect it by setting its corresponding parameter to a value near zero.

[53] The used CPU was an Intel Xeon E5-2630 at 2.6 GHz with 32 GB of RAM.

As for the **feedback** case, a similar trend as in the open-loop case could be observed. Nevertheless, it can be stated that the magnitude of both parameter and NSAE for IOC and IRL methods is smaller than in the open-loop case. One possible reason is that the linear system dynamics allow for better identification, especially in the case of the IRL which relies on a dynamics linearization (which is nevertheless time-variant, i.e. it is computed in every time step). However, this may be best explained by the LS estimation of the feedback matrices $\boldsymbol{K}_i$ which is done by means of the control and state trajectories. This estimation is, theoretically speaking, not bias-free. The noise has zero mean, but is applied to both the control and the state values. In spite of this fact, the estimation works well in practice such that a relatively accurate functional relationship between the states and the control is provided to the IOC and IRL methods. This is also reflected by the good results obtained by all methods in the analysis of basis function mismatch.

To finish this chapter, the main findings are summarized as follows:

- Approaches based on IOC offer the most precise parameter identification results in case of uncorrupted observations of Nash equilibrium trajectories. They are less robust towards measurement noise than the other methods and may be affected by a significant mismatch in the basis functions, but are the least computationally expensive of all methods. The latter property indicates that this method class is the most appropriate for a potential online application.
- Approaches based on IRL provide a good compromise between computation time and quality of identification. They are the most robust towards measurement noise among all tested methods. Moreover, they are more robust to non-adequate basis functions than the IOC method and yield similar results than the DB method in this case.
- The direct bilevel approach has been shown to lead to very good results and to be robust to noise and slight errors in the basis functions, but the computation time is greater than IOC methods (up to a factor of approximately 20 000) and IRL methods (up to a factor of approximately 15) and therefore is the least efficient among all methods.
- The robustness of all methods to measurement noise and to errors in the basis function selection is higher for the state trajectory approximations. Especially for the IOC and IRL methods, the approximation of the controls is more sensitive to violations of the assumptions the methods are based on.

After this analysis of inverse dynamic game methods in a simulation environment, the following chapter presents a first application of inverse dynamic game methods with real experimental data.

8 Application to Shared Control Systems

This chapter presents an application example for inverse dynamic games. The aim of this chapter is to provide a first evaluation of the applicability of inverse dynamic games to identify cost functions in a real scenario. In the following, a shared control scenario between two humans is considered. Shared control stems from the field of human-machine cooperation. It usually describes a situation where humans and machines simultaneously control a dynamic system.[54] Therefore, it has led to a rising number of applications including robot-assisted rehabilitation in medicine as well as all kinds of technical assistance systems for vehicle control or for various types of technical devices including construction machines, wheelchairs, etc. For the evaluations in this chapter, an experiment in which several pairs of subjects simultaneously control a steering system is employed. This scenario is modeled by means of a differential game such that cost functions describing the interaction of human pairs can be identified from measured data. The two method classes for inverse differential games presented in this thesis, IOC and IRL, shall be evaluated by means of this experiment. Furthermore, similar to Chapter 7, the results shall be compared to the results of applying the DB approach for identification.

8.1 Experimental Setup

The experimental setup which was used can be seen as a simplified scenario of the lateral control of a vehicle. This section presents all details concerning the hardware setup and the implementation of the haptic feedback. In the following, this setup will be referred to as the *cooperative steering system.*[55]

The cooperative steering system consists of four main components: two active steering wheels, two monitors with visualization windows and a real-time processing unit of dSPACE. The steering wheels are equipped with an incremental encoder of 40000 increments per full rotation for measuring the steering angles with a sampling frequency of $f_s = 100$ Hz. Furthermore, they are active due to integrated motors which can apply a torque on each of them. The

[54] The reader is referred to [ACM+18] for a formal definition of Shared Control and its multiple applications.

[55] The experiment described in this chapter has been also presented in the conference paper [IFH19], where the differential game model was shown to better explain cooperative steering behavior than an alternative state-of-the-art model (presented in [IEFH18]).

maximum torque of the motors is 15.6 Nm. One of the components of the motor torque is calculated such that the steering wheel has the dynamics of a spring-damper system. Therefore, the dynamics of the steering wheel $j \in \{1, 2\}$ are described by means of the equation

$$\Theta_{\mathrm{SW},j}\, \ddot{\varphi}_j(t) = M_j(t) - d_j\dot{\varphi}_j(t) - c_j\varphi_j(t), \tag{8.1}$$

with the spring constant c_j, damping constant d_j and the moment of inertia $\Theta_{\mathrm{SW},j}$ and where $\varphi_j(t)$ and M_j denote the steering wheel angle and the human input torque, respectively. The parameters of the steering wheels are given in Section F.1.1 of the Appendix.

In the experiment, the two steering wheels are haptically coupled. This virtual coupling is implemented in a real-time environment with the dSPACE processor unit. This unit is also used to establish the communication between all components. The haptic coupling is effectuated by calculating the required torque $M_{\mathrm{C}}(t)$ such that the angular difference between the two steering wheels is reduced to zero. This is achieved by emulating a virtual spring-damper element between both steering wheels with an automatic controller. Therefore, with the haptic coupling, a further torque exists which influences the dynamics of each steering wheel, leading to the dynamics equation

$$\Theta_{\mathrm{SW},j}\, \ddot{\varphi}_j(t) = M_j(t) - d_j\dot{\varphi}_j(t) - c_j\varphi_j(t) + M_{\mathrm{C}}(t). \tag{8.2}$$

The implementation of the controller was done in `MATLAB/Simulink` 2010b. Further details on this controller can be found in Section F.1.2 of the Appendix.

A computer interacts with the real-time system and generates two separate visualization windows on two monitors in order to give visual feedback of the current steering wheel position to each participant. This visualization was implemented by means of *OpenGL* and includes a marker (green square) which moves horizontally in the window according to the value of the steering angle. The steering wheel value range which is mapped onto the screen is $[-180°; 180°]$, where a positive angle corresponds to a counterclockwise rotation. A further element in the visualization window is the reference trajectory. The points which constitute the trajectory pass downwards through the window at a constant speed. A single point crosses the entire visualization window in 2 seconds. The vertical position of the marker is fixed at 75% of the window height. Figure 8.1 depicts all components of the experimental setup as well as an example of the visualization window and the black curtain (thick black line) which served to separate each subject's area.

8.2 Modeling

The experiment consists of a shared control task, in which pairs of participants control the cooperative steering system simultaneously. The aim of the subjects is to follow the reference trajectory shown on the monitor by means of their corresponding steering wheel. This

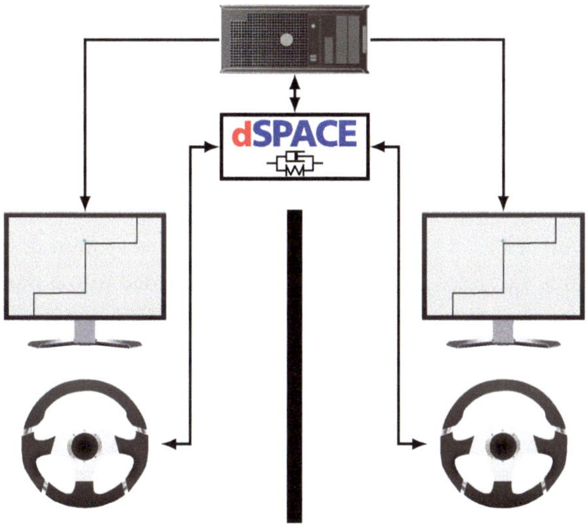

Figure 8.1: Hardware setup for the experiment

scenario is modeled by means of a differential game such that the observed data can be used to identify cost functions of each subject which explain their cooperative behavior. In the following, the differential game is formalized mathematically. Afterwards, the system dynamic equations and cost function structure are stated more precisely for the scenario at hand.

8.2.1 Shared Control Modeling via Differential Games

Consider two human players controlling a dynamic system

$$\dot{\boldsymbol{x}}(t) = \boldsymbol{A}\boldsymbol{x}(t) + \boldsymbol{B}_1\boldsymbol{u}_1(t) + \boldsymbol{B}_2\boldsymbol{u}_2(t) \tag{8.3}$$

with $\boldsymbol{x}(0) = \boldsymbol{x}_0$, where $\boldsymbol{x}(t) \in \mathbb{R}^n$ represents the system states and $\boldsymbol{u}_i(t) \in \mathbb{R}^{m_i}$ denotes the control trajectories of player i. In addition, suppose a reference signal is given, which is the output of the known linear reference model

$$\dot{\boldsymbol{z}}(t) = \boldsymbol{H}\boldsymbol{z}(t). \tag{8.4}$$

Given that the framework of feedback control is the most suitable for modeling human motor control [TJ02, Tod04], it is assumed that the human players select a feedback strategy $\boldsymbol{\gamma}_i \in \Gamma_i^{\mathrm{FB}}$ according to Definition 3.6. Furthermore, the cost function structure

$$J_i = \int_0^\infty \boldsymbol{e}(t)^\top \boldsymbol{Q}_i \boldsymbol{e}(t) + \boldsymbol{u}_i(t)^\top \boldsymbol{R}_{ii} \boldsymbol{u}_i(t)\, \mathrm{d}t, \quad i \in \{1, 2\} \tag{8.5}$$

is assumed for each player, where $\boldsymbol{e}(t) = \boldsymbol{x}(t) - \boldsymbol{z}(t)$. In this way, the cost function models the objective of both humans to track a given reference, i.e. minimize the error between the state and reference trajectories.

While the cost function (8.5) is quadratic, it is not a standard quadratic cost function since the cost function matrix $\boldsymbol{Q}_i$ is not penalizing the state variable $\boldsymbol{x}(t)$, but the state-reference deviation $\boldsymbol{e}(t)$. Therefore, the methods for inverse linear-quadratic dynamic games cannot be applied directly. Nevertheless, it is possible to introduce a new system state including both the states and the reference variables such that (8.5) is transformed into a standard quadratic cost function. This leads to extended system dynamics where the linearity property is maintained. In this way, we obtain a linear-quadratic differential game according to Definition 3.11. The details on these reformulations are presented in Section B.7 of the Appendix.

8.2.2 Cooperative Steering System Dynamics

To simplify the model of the cooperative steering system, an ideal coupling of the two steering wheels is assumed. This means that both steering wheels have the same angle φ and angular velocity $\dot{\varphi}$. With this assumption, the dynamics of the system of coupled steering wheels are given by

$$\dot{\boldsymbol{x}}(t) = \begin{bmatrix} -\frac{d_c}{\Theta_{\mathrm{sum}}} & -\frac{c_c}{\Theta_{\mathrm{sum}}} \\ 1 & 0 \end{bmatrix} \boldsymbol{x}(t) + \begin{bmatrix} \frac{1}{\Theta_{\mathrm{sum}}} \\ 0 \end{bmatrix} u_1(t) + \begin{bmatrix} \frac{1}{\Theta_{\mathrm{sum}}} \\ 0 \end{bmatrix} u_2(t) \tag{8.6}$$

where $\boldsymbol{x}(t) = \begin{bmatrix}\dot{\varphi}(t) & \varphi(t)\end{bmatrix}^\top$ and $u_i(t) = M_i(t)$ is the steering torque of human i. The variable Θ_{sum} denotes the sum of the moments of inertia of both steering wheels. All system parameters are given in Table 8.1.

Table 8.1: Cooperative steering system model parameters

Parameter	Value		Description
Θ_{sum}	0.094	$\mathrm{kg\,m^2}$	Rotational inertia of the coupled steering wheels
c_c	1.146	Nm/rad	Spring constant
d_c	0.859	Nm · s/rad	Damping constant

8.2.3 Cost Functions

The cost function structure is given by (8.5). Furthermore, diagonal matrices $\boldsymbol{Q}_i = \mathrm{diag}(q_i^{(1)}, q_i^{(2)})$ are assumed such that off-diagonal parameters are neglected. This is a common procedure in optimal control theory since off-diagonal matrix elements represent mixed terms in the cost function which are usually not interpretable [BH75]. The state reference is given by $\boldsymbol{z}(t) = \begin{bmatrix}\dot{\varphi}_{\mathrm{ref}}(t) & \varphi_{\mathrm{ref}}(t)\end{bmatrix}^\top$, representing the reference values for the steering angle velocity and the steering angle, which is visible on the monitor. It is assumed that the participants do not aim to follow a particular reference trajectory of the steering velocity since none was specified, neither visually nor verbally. Conversely, the reference trajectory of the steering angle $\varphi_{\text{ref}}(t)$ corresponds to the one visible on the monitor and is equal for both participants.

8.3 Data Acquisition and Preparation

In order to apply inverse dynamic game methods, a set of state and control trajectories is needed. As mentioned previously in Section 8.1, a sensor for measuring the angle $\varphi_j(t)$ of each steering wheel is available. The steering angle velocity $\dot{\varphi}_j(t)$ and the acceleration $\ddot{\varphi}_j(t)$ are determined offline by a numerical differentiation and a subsequent smoothing process via a cubic spline interpolation (MATLAB function `csaps` with parameter $p = 0.99995$). The steering torque of each human $u_i(t) = M_i(t)$ is then calculated by means of (8.2), i.e. the system dynamics equation of each steering wheel. Due to the ideal coupling of the steering wheels, the steering wheel angle $\varphi(t)$ and angular velocity $\dot{\varphi}(t)$ of the cooperative steering system are set equal to the mean value of both steering wheel angles and velocities, respectively.

8.4 Experimental Protocol

Fifty-two subjects (age 25 ± 2.27) participated in the experiment in pairs. They did not have the possibility to make any eye-contact and were told to refrain from speaking during the experiment. They were aware that they were completing the task with a partner. Each subject pair was told to track the reference trajectory as well as they could.

Each pair of subjects completed an approximately two minutes long run which consisted of

- An approximately one minute long initial part (P1) which allowed the participants to become familiar with the haptically coupled system,
- A 4 seconds long middle part (P2) which was used for identification and validation,
- A 32 seconds long final part (P3) which was not used for analysis.

The first part P1 included splines and step functions as visible reference trajectories for the steering angle. On the other hand, P2 consisted of only step functions. Step functions were used for evaluation since these represent goal-oriented or point-to-point movements, also known as reaching movements. This kind of movements are often considered in studies concerning human motor behavior both from a neuroscience and biology perspective [FH91, Kal09, KM11] as well as from a control theoretical perspective [ARARU+11, CS17].

The reference trajectory of P2 describes 4 point-to-point movements defined by the fixed positions (120°, 0°, -120°, 0°, 120°). Finally, P3 included similarly to P1 step functions as well as splines. The subjects were unaware of this scenario subdivision and all related details.

8.5 Evaluation Procedure

As described in Section 8.2.1, the shared control scenario is modeled as a linear-quadratic differential game with feedback strategies. Therefore, the methods for inverse feedback dynamic games (the same as in Section 7.5) are applied for cost function identification. In the following, they are also referred to as the IOC, IRL and DB methods. All methods were given the same system dynamics and cost function structure. The data obtained from the middle part of the test run (P2) was used for estimating the cost function parameters of both participants with each of the aforementioned methods.

Contrary to the simulations presented in Chapter 7, no ground truth cost function parameters $\boldsymbol{\theta}^* = (\boldsymbol{\theta}_1^*, \boldsymbol{\theta}_2^*)$ are available in a real application. Therefore, the only way to evaluate the identification results is by using the estimated cost functions to generate estimated trajectories $\hat{\boldsymbol{x}}(t)$, $\hat{\boldsymbol{u}}_1(t)$, and $\hat{\boldsymbol{u}}_2(t)$ and compare them with the measured trajectories $\tilde{\boldsymbol{x}}(t)$, $\tilde{\boldsymbol{u}}_1(t)$ and $\tilde{\boldsymbol{u}}_2(t)$. This comparison is done by means of the NSAE for states and controls introduced in Section 7.3.2. The 52 participants formed 26 pairs of subjects and therefore, 26 data sets were available for analysis. These 26 sets of trajectories lead each to an estimation of the cost function parameters. Therefore, we obtain the parameters $\hat{\boldsymbol{\theta}}^{(s)}$, $s \in \{1, \ldots, 26\}$ for each of the methods IOC, IRL and DB. Afterwards, each set of identified parameter vectors consisting of $\hat{\boldsymbol{\theta}}^{(s)}_{\mathrm{IOC}}$, $\hat{\boldsymbol{\theta}}^{(s)}_{\mathrm{IRL}}$ and $\hat{\boldsymbol{\theta}}^{(s)}_{\mathrm{DB}}$ is used to solve for the Nash equilibrium trajectories $\hat{\boldsymbol{x}}^{(s)}(t)$, $\hat{\boldsymbol{u}}_1^{(s)}(t)$ and $\hat{\boldsymbol{u}}_2^{(s)}(t)$, $s \in \{1, \ldots, 26\}$. This is done by applying the reformulations of Section B.7 to obtain a standard LQ differential game and using Theorem 3.7 afterwards. The Nash trajectories are compared to the observed trajectories $\tilde{\boldsymbol{x}}(t)$, $\tilde{\boldsymbol{u}}_1(t)$ and $\tilde{\boldsymbol{u}}_2(t)$ by computing the corresponding NSAE as described in Section 7.3.2. Figure 8.2 summarizes the evaluation procedure applied in this chapter.

8.6 Results

The NSAE of states and controls was calculated for all data sets and all corresponding identification results. All values are given in the Section F.2 of the Appendix. Due to the small data set, the median values $\delta^{\boldsymbol{x}}_{\mathrm{median}}$ of the errors are considered instead of the mean values. The median values and the standard deviations $\delta^{\boldsymbol{x}}_{\mathrm{SD}}$ of the errors for all used inverse dynamic methods are given in Table 8.2. The statistical results are summarized and depicted in Figure 8.3.

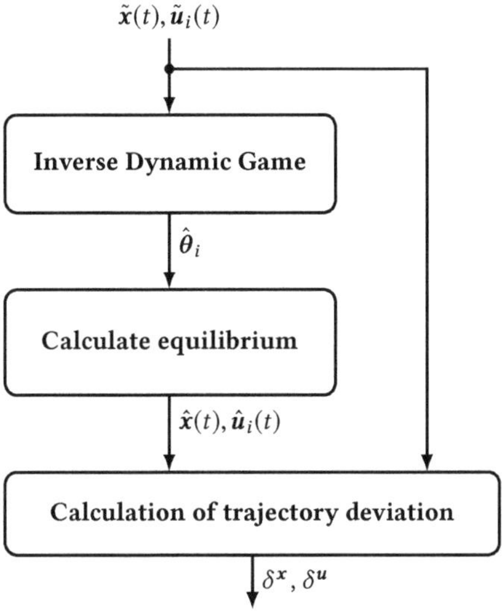

Figure 8.2: Evaluation procedure for the identification in a real shared control scenario

Table 8.2: Mean value and standard deviation of NSAE obtained from identification with IOC, IRL and DB methods

	δ^x		δ^u	
	δ^x_{median}	δ^x_{SD}	δ^u_{median}	δ^u_{SD}
IOC	127.429	58.632	166.578	52.904
IRL	101.236	35.611	173.202	49.356
DB	89.672	19.372	143.952	22.867

The first noticeable characteristic of the results is the considerably higher magnitude of the error compared to the magnitudes seen in Chapter 7. In general, it can be discerned that the DB approach led to smaller mean values and variances of errors than the IRL and IOC based approaches. The IRL method performed better than the IOC method in terms of the state trajectory approximation. Nevertheless, the mean values of the NSAE of the controls are very similar. The range and standard deviation of the errors shown in Figure 8.3 are smaller for the DB method compared to IOC and IRL based approaches. In order to test the statistical significance of these errors, a Wilcoxon signed rank test[56] was conducted on the data sets of δ^x, δ^u. The test results confirmed that all differences are statistically significant with a significance level of $\alpha = 0.01$. Nevertheless, the control errors of the IOC and IRL methods are an exception. The signed rank test confirmed that their difference is not statistically significant. Detailed results with p-values are provided in Section F.2.1 of the Appendix.

[56] A Wilcoxon signed rank test (see e.g. [SC88]) is a statistical test where, contrary to more widespread statistical test methods as e.g. student's t-test, it is not assumed that the data follows a normal distribution. This assumption was avoided here due to the relatively small data population.

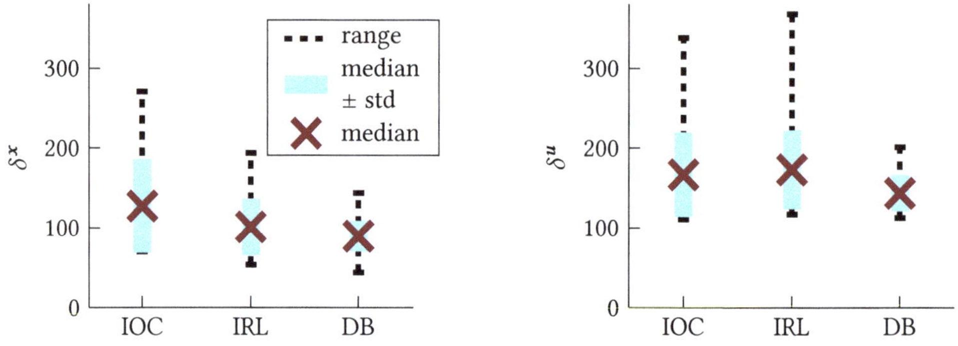

(a) Sum of absolute state errors

(b) Sum of absolute control errors

Figure 8.3: Statistical results of the cost function identification in the experiment

In order to further illustrate the identification results, the measured data and the estimated trajectories $\boldsymbol{x}^{(s)}(t)$ and $\boldsymbol{u}_i^{(s)}(t)$ for some representative subject pairs $s \in \{1, \ldots, 26\}$ are shown in the following. Figure 8.4 shows the data and identification results of subject pair 1. This data set yielded the smallest error for all methods. It can be recognized that the states are approximated the best by the DB approach, followed by the IRL method. The control trajectories cannot be exactly described by the dynamic game with the estimated parameters $\hat{\boldsymbol{\theta}}$ any method. Only the qualitative course can be described and several changes in the torque cannot be accounted for.

The following identification result in Figure 8.5 corresponds to subject pair 2. The DB and IRL method yield the best results regarding state trajectory approximation. Nevertheless, the error is higher than in the results shown in Figure 8.4. In the case of the control trajectories, it is noticeable that the IRL approach fails to identify the control actions of the first subject, but estimates the control of the second subject as higher. This leads to the same state trajectories as the DB approach. The estimation of a control trajectory as (nearly) a constant is an effect which was observed for some data sets, not only for the IRL method, but also for the IOC and DB method. This effect can be seen e.g. in the results of subject pair 22 depicted in Figure 8.6. The DB approach is able to describe the control trajectories better, but on the other hand, the IOC and IRL methods are able to approximate the state trajectories slightly better than the DB method for this data set.

8.7 Computation Time

Analogously to Chapter 7, the computation time required for the solution of inverse dynamic games is analyzed.[57] The mean of the computation times was calculated for each of the

[57] The used CPU was an Intel Core i7-6600U at 2.6 GHz with 12 GB of RAM.

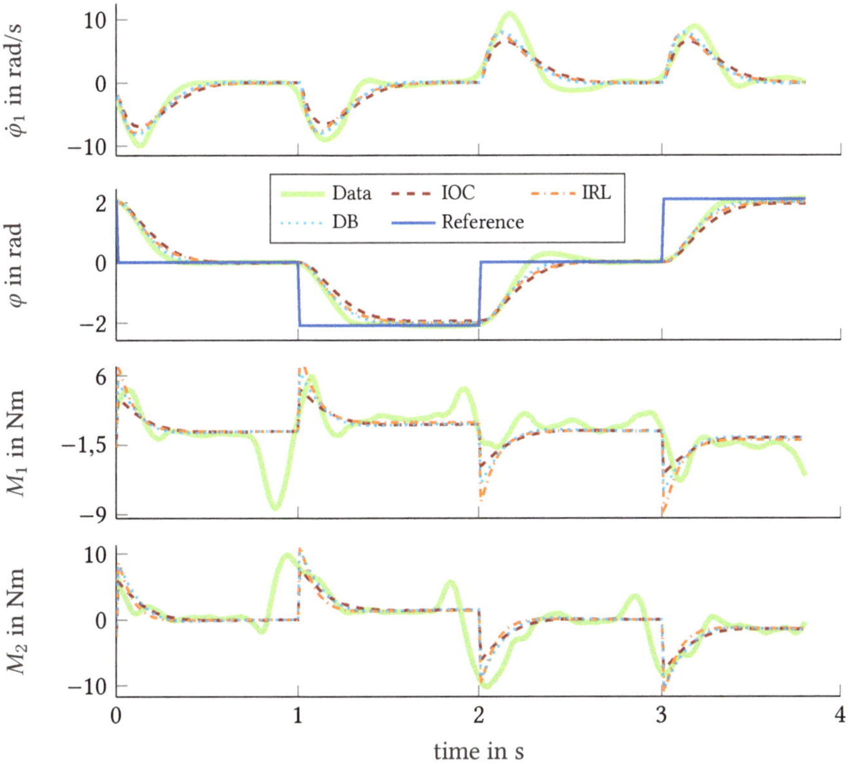

Figure 8.4: Identification results of subject pair 1

method classes considered. The values are listed in Table 8.3. It can be observed that the results of Section 7.7 are replicated. The DB approach needs the most computation time, followed by the IRL and IOC method. The IOC and IRL approaches need 0.01 % and 1.57 % of the DB method's required computation time, respectively.

Table 8.3: Mean computation time for identification of both cost functions of a subject pair in the cooperative steering experiment.

Method	T_{CPU}
IOC	0.04 s
IRL	4.6 s
DB	291.75 s

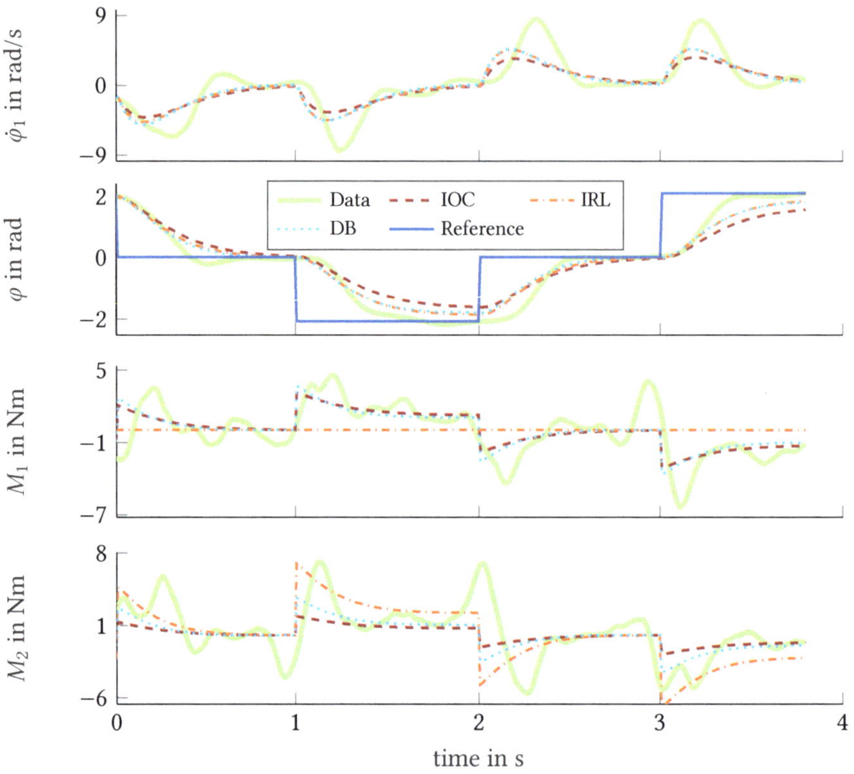

Figure 8.5: Identification results of subject pair 2

8.8 Discussion

This section is devoted to a discussion of the results of the previous sections. The results are analyzed and the limitations of the methods and the experiment are reviewed.

Overall, it can be stated that the inverse feedback dynamic game method based on the DB approach performs better than its IRL and IOC based counterparts in terms of trajectory approximation. This is shown by the mean values of the errors $\delta^{\boldsymbol{x}}_{\text{DB, mean}} < \delta^{\boldsymbol{x}}_{\text{IRL, mean}} < \delta^{\boldsymbol{x}}_{\text{IOC, mean}}$ of both states and controls in Table 8.2. Furthermore, the standard deviations $\delta^{\boldsymbol{x}}_{\text{SD}}$ and $\delta^{\boldsymbol{u}}_{\text{SD}}$ are the smallest for the DB approach, indicating that this method led to more consistent results.

The better results of the DB approach can be similarly explained as in the simulation results of Chapter 7. The underlying optimization problem in the DB method directly minimizes the error between observed and estimated trajectories. In turn, the IRL method does this indirectly by means of an implicit requirement included in the likelihood function. In a very different approach, the IOC method aims to minimize the violation of Nash equilibrium conditions and does not consider the error between trajectories in the process.

In general terms, the methods appear to be able to describe the state trajectories better than the control trajectories. However, there were several data sets for which the state trajectories could not be explained adequately by the cost functions with identified parameters, regardless of the selected inverse dynamic game method. The question arises as to which reasons this effect might have.

One potential source of error is an inexact modeling of the cooperative steering system. In particular, the assumption of an ideal coupling of the steering wheels may have been too strong for the used system, such that the description by means of (8.6) is not accurate enough. It is conceivable that this inaccuracy is higher the more dynamic the interaction is, i.e. when the partners act very differently and change the direction of the torque very often. Besides this fact, the subject pairs were observed to have partially disobeyed the instructions of the experiment. For example, in Figure 8.6, the time span between 1 s and 2 s shows that player 1 applied a torque contrary to the one which is needed to bring the steering angle towards the reference value. This behavior had to be compensated for by player 2. Such behavior contradicts the rationality implied by a model based on differential games and thus cannot be accounted for.

Overall, the results suggest that the players may not act exactly optimally and thus the interaction may sometimes not be exactly represented by a Nash equilibrium. If the trajectories do not represent a Nash equilibrium, then worse results of the IOC and IRL methods are potentially obtained, given the fact that they rely on the estimation of a Nash equilibrium control law from these trajectories. For example, the IOC method first calculates an estimation $\hat{\boldsymbol{K}}_i$ of the linear control law which best describes the relation between measured controls and states; afterwards, cost function parameters are determined which correspond to the identified control matrix. However, these control matrices $\hat{\boldsymbol{K}}_i$ which are optimal in a least-squares sense (cf. Section 5.4.2) do not necessarily correspond to a Nash equilibrium. Consequently, the cost functions with parameters $\hat{\boldsymbol{\theta}}_i$ describe a Nash equilibrium which is the "closest" to $\hat{\boldsymbol{K}}_i$ in the sense that the violation of the Riccati equations is minimal. To illustrate this, consider the value of the residual $||\hat{\boldsymbol{M}}_i\hat{\boldsymbol{\theta}}_i||$, where $\hat{\boldsymbol{M}}_i$ is calculated by means of the $\hat{\boldsymbol{K}} = (\hat{\boldsymbol{K}}_1, ..., \hat{\boldsymbol{K}}_N)$ identified via the LS method (see (5.36)). This describes the extent up to which identified parameters $\hat{\boldsymbol{\theta}}_i$, together with $\hat{\boldsymbol{K}}_i$, violate the necessary and sufficient conditions for Nash equilibria. Therefore, it can be seen as a measure of the "non-Nash" character of the estimated $\hat{\boldsymbol{K}}$[58]. Figure 8.7 shows that some of the identified $\hat{\boldsymbol{K}}$ are approximately a Nash equilibrium, but some others present less Nash character. In particular, the good results of Figure 8.4 can be associated to a low value of the residual. Nevertheless, it could be observed that the residual value does not allow forseeing the quality of the trajectory approximation results.

58 Note that $||\hat{\boldsymbol{M}}_i\hat{\boldsymbol{\theta}}_i|| \neq 0$ is possible while $||\hat{\boldsymbol{M}}'_i\hat{\boldsymbol{\theta}}_i|| \approx 0$. $\hat{\boldsymbol{M}}'_i$ is calculated with $\hat{\boldsymbol{K}}'$ which arise from the solution of the differential game corresponding to the identified parameters $\hat{\boldsymbol{\theta}}$. The latter lead to a Nash equilibrium according to the necessary and sufficient conditions used for determining the trajectories $\hat{\boldsymbol{u}}_i(t)$ and $\hat{\boldsymbol{x}}(t)$.

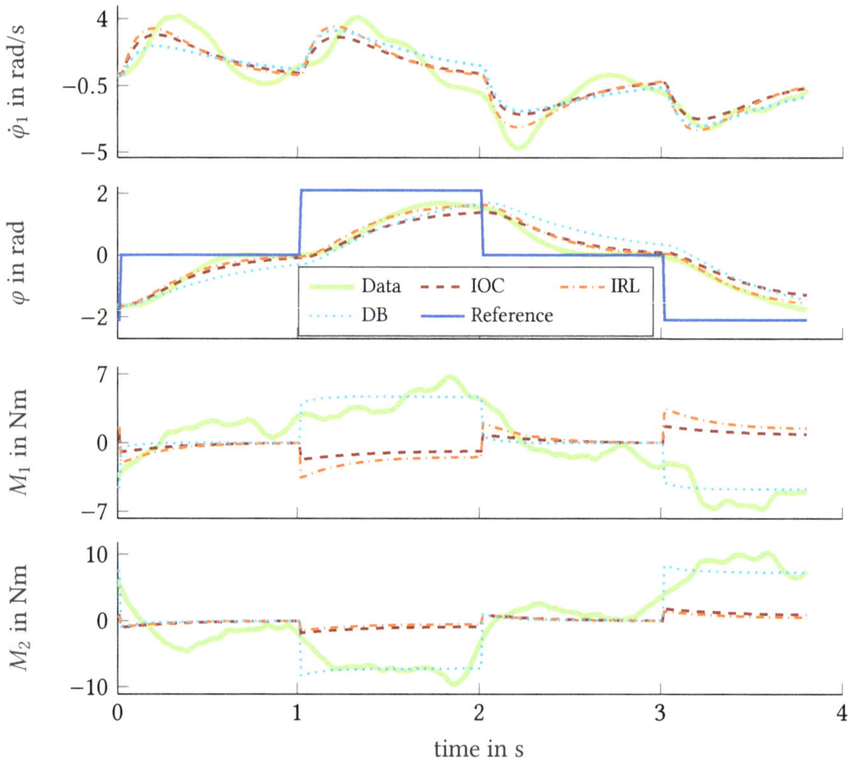

Figure 8.6: Identification results of subject pair 22

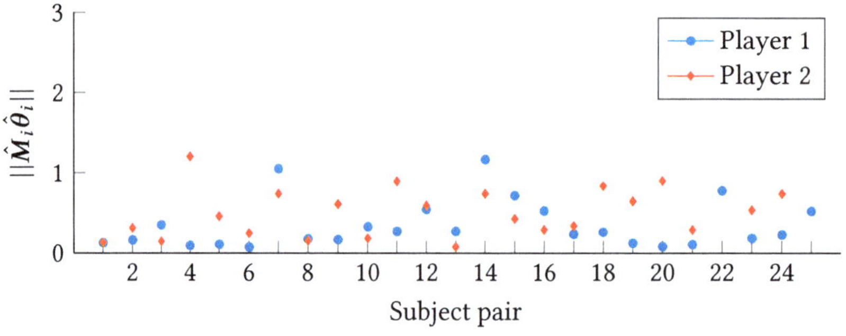

Figure 8.7: Residual values of the identified control law and parameters for all subject pairs. Here, the outlier $||\hat{M}_2\hat{\theta}_2|| = 44.84$ for subject pair 22 is not depicted in favor of better visibility of the other values.

Another problem arises if the estimated $\hat{K}_i$ yields higher values of the objective function of the least-square estimation functional $||u_i + K_i x||$ (cf. (5.36)), i.e. the linear feedback is unable to reproduce the relationship between $u_i(t)$ and $x(t)$. A consequence would be a detriment of the approximation capabilities of the inverse dynamic game methods based on IOC and IRL since they rely on this feedback law estimation to include the influence of the other player's controls on the system dynamics.

Finally, the mean computation times presented in Table 8.3 show that the IOC method would be the most appropriate method in terms of a potential online application such that cost function parameters are constantly updated as new data points are available. The IRL approach may also serve for such a purpose with more efficient coding. On the other hand, the computation time of the DB approach confirm that it is not suitable for an online application. Cost function parameters may change over time due to different effects, e.g. fatigue or even sudden events. These alterations cannot be quickly detected by the DB method, but rather by the alternative methods developed in this thesis.

8.9 Concluding Remarks

In this chapter, an application example for inverse dynamic games was presented. A cooperative steering experiment was conducted where pairs of subject interact haptically to cooperatively complete a control task. The results indicate that it is possible to describe cooperative system behavior by means of dynamic games, and that inverse dynamic game methods can be used to identify cost functions which explain the observed behavior.

The results showed the following insights:

- All methods are influenced by dynamic system model inacurracy, irrational behavior with respect to the control task, and the violation of the assumption of Nash equilibrium trajectories. The IOC and IRL methods are the most affected by this violation.
- The IOC method is confirmed as the most promising method for the online estimation of cost function parameters in real applications due to computation times of fractions of a second.
- The IRL method performs better than the IOC method but the estimation demands more computation time. It still is less computationally demanding than the DB method but has a lower performance.
- The DB method is the most robust towards all kinds of perturbations, but at the cost of a high computational burden. In the evaluations conducted in this chapter, the computation time was over 60 times and 7000 times bigger than the ones achieved by the IRL and IOC methods, respectively.

The system used for the experiment and its dynamic model resulted to be too inaccurate to make reliable conclusions concerning cooperative behavior of human in haptic interaction. The results of this experiment suggest that the assumption of a Nash equilibrium in haptic interaction may be reasonable in certain situations. In order to give answers to these questions, which are also interesting for other scientific communities, more studies and experiments have to be conducted. Nevertheless, the methods presented in this thesis showed the potential of application to these purposes.

9 Conclusion

As technical systems become more intelligent, they are also required to be able to interact with other technical systems and humans. The theory of dynamic games provides a useful mathematical framework for describing the interaction between several players with possibly conflicting interests. A large body of work exists concerning the calculation of the outcome of the dynamic game from known objectives of all players. On the contrary, the inverse problem of dynamic games, which consists in finding the cost functions each player minimized which led to the observed behavior, has received limited attention. This thesis contributes to this line of research by developing methods for the solution of N-player inverse dynamic games with both open-loop and feedback structures and with two different classes of methods, assuming that the interaction between players led to an open-loop or a feedback Nash equilibrium. Following the line of a large number of studies in the identification of cost function in a single-player case, the structure of the cost functions is fixed by assuming a linear combination of basis functions such that the problem is reduced to finding cost function parameters for each player. In addition, the results give a substantial insight on the properties of inverse optimal control and inverse dynamic game problems.

The first method class proposed in this thesis is given by a residual-based IOC method and exploits necessary and sufficient conditions for Nash equilibria which are based on control-theoretical techniques. In the open-loop case, the reformulations of these conditions allow to pose the problem of identifying cost function parameters as an unconstrained quadratic program. Furthermore, sufficient conditions are given to test for the uniqueness of the cost function parameters up to a multiplying constant. For a feedback structure, the use of the same techniques is possible. Nevertheless, the knowledge of the feedback law becomes necessary. Identifying the feedback law is feasible for the main class of dynamic games given by infinite-horizon linear-quadratic dynamic games with an infinite horizon. Therefore, the inverse problem of dynamic games was thoroughly analyzed for this particular class of games. By exploiting the necessary and sufficient conditions for Nash equilibria given by algebraic Riccati equations, explicit solution sets describing all possible cost function parameters which correspond to the same Nash equilibrium were established. Furthermore, a quadratic program was formulated to efficiently find a solution of the inverse dynamic game. An analysis of the properties of this quadratic program yields necessary and sufficient conditions for the uniqueness of the inverse LQ dynamic game solutions.

The second method class which was proposed is an IRL approach, where a probability density function is stated as a likelihood function which depends on the cost function parameters of each player. The likelihood function, found by means of the principle of Maximum Entropy, implicitely includes the requirement that the expected costs of the trajectories sampled from a density function with the estimated parameters correspond to the costs of the observed trajectories. The cost function parameters are determined via a Maximum-Likelihood estimation. For this approach, it was proved that by maximizing the likelihood function we obtain equal expected costs of trajectories generated by the probability density function with ground truth parameters and the one with the estimated parameters.

Having proposed two major classes of inverse dynamic game methods for each of the two information structures considered, i.e. open-loop and feedback, a systematic evaluation was conducted where each method was tested using Nash equilibrium trajectories of a test system. Until now, such a study was missing in literature, even for the single-player case. For inverse dynamic games with open-loop strategies, a two-player game with a nonlinear ball-on-beam dynamic system was considered. The evaluation in the case of a feedback Nash equilibrium was done using a three-player linear-quadratic dynamic game. Both cases included a comparison of the performance of IOC and IRL based methods as well as a direct bilevel (DB) approach analogous to the widespread state-of-the-art single-player inverse dynamic game method of Mombaur et al. [MTL10]. The main findings confirm previous evidence that bilevel methods generally need a high computational effort, since they demand the solution of several dynamic games, i.e. determining Nash equilibria from current candidate cost function parameters. The IOC method outperformed IRL and the DB method in the case of perfect measurements. However, it was shown that the DB and IRL methods are similar to each other and more robust towards measurement noise than IOC methods, since the results of the latter deteriorate with higher measurement noise. Nevertheless, if the measurement noise is low, IOC methods can yield even better results than the DB approach, as it could be observed that the IOC method needs between 0.005% and 0.01% of the DB method's computational time. In addition, the inverse dynamic game methods which exploit the estimation of the feedback Nash equilibrium control laws were shown to be more robust towards measurement noise. As for potential errors in the basis functions, the IRL method offers the ability of detecting irrelevant basis functions with less computational effort than the DB method. The IOC methods show a higher dependency on meaningful basis functions.

Finally, an application example of cooperative system identification was presented, where the aim was to identify cost functions which explain cooperative behavior of humans while completing together a control task and interacting haptically in the process. The results confirmed the trends observed in the simulations, showing that the DB method is the most robust method, followed by IRL and IOC methods. Nevertheless, some data sets could not be described properly by any of the methods. The results indicate that an accurate dynamic system model is of utmost importance for the use of these methods. With a model which better describes the dynamic system both humans interact through, it is conceivable that the developed methods based on IRL and IOC yield a good performance with a reasonable required computational time (of seconds or even milliseconds), thus allowing for their use in real applications where an online estimation of cost function parameters is of interest.

To summarize, this thesis makes a contribution to the theory of inverse problems in optimal control and dynamic game theory. The results not only provide new methods for solving this class of problems, but also shed new light onto their properties. In particular, the novel necessary and sufficient conditions for unique solutions of inverse dynamic games, as well of the unbiasedness of the estimation in an IRL setting, are also valid for the single-player case. The methods open new possibilites for applications regarding the description of multi-agent or cooperative system behavior, e.g. for the identification of human behavior during the interaction with a machine or of biological systems in general, leading to the possibility of employing a learning-by-demonstration approach in a multi-agent setting.

A Infinite Dynamic Games in Discrete Time

This section gives an overview of the relevant definitions and theorems for discrete-time dynamic games which are considered in Chapter 6 of this thesis. The definitions and theorems are analogous to the ones in continuous time. Therefore, each of them has a corresponding counterpart which can be found in Chapter 3. The following selection is based on the books [BO99, HKZ12].

A.1 Basic Definitions

A discrete-time dynamic game involves N players taking actions in several discrete time steps. Since their possible actions are infinite, typical description forms as *payoff matrices* or *game trees* are not possible (see e.g [BO99, Chapter 3]). Instead, the evolution of their decision process is described by means of a dynamic system in discrete-time which is defined as follows.

Definition A.1 (Dynamic System in Discrete-Time State Space Representation) *A dynamic system is defined by a difference equation and an initial condition given by*

$$\boldsymbol{x}^{(k+1)} = \boldsymbol{f}_D^{(k)}\left(\boldsymbol{x}^{(k)}, \boldsymbol{u}_1^{(k)}, \ldots, \boldsymbol{u}_N^{(k)}\right) \tag{A.1a}$$

$$\boldsymbol{x}^{(1)} = \boldsymbol{x}_1 \tag{A.1b}$$

where $\boldsymbol{x}^{(k)} \in \mathbb{R}^n$ and $\boldsymbol{u}_i^{(k)} \in \mathbb{R}^{m_i}$ denote the system state vector and the control vector of player i at time step $k \in \{1, 2, ..., k_E\} =: \mathcal{K}$, respectively.

Each player $i \in \mathcal{P}$ acts upon the system in Definition A.1 by applying a sequence of inputs or controls $\boldsymbol{u}_i^{(k)}, \forall k \in \mathcal{K}$ which belongs to an (here infinite) action space $\mathcal{U}_i$. Analogously to the continuous-time case, each player decides on a particular strategy $\boldsymbol{\gamma}_i^{(k)}$ from the space Γ_i. The control decision is based on the information available to them which is represented by a set-valued function $\eta_i^{(k)}$. This function is generally defined for each player $i \in \mathcal{P}$ and all time steps $k \in \mathcal{K}$ as a subset of

$$\mathcal{I}_i = \left\{\{\boldsymbol{y}_i^{(j)}\}, \{\boldsymbol{u}_i^{(j)}\}\right\}_{\substack{i \in \mathcal{P} \\ j=1,\ldots,k}}, \tag{A.2}$$

where $\boldsymbol{y}_i^{(k)} = \boldsymbol{h}_i^{(k)}(\boldsymbol{x}^{(k)})$ denotes the observed values of the state $\boldsymbol{x}^{(k)}$ according to a function $\boldsymbol{h}_i^{(k)}$. Consequently, the control value at step k results from $\boldsymbol{\gamma}_i(\eta_i^{(k)}) = \boldsymbol{u}_i^{(k)}$, $\boldsymbol{\gamma}_i \in \Gamma_i$.

Each player selects its strategy according to an individual stage-additive cost function of the form

$$J_i = \sum_{k=1}^{k_E} g_{D,i}\left(\boldsymbol{x}^{(k)}, \boldsymbol{u}_1^{(k)}, ..., \boldsymbol{u}_N^{(k)}\right). \tag{A.3}$$

To summarize, a definition of the discrete-time infinite dynamic game is given.

Definition A.2 (Non-Cooperative Discrete-Time Dynamic Game)
A non-cooperative discrete-time dynamic game is defined by

- *A set of players* $\mathcal{P} = \{1, ..., N\}$
- *A set* $\mathcal{K} = \{1, ..., k_E\}$ *including the stages of the game*
- *An infinite action set* $\mathcal{U}_i$, $i \in \mathcal{P}$
- *A set-valued function* $\eta_i^{(k)}$ *describing the state information of player* $i \in \mathcal{P}$ *at time step* k
- *A system given by Definition A.1*
- *A set of stage-additive cost functions* $\mathcal{J} = \{J_1, ..., J_N\}$, $i \in \mathcal{P}$.

The elements and the definition strongly resemble those introduced in Chapter 3. In fact, in system-theoretical terms, if a time difference between each level of play (e.g. k and $k+1$) in a discrete-time dynamic game can be stated and this difference tends towards zero, the game may be considered an approximation of a corresponding continuous-time differential game (quasi-continuous analysis). Indeed, this fact was exploited in order to apply the IRL-based inverse dynamic game methods of Chapter 6 to continuous-time models, e.g. the physically interpretable model of the ball-on-beam system. Furthermore, this allows the comparison of the methods presented in this thesis.

A.2 Information Structures

In the following, a definition of the information structures analogous to the ones in Definition 3.4 is given.

Definition A.3 (Information Structure of the Players in Discrete-Time Dynamic Games)
The information structure of player i is said to be

(i) *open-loop (OL) pattern if* $\eta_i^{(k)} = \boldsymbol{x}^{(1)},\ k \in \mathcal{K}$.

(ii) *memoryless perfect state (MPS) pattern if* $\eta_i^{(k)} = \left\{\boldsymbol{x}^{(1)}, \boldsymbol{x}^{(k)}\right\},\ k \in \mathcal{K}$.

(iii) *feedback (FB) pattern if* $\eta_i^{(k)} = \{\boldsymbol{x}^{(k)}\},\ k \in \mathcal{K}$.

A.3 Strategies

Similar to Section 3.4, the following definitions describe open-loop and feedback strategies in discrete-time dynamic games.

Definition A.4 (Open-Loop Strategy in Discrete-Time Dynamic Games)
An open-loop strategy $\boldsymbol{\gamma}_i^{(k)}$ *for player* $i \in \mathcal{P}$ *selects a control action according to*

$$\boldsymbol{u}_i^{(k)} = \boldsymbol{\gamma}_i^{(k)}(\boldsymbol{x}_1), \quad \forall \boldsymbol{x}_1 \in \mathbb{R}^n,\ k \in \mathcal{K}. \tag{A.4}$$

The set of all such possible strategies is denoted by Γ_i^{OL}.

Definition A.5 (Feedback Strategy in Discrete-Time Dynamic Games)
An feedback strategy $\boldsymbol{\gamma}_i^{(k)}$ *for player* $i \in \mathcal{P}$ *selects a control action according to*

$$\boldsymbol{u}_i^{(k)} = \boldsymbol{\gamma}_i^{(k)}(\boldsymbol{x}^{(k)}), \quad k \in \mathcal{K}. \tag{A.5}$$

The set of all such possible strategies is denoted by Γ_i^{FB}.

A.4 Conditions for Nash Equilibria and Pareto Efficient Solutions in Discrete-Time Dynamic Games

The definition of the solution concepts, i.e. Nash equilibrium, Stackelberg and Pareto efficient solutions, are identical to the ones given in Section 3.5. The only difference is the definition of the strategies $\boldsymbol{\gamma}_i$ which are defined for discrete-time dynamic games by Definitions A.4 and A.5. Therefore, the definitions are not rewritten here. Nevertheless, in the following, analogous results to Theorems 3.1 – 3.3 are given. These serve as a basis for the calculation of solutions of discrete-time dynamic games.

Nash Equilibrium

The following theorems are based on the discrete-time Hamiltonian function

$$\begin{aligned} H_i^{(k)}(\boldsymbol{\psi}_i^{(k+1)}, \boldsymbol{u}_i^{(k)}, \boldsymbol{u}_{\neg i}^{(k)}) := g_{D,i}^{(k)}(\boldsymbol{x}^{(k)}, \boldsymbol{u}_i^{(k)}, \boldsymbol{u}_{\neg i}^{(k)}) + \\ \boldsymbol{\psi}_i^{(k+1)\top} \boldsymbol{f}_D^{(k)}(\boldsymbol{x}^{(k)}, \boldsymbol{u}_i^{(k)}, \boldsymbol{u}_{\neg i}^{(k)}), \quad k \in \mathcal{K}, \, i \in \mathcal{P}. \end{aligned} \tag{A.6}$$

Furthermore, the shorthand notations

$$\boldsymbol{f}_D^{(k)*} = \boldsymbol{f}_D^{(k)}(\boldsymbol{x}^{(k)*}, \boldsymbol{u}_1^{(k)*}, \ldots, \boldsymbol{u}_N^{(k)*}) \tag{A.7}$$

$$g_{D,i}^{(k)*} = g_{D,i}^{(k)}(\boldsymbol{x}^{(k)*}, \boldsymbol{u}_1^{(k)*}, \ldots, \boldsymbol{u}_N^{(k)*}) \tag{A.8}$$

are introduced.

The following theorem is the discrete-time counterpart of Theorem 3.1.

Theorem A.1 (Necessary Conditions for Open-Loop Nash Equilibria in Discrete-Time Dynamic Games)
For an N-player discrete-time infinite dynamic game, let $\boldsymbol{f}_D^{(k)}(\boldsymbol{x}^{(k)}, \boldsymbol{u}_i^{(k)}, \boldsymbol{u}_{\neg i}^{(k)})$ be convex and $g_{D,i}\left(\boldsymbol{x}^{(k)}, \boldsymbol{u}_i^{(k)}, \boldsymbol{u}_{\neg i}^{(k)}\right)$ be continuously differentiable on $\mathbb{R}^n$ for all $k \in \mathcal{K}$, $i \in \mathcal{P}$. Then, if $(\boldsymbol{\gamma}_1^*(\boldsymbol{x}_1), \ldots, \boldsymbol{\gamma}_N^*(\boldsymbol{x}_1))$ with $\boldsymbol{\gamma}_i^*(\boldsymbol{x}_i) = \underline{\boldsymbol{u}}_i^*$ provides an open-loop Nash equilibrium solution with $\underline{\boldsymbol{x}}^*$ as the corresponding state trajectory, there exists a finite sequence of costate functions $(\boldsymbol{\psi}_{D,i}^{(1)}, \ldots, \boldsymbol{\psi}_{D,i}^{(k_E)})$, $i \in \mathcal{P}$ such that the following relations are satisfied:

$$\boldsymbol{x}^{(k+1)} = \boldsymbol{f}_D^{(k)*}, \quad \boldsymbol{x}^{(1)*} = \boldsymbol{x}_1 \tag{A.9a}$$

$$\boldsymbol{u}_i^{(k)*} = \underset{\boldsymbol{u}_i^{(k)}}{\arg\min}\, H_i^{(k)}\left(\boldsymbol{\psi}_{D,i}^{(k+1)}, \boldsymbol{x}^{(k)*}, \boldsymbol{u}_i^{(k)}, \boldsymbol{u}_{\neg i}^{(k)*}\right) \tag{A.9b}$$

$$\boldsymbol{\psi}_{D,i}^{(k)} = \nabla_{\boldsymbol{x}^{(k)}} H_i^{(k)}\left(\boldsymbol{\psi}_{D,i}^{(k+1)}, \boldsymbol{x}^{(k)*}, \boldsymbol{u}_i^{(k)*}, \boldsymbol{u}_{\neg i}^{(k)*}\right) \tag{A.9c}$$

$$\boldsymbol{\psi}_i^{(K)} = \boldsymbol{0}, \tag{A.9d}$$

where $\nabla_{\boldsymbol{x}^{(k)}}$ denotes a partial derivative with respect to the states $\boldsymbol{x}^{(k)}$.

Proof:
See e.g. the proof of Theorem 6.1 of [BO99]. □

Before presenting the theorem which represents necessary and sufficient conditions for feedback Nash equilibria, the discrete-time value function is defined.

Definition A.6 (Value Function)
Consider a player $i \in \mathcal{P}$. Let the optimal strategies of the other players $\boldsymbol{\gamma}^*_{\neg i}$ associated to an N-player non-cooperative discrete-time infinite dynamic game be given. The value function $V_i : \mathbb{R}^n \times \mathcal{K} \mapsto \mathbb{R}$ of player i is defined by

$$V_i(\boldsymbol{x}, k) = \min_{\boldsymbol{\gamma}_i^{(k)}, \dots, \boldsymbol{\gamma}_i^{(k_E)}} \sum_{j=k}^{k_E} g_{D,i}\left(\boldsymbol{x}^{(j)}, \boldsymbol{\gamma}_i^{(j)}, \boldsymbol{\gamma}_{\neg i}^{(j)*}\right), \tag{A.10}$$

where $\boldsymbol{x}^{(k)} = \boldsymbol{x}$.

The following theorem is the discrete-time counterpart of Theorem 3.2.

Theorem A.2 (Necessary and Sufficient Conditions for Feedback Nash Equilibria in Discrete-Time Dynamic Games)
For an N-player discrete-time dynamic game, an N-tuple of feedback strategies $(\boldsymbol{\gamma}_1^{(k)*}, \dots, \boldsymbol{\gamma}_N^{(k)*})$ provides a feedback Nash equilibrium (FNE) solution if, and only if, there exist value functions V_i according to Definition A.6 such that the following recursive relations are satisfied for all players $i \in \mathcal{P}$:

$$\begin{aligned} V_i(\boldsymbol{x}, k) &= \min_{\boldsymbol{u}_i^{(k)}} \left[\tilde{g}_{D,i}^{(k)*}(\boldsymbol{x}, \boldsymbol{u}_i^{(k)}) + V_i\left(\tilde{\boldsymbol{f}}_{D,i}^{(k)*}(\boldsymbol{x}, \boldsymbol{u}_i^{(k)}), k+1\right)\right] \\ &= \tilde{g}_{D,i}^{(k)*}(\boldsymbol{x}, \boldsymbol{u}_i^{(k)*}) + V_i\left(\tilde{\boldsymbol{f}}_{D,i}^{(k)*}(\boldsymbol{x}, \boldsymbol{u}_i^{(k)}), k+1\right); \quad V_i(\boldsymbol{x}, k_E) = \mathbf{0}, \end{aligned} \tag{A.11}$$

where

$$\begin{aligned} \tilde{\boldsymbol{f}}_{D,i}^{(k)*}(\boldsymbol{x}, \boldsymbol{u}_i^{(k)}) &= \boldsymbol{f}_D(\boldsymbol{x}, \boldsymbol{\gamma}_{\neg i}^{(k)*}(\boldsymbol{x}), \boldsymbol{u}_i^{(k)}), \\ \tilde{g}_{D,i}^{(k)*}(\boldsymbol{x}, \boldsymbol{u}_i^{(k)}) &= g_{D,i}^{(k)}(\boldsymbol{x}, \boldsymbol{\gamma}_{\neg i}^{(k)*}(\boldsymbol{x}), \boldsymbol{u}_i^{(k)}). \end{aligned} \tag{A.12}$$

The corresponding Nash equilibrium cost for player i is $V_i(\boldsymbol{x}_1, 1)$.

Proof:
See the proof of Theorem 6.6 of [BO99]. □

Theorem A.2 gives not only sufficient conditions for FNE (cf. Theorem 3.2), but also necessary conditions. Its core consists of the N Bellman equations (A.11) which, analogous to the single-player case, follow from the principle of optimality stated by Bellman [Bel66].[59] For dynamic games, the Bellman equations imply that the N inequalities corresponding to the definition of the Nash equilibrium must hold true for all possible local games (with $\boldsymbol{\gamma}_i^{(k)} \in \Gamma_i^{\mathrm{FB}}$) defined at each possible initial point $\boldsymbol{x}^{(k)}$, $k \in \mathcal{K}$, thus leading to the strong time consistency property of the FNE.

Pareto Efficient Solutions

The following theorem presents necessary and sufficient conditions for Pareto efficient solutions in discrete-time dynamic games. It constitutes the counterpart of Theorem 3.3.

Theorem A.3 (Necessary and Sufficient Conditions for Pareto Efficient Solutions in Discrete-Time Dynamic Games)
Let $\tau_i > 0$, for all $i \in \mathcal{P}$, satisfy

$$\sum_{i=1}^{N} \tau_i = 1. \tag{A.13}$$

Now consider an N-player differential game. If $\boldsymbol{\gamma}^P = \{\boldsymbol{\gamma}_1^P, \dots, \boldsymbol{\gamma}_N^P\}$ is such that

$$\boldsymbol{\gamma}^P = \arg\min_{\boldsymbol{\gamma}} \sum_{i=1}^{N} \tau_i J_i(\boldsymbol{\gamma}) \tag{A.14a}$$

$$\text{w.r.t}$$

$$\boldsymbol{x}^{(k+1)} = \boldsymbol{f}_D(\boldsymbol{x}^{(k)}, \boldsymbol{u}_1^{(k)}, \dots \boldsymbol{u}_N^{(k)}) \tag{A.14b}$$

$$\boldsymbol{x}(1) = \boldsymbol{x}_1 \tag{A.14c}$$

then $\boldsymbol{\gamma}^P$ is a Pareto efficient solution (PES). Moreover, if the strategy spaces Γ_i are convex and J_i are convex in $\boldsymbol{u}_i^{(k)}$ for all $i \in \mathcal{P}$, $k \in \mathcal{K}$, then for all Pareto-efficient $\boldsymbol{\gamma}^P$ there exist τ such that $\boldsymbol{\gamma}^P$ solves the optimization problem in (A.14).

Proof:
The theorem is stated analogously to Theorem 3.3. According to [LZ18], both the sufficiency (first theorem assertion) and the necessary part which are taken from the continuous-time result are valid for the discrete-time case. □

[59] The principle of optimality as stated in [Bel66] reads: "An optimal policy has the property that whatever the initial state and initial decision are, the remaining decisions must constitute an optimal policy with regard to the state resulting from the first decision". This result was used to derive the Bellman equation in single-player optimal control (see e.g. [Kir04, Chapter 3]).

Similar to Theorem 3.3, the optimization problem (A.14) allows the use of the discrete-time minimum principle to solve for the PES. Further results concerning the necessary and sufficient conditions, in terms of the minimum principle corresponding to the problem defined by (A.14), are presented in [LZ18].

A.5 Discrete-Time Linear-Quadratic Dynamic Games

Analogously to LQ differential games, discrete-time LQ dynamic games are defined as follows.

Definition A.7 (Linear-Quadratic Dynamic Game)
A linear-quadratic dynamic game is defined by the same elements as Definition A.2. The system dynamics are linear, i.e. are defined by

$$\boldsymbol{x}^{(k+1)} = \boldsymbol{A}_D \boldsymbol{x}^{(k)} + \sum_{j=1}^{N} \boldsymbol{B}_{D,j} \boldsymbol{u}_j^{(k)} \tag{A.15}$$

where $\boldsymbol{x} \in \mathbb{R}^n$, $\boldsymbol{u} \in \mathbb{R}^{m_i}$. *The cost functions are quadratic, i.e.*

$$J_i = \frac{1}{2} \sum_{k=1}^{k_E} \left(\boldsymbol{x}^{(k)\top} \boldsymbol{Q}_i \boldsymbol{x}^{(k)} + \sum_{j=1}^{N} \boldsymbol{u}_j^{(k)\top} \boldsymbol{R}_{ij} \boldsymbol{u}_j^{(k)} \right). \tag{A.16}$$

where $\boldsymbol{Q}_i, \boldsymbol{R}_{ij}$ *are symmetric for all* $i, j \in \mathcal{P}$ *and* $\boldsymbol{R}_{ii} \succ \boldsymbol{0}$.

The positive semidefiniteness of $\boldsymbol{Q}_i$ and $\boldsymbol{R}_{ij}$, $i, j \in \mathcal{P}$, $i \neq j$ can be sometimes required in order to state necessary and sufficient conditions for Nash equilibria in open-loop and feedback information structures by means of discrete-time coupled Riccati equations. These equations are also derived from the discrete-time minimum principle, i.e. Theorem A.1, and the coupled HJB equations, i.e. Theorem A.2, respectively. In this thesis, a quasi-continuous analysis was considered such that the trajectories of states and controls in LQ dynamic games were generated by the continuous-time RDEs. Therefore, the discrete-time Riccati equations are not explicitely given here. The reader is referred to

- [BO99, Theorem 6.2] for discrete-time Riccati equations in LQ open-loop dynamic games
- [BO99, Corollary 6.1] for discrete-time Riccati equations in LQ feedback dynamic games
- [BO99, Proposition 6.3] for discrete-time Riccati equations in infinite-horizon LQ feedback dynamic games.

B Mathematical Supplements

In this section, further mathematical details are given which complement various sections of this thesis.

B.1 Proof of Theorem 3.4

To the best of the author's knowledge, the precise formulation of Theorem 3.4 is not available in literature. Similar results can be found in [BO99, Theorem 6.12]. However, a formulation similar to the results in [Eng05] was chosen in this thesis in favor of simplicity.

Proof:

[Eng05, Theorem 7.2] states that an OLNE exists if the coupled RDEs (3.60) with conditions (3.61) have a solution $\boldsymbol{P}_i$, $i \in \mathcal{P}$ and additionally, a symmetric solution $\bar{\boldsymbol{P}}_i(t)$ to the non-coupled RDE

$$\dot{\bar{\boldsymbol{P}}}_i(t) = -\boldsymbol{A}^\top \bar{\boldsymbol{P}}_i(t) - \bar{\boldsymbol{P}}_i(t)\boldsymbol{A} + \bar{\boldsymbol{P}}_i(t)\boldsymbol{S}_i\bar{\boldsymbol{P}}_i(t) - \boldsymbol{Q}_i(T) \tag{B.1}$$

exists on $[0, T]$ for all players $i \in \mathcal{P}$. Under the theorem conditions $\boldsymbol{Q}_i \succeq \boldsymbol{0}$ and $\boldsymbol{Q}_{i,T} \succeq \boldsymbol{0}$, $i \in \mathcal{P}$, results of the theory of differential equations can be leveraged to state that the solutions $\bar{\boldsymbol{P}}_i(t)$ of (B.1) are guaranteed to exist (cf. proof of [BO99, Proposition 5.3]. The theorem assertion follows. □

B.2 Equivalence of Cost Functions

Inverse optimal control and inverse dynamic game problems have an inherent ill-posedness property. We give in this section definitions of the equivalence of cost functions in an optimal control and dynamic game scenario.

B.2.1 Optimal Control

In an optimal control problem, where optimal control trajectories $\boldsymbol{u}^*(t)$ which minimize a cost function J are sought, more than one cost function exists which would lead to the same optimal control $\boldsymbol{u}^*(t)$. Consequently, if the system dynamics are unchanged, they lead to the same state trajectories $\boldsymbol{x}^*(t)$. Mathematically, this means that even if

$$J^{(1)}(\boldsymbol{u}(t)) \neq J^{(2)}(\boldsymbol{u}(t)), \tag{B.2}$$

it is still possible to obtain

$$\arg\min_{\boldsymbol{u}(t)} J^{(1)}(\boldsymbol{u}(t)) = \arg\min_{\boldsymbol{u}(t)} J^{(2)}(\boldsymbol{u}(t)). \tag{B.3}$$

For example, it is a well-known fact that (B.3) holds for $J^{(2)}(\boldsymbol{u}(t)) = cJ^{(1)}(\boldsymbol{u}(t))$, $c \in \mathbb{R}^+$. Nevertheless, according to [NF04], the illposedness of a general inverse LQ optimal control problem may transcend the ill-posedness due to a positive real constant. Therefore, it is conceivable that this property is still present in a general inverse (non-LQ) optimal control problem. To define when two cost functions are equivalent, we introduce the following definition.

Definition B.1 (Equivalence of Cost Functions in an Optimal Control Problem)
Two cost functions $J^{(1)}$ and $J^{(2)}$ are equivalent if and only if

$$\mathcal{S}^{(1)} = \mathcal{S}^{(2)} \tag{B.4}$$

where $\mathcal{S}^{(j)}$, $j \in \{1, 2\}$, denotes the set of solutions for cost function $J^{(j)}$, i.e.

$$\mathcal{S}^{(j)} = \left\{ \boldsymbol{u}(t) \mid \boldsymbol{u}(t) = \arg\min_{\boldsymbol{u}(t)} J^{(j)}(\boldsymbol{u}(t)) \right\}. \tag{B.5}$$

B.2.2 Differential Game

An N-player differential game can be considered a generalization of an optimal control problem. Consequently, the ill-posedness issues discussed in the last section are valid in this more general case as well. Analogously to Definition B.1, it is possible to define two equivalent cost functions of a specific player i in a differential game with N players.

Definition B.2 (Equivalence of Cost Functions in a Differential Game)
Two cost functions $J_i^{(1)}$ *and* $J_i^{(2)}$ *are equivalent if and only if*

$$\mathcal{S}_i^{(1)} = \mathcal{S}_i^{(2)} \tag{B.6}$$

where $\mathcal{S}_i^{(j)}, j \in \{1, 2\}$ *denotes the set of solutions of cost function* $J_i^{(j)}$, *i.e.*

$$\mathcal{S}_i^{(j)} = \left\{ \boldsymbol{u}_i(t) \mid \boldsymbol{u}_i(t) = \underset{\boldsymbol{u}_i(t)}{\arg\min}\, J_i^{(j)}(\boldsymbol{u}_i(t), \boldsymbol{u}_{\neg i}^*(t)) \right\}. \tag{B.7}$$

This definition can be interpreted as follows. Let $J_{\neg i}$ represent $N - 1$ cost functions except the cost function of player i. If these cost functions are fixed, then according to Definition B.2, two cost functions for player i are equivalent if and only if, together with $J_{\neg i}$, they lead to the same Nash equilibrium.

B.3 Calculation of Open-Loop Nash Equilibria With the Minimum Principle

Section 3.6.1 presented Theorem 3.1 as necessary conditions for OLNE which consist of several coupled differential equations. Under certain restrictions, these can be used to state a two-point boundary value problem (TPBVP) to solve for Nash equilibrium state trajectories $\boldsymbol{x}^*(t)$. The following lemma represents a useful result for this purpose.

Lemma B.1
Consider an N*-player differential game where the system dynamics are affine in the controls, i.e.*

$$\dot{\boldsymbol{x}}(t) = \boldsymbol{f}(\boldsymbol{x}(t), \boldsymbol{u}_1(t), \ldots, \boldsymbol{u}_N(t), t) = \boldsymbol{f}_{\boldsymbol{x}}(\boldsymbol{x}(t), t) + \sum_{i=1}^{N} G_i(\boldsymbol{x}, t)\boldsymbol{u}_i(t) \tag{B.8}$$

and the running costs g_i *of the cost function* J_i *in* (3.3) *are given by*

$$g_i(\boldsymbol{x}(t), \boldsymbol{u}_1(t), \ldots, \boldsymbol{u}_N(t)) = g_{i,1}(\boldsymbol{x}(t), \boldsymbol{u}_1(t)) + \ldots + g_{i,N}(\boldsymbol{x}, \boldsymbol{u}_N(t)), \quad \forall i \in \mathcal{P}. \tag{B.9}$$

Furthermore, assume that the functions $\boldsymbol{u}_j \mapsto g_{i,j}(\boldsymbol{x}(t), \boldsymbol{u}_j(t))$ are strictly convex for all $i, j \in \mathcal{P}$ and that $g_{i,i}$ has superlinear growth, i.e.

$$\lim_{||\boldsymbol{u}_i|| \to \infty} \frac{g_{i,i}(\boldsymbol{x}, \boldsymbol{u}_i)}{\boldsymbol{u}_i} \to \infty \tag{B.10}$$

Then, for every $(\boldsymbol{x}, t) \in \mathbb{R}^n \times [0, T]$ and every tuple $(\boldsymbol{\psi}_1, ..., \boldsymbol{\psi}_N) \in \mathbb{R}^n \times ... \times \mathbb{R}^n$, the minimization problem

$$\boldsymbol{u}_i^*(t) = \arg\min_{\boldsymbol{u}_i} \{\boldsymbol{\psi}_i^\top G_i(\boldsymbol{x}, t)\boldsymbol{u}_i(t) + g_{i,i}(\boldsymbol{x}(t), \boldsymbol{u}_i(t))\} \tag{B.11}$$

has a unique solution.

Proof:
The proof is analogous to the proof of Lemma 4.1 in [Bre11] in a two-player case.. □

The implications of Lemma B.1 are explained in the following. By using the n algebraic equations defined by (3.17) and the results of Lemma B.1, $\boldsymbol{u}_i^*(t)$ can be written as the unique map

$$\boldsymbol{u}_i^*(t) = \boldsymbol{\eta}_i^*(\boldsymbol{x}(t), \boldsymbol{\psi}_i(t), t). \tag{B.12}$$

By inserting (B.12) in (3.16a) and (3.16c), we obtain a system of coupled non-linear differential equations consisting of

$$\dot{\boldsymbol{x}}(t) = \boldsymbol{f}\left(\boldsymbol{x}(t), \boldsymbol{\eta}_i^*(t), \boldsymbol{\eta}_{\neg i}^*(t), t\right) \tag{B.13}$$

$$\dot{\boldsymbol{\psi}}_i(t) = -\nabla_{\boldsymbol{x}} H_i(\boldsymbol{\psi}_i(t), \boldsymbol{x}(t), \boldsymbol{\eta}_i^*(t), \boldsymbol{\eta}_{\neg i}^*(t), t), \tag{B.14}$$

where $\boldsymbol{\eta}_i^*(t)$ and $\boldsymbol{\eta}_{\neg i}^*(t)$ is used as a short notation for $= \boldsymbol{\eta}_i^*(\boldsymbol{x}(t), \boldsymbol{\psi}_i(t))$ and $\boldsymbol{\eta}_{\neg i}^*(\boldsymbol{x}(t), \boldsymbol{\psi}_{\neg i}(t))$, respectively, and the boundary conditions

$$\boldsymbol{x}^*(0) = \boldsymbol{x}_0 \tag{B.15a}$$

$$\boldsymbol{\psi}_i(T) = \nabla_{\boldsymbol{x}} h_i(\boldsymbol{x}(T)). \tag{B.15b}$$

The TPBVP arising from (B.13), the differential equations (B.14) for each $i \in \mathcal{P}$ and boundary conditions (B.15) can be solved using numerical methods, e.g. shooting methods or collocation methods [AMR95, Chapter 4]. The solution of this TPBVP describes an OLNE.

B.4 Open-Loop Nash Equilibrium of the Ball-on-Beam System

In this section, details on the computation of the OLNE for the differential game with the ball-on-beam system considered in Section 7.4 are provided. In the following, time dependencies shall be omitted for brevity. Furthermore, all equations with the index i refer to player $i \in \{1, 2\}$. The ball-on-beam system dynamics are given by

$$\dot{\boldsymbol{x}} = \begin{bmatrix} x_2 \\ \frac{m_b r_b^2 (x_1 x_4^2 - g_e \sin(x_3))}{\Theta_b + m_b r_b^2} \\ x_4 \\ \frac{-2 m_b x_1 x_2 x_4 - m_b g_e x_1 \cos(x_3) + u_1 + u_2}{m_b x_1^2 + \Theta_w} \end{bmatrix} \tag{B.16}$$

and the cost functions are defined as

$$J_i = \int_0^T \boldsymbol{\theta}_i^\top \boldsymbol{\phi}_i \, \mathrm{d}t$$

with the parameter vector $\boldsymbol{\theta}_i \in \mathbb{R}^{5 \times 1}$ and the basis function vector

$$\boldsymbol{\phi}_i = \begin{bmatrix} x_1^2 & x_2^2 & x_3^2 & x_4^2 & u_i^2 \end{bmatrix}^\top . \tag{B.17}$$

The corresponding Hamiltonian is

$$H_i = \boldsymbol{\theta}_i^\top \boldsymbol{\phi}_i + \boldsymbol{\psi}_i^\top \boldsymbol{f}(\boldsymbol{x}, \boldsymbol{u}_i, \boldsymbol{u}_{\neg i}). \tag{B.18}$$

Using (3.17), we obtain for each players' controls

$$u_i^* = \eta_i(\boldsymbol{x}, \boldsymbol{\psi}_i) = -\frac{\psi_{i,4}}{2\theta_{i,5}(m_b x_1^2 + \Theta_w)}. \tag{B.19}$$

Next, we apply (3.16c) to obtain

$$\dot{\boldsymbol{\psi}}_i = - \begin{bmatrix} 2\theta_{i,1} x_1 + \psi_{i,2} (\nabla_{\boldsymbol{x}} \boldsymbol{f})_{(2,1)} + \psi_{i,4} (\nabla_{\boldsymbol{x}} \boldsymbol{f})_{(4,1)} \\ 2\theta_{i,2} x_2 + \psi_{i,1} (\nabla_{\boldsymbol{x}} \boldsymbol{f})_{(1,2)} + \psi_{i,4} (\nabla_{\boldsymbol{x}} \boldsymbol{f})_{(4,2)} \\ 2\theta_{i,3} x_3 + \psi_{i,2} (\nabla_{\boldsymbol{x}} \boldsymbol{f})_{(2,3)} + \psi_{i,4} (\nabla_{\boldsymbol{x}} \boldsymbol{f})_{(4,3)} \\ 2\theta_{i,4} x_4 + \psi_{i,2} (\nabla_{\boldsymbol{x}} \boldsymbol{f})_{(2,4)} + \psi_{i,3} (\nabla_{\boldsymbol{x}} \boldsymbol{f})_{(3,4)} + \psi_{i,4} (\nabla_{\boldsymbol{x}} \boldsymbol{f})_{(4,4)} \end{bmatrix}, \tag{B.20}$$

where $(\nabla_{\boldsymbol{x}} f)_{(r,c)}$, $r, c \in \{1, \ldots, 4\}$ denote the elements of the matrix of partial derivatives

$$\nabla_{\boldsymbol{x}} f = \begin{bmatrix} 0 & 1 & 0 & 0 \\ \frac{m_b r_b^2 x_4^2}{m_b r_b^2 + \Theta_b} & 0 & \frac{-g_e m_b r_b^2 \cos(x_3)}{m_b r_b^2 + \Theta_b} & \frac{2 m_b r_b^2 x_1 x_4}{m_b r_b^2 + \Theta_b} \\ 0 & 0 & 0 & 1 \\ D & -\frac{2 m_b x_1 x_4}{Z} & \frac{g_e m_b x_1 \sin(x_3)}{Z} & -\frac{2 m_b x_1 x_2}{Z} \end{bmatrix} \tag{B.21}$$

with

$$\begin{aligned} D &= \frac{-2 m_b x_2 x_4 - g_e m_b \cos(x_3)}{Z} - \frac{2 m_b x_1 (u_1^* + u_2^* - 2 m_b x_1 x_2 x_4 - g_e m_b x_1 \cos(x_3))}{Z^2} \\ Z &= m_b x_1^2 + \Theta_w. \end{aligned}$$

Following the procedure described in Section B.3, we insert (B.19) in (B.16) and obtain the system dynamics

$$\dot{\boldsymbol{x}} = \begin{bmatrix} x_2 \\ \frac{m_b r_b^2 (x_1 x_4^2 - g_e \sin(x_3))}{\Theta_b + m_b r_b^2} \\ x_4 \\ f_4^\eta \end{bmatrix}, \tag{B.22}$$

where

$$f_4^\eta = \frac{\left(-4\theta_{1,(5)}\theta_{2,(5)} x_2 x_4 - 2\theta_{1,(5)}\theta_{2,(5)} g_e \cos(x_3)\right) m_b x_1 Z - \psi_{1,(4)}\theta_{2,(5)} - \psi_{2,(4)}\theta_{1,(5)}}{2\theta_{1,(5)}\theta_{2,(5)} Z^2}. \tag{B.23}$$

Furthermore, we insert (B.19) in (B.20) and obtain the same costate differential equation, yet with

$$(\nabla_{\boldsymbol{x}} f)_{(4,1)} = D^\eta = \frac{-2 m_b x_2 x_4 - g_e m_b \cos(x_3)}{Z} - \frac{2 m_b x_1 f_4^\eta}{Z}. \tag{B.24}$$

The system dynamics (B.22) and the differential equations of $\boldsymbol{\psi}_1$ and $\boldsymbol{\psi}_2$ defined by (B.20) and (B.24) constitute a TPBVP which can be solved numerically. In this thesis, the MATLAB function `bvp4c` is used which applies a collocation method (see [SKR00]).

B.5 Approximations for the Maximum Entropy Probability Density Function

This section presents the steps needed for the approximation result of the probability density function given in (6.47). For brevity, the subscript i is omitted from all variables related to player i in the following. Likewise, for the following derivations are based on the assumption that one single demonstration ($n_t = 1$) lies at hand such that the subscript l can also be neglected.

Inserting (6.44) in (6.24) results in

$$\begin{aligned}
\mathrm{p}\left(\tilde{\zeta}\middle|\theta\right) &= \mathrm{p}\left(\underline{\tilde{u}}\middle|\underline{\tilde{u}}_{\neg i}, x^{(1)}, \theta\right) \\
&= \mathrm{e}^{-J(\underline{\tilde{u}})}\left[\int_{-\infty}^{\infty}\mathrm{e}^{-J(\underline{u})}\mathrm{d}\underline{u}\right]^{-1} \\
&\approx \mathrm{e}^{-J(\underline{\tilde{u}})}\left[\int_{-\infty}^{\infty}\mathrm{e}^{\left\{-J(\underline{\tilde{u}})-(\underline{u}-\underline{\tilde{u}})^{\top}g-\frac{1}{2}(\underline{u}-\underline{\tilde{u}})^{\top}G(\underline{u}-\underline{\tilde{u}})\right\}}\mathrm{d}\underline{u}\right]^{-1} \\
&= \left[\int_{-\infty}^{\infty}\mathrm{e}^{\left\{\frac{1}{2}g^{\top}G^{-1}g-\frac{1}{2}\left((\underline{u}-\underline{\tilde{u}})^{\top}G(\underline{u}-\underline{\tilde{u}})+g^{\top}(\underline{u}-\underline{\tilde{u}})+(\underline{u}-\underline{\tilde{u}})^{\top}g+g^{\top}G^{-1}g\right)\right\}}\mathrm{d}\underline{u}\right]^{-1} \\
&= \left[\int_{-\infty}^{\infty}\mathrm{e}^{\left\{\frac{1}{2}g^{\top}G^{-1}g-\frac{1}{2}(G(\underline{u}-\underline{\tilde{u}})+g)^{\top}G^{-1}(G(\underline{u}-\underline{\tilde{u}})+g)\right\}}\mathrm{d}\underline{u}\right]^{-1}.
\end{aligned} \tag{B.25}$$

We note that the relation

$$\begin{aligned}
&\left(G\left(\underline{u}-\underline{\tilde{u}}\right)+g\right)^{\top}G^{-1}\left(G\left(\underline{u}-\underline{\tilde{u}}\right)+g\right) \\
&\quad= \left(G\underline{u}-G\underline{\tilde{u}}+g\right)^{\top}G^{-1}\left(G\underline{u}-G\underline{\tilde{u}}+g\right) \\
&\quad= \left(\underline{u}^{\top}G^{\top}G^{-1}-\underline{\tilde{u}}^{\top}G^{\top}G^{-1}+g^{\top}G^{-1}\right)G\left(\underline{u}-\underline{\tilde{u}}+G^{-1}g\right) \\
&\quad= \left(\underline{u}^{\top}+\left(g^{\top}G^{-1}-\underline{\tilde{u}}^{\top}\right)\right)G\left(\underline{u}+G^{-1}g-\underline{\tilde{u}}\right) \\
&\quad= \left(\underline{u}+\left(G^{-1}g-\underline{\tilde{u}}\right)\right)^{\top}G\left(\underline{u}+\left(G^{-1}g-\underline{\tilde{u}}\right)\right),
\end{aligned} \tag{B.26}$$

holds due to the symmetry of the second derivative G of the cost function. By appling (B.26) in (B.25), the right hand side results in

$$\mathrm{e}^{\left\{-\frac{1}{2}g^{\top}G^{-1}g\right\}}\left[\int_{-\infty}^{\infty}\mathrm{e}^{\left\{-\frac{1}{2}(\underline{u}+(G^{-1}g-\underline{\tilde{u}}))^{\top}G(\underline{u}+(G^{-1}g-\underline{\tilde{u}}))\right\}}\mathrm{d}\underline{u}\right]^{-1}. \tag{B.27}$$

Finally, since

$$\int_{-\infty}^{\infty}\frac{1}{\sqrt{(2\pi)^{\dim(y)}\left|\Sigma_{y}\right|}}\mathrm{e}^{\left\{-\frac{1}{2}(y-\mu_{y})^{\top}\Sigma_{y}^{-1}(y-\mu_{y})\right\}}\mathrm{d}y \overset{!}{=} 1 \tag{B.28}$$

holds for a multidimensional Gaussian distribution with the mean μ_y and the covariance matrix Σ_y, we may rewrite (B.27) and obtain the approximated probability density function

$$\mathrm{p}\left(\zeta_i\middle|\theta\right) \approx \mathrm{e}^{-\left\{\frac{1}{2}g^{\top}G^{-1}g\right\}}\det(G)^{\frac{1}{2}}\left(2\pi\right)^{-\frac{1}{2}\dim(\underline{u})}, \tag{B.29}$$

where $\dim(\underline{u}) = mk_E$ denotes the dimension of $\underline{u}$. From (B.29), the approximated log-likelihood function results in

$$\begin{aligned}\ln \mathcal{L}\left(\tilde{\zeta} \middle| \theta\right) &= \ln\left(\mathrm{p}\left(\underline{u} \middle| \underline{\tilde{u}}_{\neg i}, x^{(1)}, \theta\right)\right) \\ &\approx -\frac{1}{2} g^{\top} G^{-1} g + \frac{1}{2} \ln\left(\det(G)\right) - \frac{1}{2} \dim\left(\underline{u}\right) \ln\left(2\pi\right).\end{aligned} \tag{B.30}$$

B.6 Implementation of the Direct Bilevel Approach

The DB approach used for comparison in this thesis is based on the minimization of the cost functional (7.1) which depends on the current candidate trajectories $u_{\theta,j}(t)$ and $x_{\theta}(t)$. These trajectories must be Nash equilibrium trajectories under an arbitrary parametrization of the cost functions θ. The solution of a forward dynamic game with the parameters θ is therefore nested inside the objective function in (7.1). Consequently, each of the objective function evaluations will include the solution of a forward dynamic game to determine an OLNE or a FNE, depending on the considered case. We note that the search for θ might lead to cost function parameter candidates for which a Nash equilibrium does not necessarily exist. Proving the existence of Nash equilibria is in general not trivial. For example, in the case of linear-quadratic differential games, the existence of Nash equilibria depends on the existence of the solution to the coupled Riccati differential equations, yet its existence has only been proved under strong assumptions. Furthermore, the proofs are not very useful for practical implementation. Therefore, existence of Nash equilibria cannot be ensured by introducing optimization constraints. Nevertheless, probably inspired by the optimal control case (cf. assumptions in the results summarized in [Kuč73]), literature on (linear-quadratic) dynamic games usually introduce constraints of the kind

$$C = \left\{\theta_i \mid \theta_{i,(j)} \geq 0,\ \forall i \in \mathcal{P}, j \in \{1, ..., M_i\}\right\}. \tag{B.31}$$

This constraint set was implemented in the minimization of the objective function for the DB approach. The occurence of succesful calculations of Nash equilibria was indeed increased with this set. Nevertheless, it was not enough to completely avoid failure. Therefore, the objective function was augmented by a resetting procedure of the candidate trajectories (potentially leading to greater costs) which became active if the forward problem, i.e. the numerical solution of the corresponding RDEs or the TPBVPs did not converge.

The algorithm describing the cost functional to be evaluated in each iteration of the optimization problem is listed below.

Algorithm 5 Cost Functional for the Direct Bilevel Approach in Inverse Differential Games.

Input: Parameter candidates $\boldsymbol{\theta}$, observed trajectory set $\mathcal{D}$, dynamics f, basis functions $\boldsymbol{\phi}_i$.
Output: Sum of squared errors J_{DB}

1: Attempt calculation of Nash equilibrium trajectories $\boldsymbol{x}_\theta(t)$ and $\boldsymbol{u}_{\theta,j}(t)$.
2: **if** Calculation fails **then**
3: Set $\boldsymbol{x}_\theta(t) = \mathbf{0}$ and $\boldsymbol{u}_{\theta,j}(t) = \mathbf{0}$, $\forall j \in \mathcal{P}$.
4: **end if**
5: Calculate sum of squared errors between candidate trajectories and observed trajectories J_{DB}.
6: **return** J_{DB}.

Therefore, the DB method used for the simulation results of Chapter 7 consists of the minimization of the cost functional described by Algorithm 5 subject to the constraints (B.31).

B.7 Solutions of the LQ Tracking Problem in the Cooperative Steering Model

This section presents reformulations of the LQ tracking problem arising in Section 8.2.1 to a standard LQ problem which allows an easier solution of the differential game. First, the general approach is presented. It is based on the reformulation proposed for the single-player case in [ML14]. Afterwards, the reformulations specific to the problem of Section 8.2.1 are shown. In the remainder of this section, time dependencies of all variables will be omitted for better readability.

B.7.1 General Reformulation to a Standard LQ Problem

To begin the reformulation, the state variable $X = \begin{bmatrix} \boldsymbol{x} & \boldsymbol{z} \end{bmatrix}^\top$ is introduced which combines the system states and the corresponding reference trajectories. With this new state, we define an extended system including the original system dynamics as well as the reference model dynamics:

$$\begin{aligned} \dot{X} &= \tilde{A}X + \tilde{B}_1 \boldsymbol{u}_1 + \tilde{B}_2 \boldsymbol{u}_2 \\ \text{with} \quad \tilde{A} &= \begin{bmatrix} A & \mathbf{0} \\ \mathbf{0} & H \end{bmatrix}, \quad \tilde{B}_i = \begin{bmatrix} B_i \\ \mathbf{0} \end{bmatrix}, \quad i \in \{1, 2\}. \end{aligned} \tag{B.32}$$

Due to the infinite horizon, the cost function (8.5) can only be applied if H is Hurwitz. This is a considerable restriction, since application-relevant reference signals, e.g. sinusoidal and step functions, will not lead to a Hurwitz reference system matrix. In order to circumvent this problem, we introduce a discount factor β such that $0 < \beta < 1$ in the cost function, thus avoiding infinite costs.

We note that the tracking error $\boldsymbol{e}$ can be written as $\boldsymbol{e} = \boldsymbol{T}\boldsymbol{X}$, where $\boldsymbol{T} = \begin{bmatrix}\boldsymbol{I}_n & -\boldsymbol{I}_n\end{bmatrix}$ and $\boldsymbol{I}_n$ is an n-dimensional identity matrix. With this transformation matrix and with the discount factor β, we rewrite (8.5) as

$$
\begin{aligned}
J_i &= \int_0^\infty \exp(-\beta t)\boldsymbol{X}^\top \boldsymbol{T}^\top \boldsymbol{Q}_i \boldsymbol{T}\boldsymbol{X} + \boldsymbol{u}_i^\top R_{ii} \boldsymbol{u}_i \,\mathrm{d}t \\
&= \int_0^\infty \exp(-\beta t)\boldsymbol{X}^\top \tilde{\boldsymbol{Q}}_i \boldsymbol{X} + \boldsymbol{u}_i^\top R_{ii} \boldsymbol{u}_i \,\mathrm{d}t
\end{aligned} \tag{B.33}
$$

where

$$
\tilde{\boldsymbol{Q}}_i = \boldsymbol{T}^\top \boldsymbol{Q}_i \boldsymbol{T} = \begin{bmatrix} \boldsymbol{Q}_i & -\boldsymbol{Q}_i \\ -\boldsymbol{Q}_i & \boldsymbol{Q}_i \end{bmatrix}. \tag{B.34}
$$

According to Modares and Lewis [ML14], the optimal control problem consisting of the system dynamics (B.32) and the cost function (B.33) for any $i \in \{1, 2\}$ to be minimized can be reformulated as an optimal control problem with a cost function without any discounting factor β, but with the new system matrix $\tilde{\boldsymbol{A}} - 0.5\beta \boldsymbol{I}_n$ instead of (B.32). This is necessary in order to ease the calculation of the solution and to prove its existence. In their paper [ML14], Modares and Lewis state that the solution exists if the matrix $\tilde{\boldsymbol{A}} - 0.5\beta \boldsymbol{I}_n$ is Hurwitz.

B.7.2 Transformed System Dynamics and Cost Functions of the Cooperative Steering System

Given that we apply constant reference values, $\boldsymbol{H} = \boldsymbol{0}$ holds for the reference system matrix in (8.4). Moreover, the velocity reference signal is zero. Therefore, we neglect this term before applying the aforementioned transformation. In this way, we obtain system dynamics of the form (B.32) with the extended state $\boldsymbol{X} = \begin{bmatrix}\dot{\varphi} & \varphi & \varphi_{\mathrm{ref}}\end{bmatrix}^\top$. This leads to a transformed system (B.32) with

$$
\tilde{\boldsymbol{A}} = \begin{bmatrix} -\frac{d_c}{\Theta_{\mathrm{sum}}} & -\frac{c_c}{\Theta_{\mathrm{sum}}} & 0 \\ 1 & 0 & 0 \\ 0 & 0 & 0 \end{bmatrix}, \quad \tilde{\boldsymbol{B}}_1 = \tilde{\boldsymbol{B}}_2 = \begin{bmatrix} \frac{1}{\Theta_{\mathrm{sum}}} \\ 0 \\ 0 \end{bmatrix}. \tag{B.35}
$$

Furthermore, the transformation matrix is given by

$$
\boldsymbol{T} = \begin{bmatrix} 1 & 0 & 0 \\ 0 & 1 & -1 \end{bmatrix}. \tag{B.36}
$$

Since our steering wheel system is stabilizable and the reference system with $\boldsymbol{H} = \boldsymbol{0}$ is marginally stable, any $\beta > 0$ suffices to make the extended system stabilizable and, consequently, to make the transformation applicable. We choose a small value of $\beta = 0.01$, leading to a modified cost function

$$J_i = \int_0^\infty \exp(-\beta t)\boldsymbol{X}^\top \tilde{\boldsymbol{Q}}_i \boldsymbol{X} + R_{ii} u_i^2 \, \mathrm{d}t, \tag{B.37}$$

where

$$\tilde{\boldsymbol{Q}}_i = \boldsymbol{T}^\top \boldsymbol{Q}_i \boldsymbol{T} = \begin{bmatrix} q_1 & 0 & 0 \\ 0 & q_2 & -q_2 \\ 0 & -q_2 & q_2 \end{bmatrix}. \tag{B.38}$$

Finally, we obtain a standard LQ differential game consisting of the system dynamics matrices $(\tilde{\boldsymbol{A}} - 0.5\beta \boldsymbol{I}_n, \tilde{\boldsymbol{B}}_1, \tilde{\boldsymbol{B}}_2)$ and the cost functions

$$J_i = \int_0^\infty \boldsymbol{X}^\top \tilde{\boldsymbol{Q}}_i \boldsymbol{X} + R_{ii} u_i^2 \, \mathrm{d}t. \tag{B.39}$$

For the solution of the inverse LQ dynamic game, parameter constraints are introduced in the corresponding optimization problems (constituting the IOC, IRL and DB approaches) such that the structure of the cost function matrix in (B.38) is ensured.

C Supplementary Results on the Solution Sets for Inverse Linear-Quadratic Differential Games

The following results complement the results of Section 5.3 to illustrate how the properties of an inverse LQ differential game are altered depending on the number of states, controls and players.

All results are based on the general structure of a quadratic cost function given by

$$J_i(\boldsymbol{x}_0, \boldsymbol{K}, \boldsymbol{Q}_i, \boldsymbol{R}_{ij}) = \frac{1}{2} \int_0^\infty \boldsymbol{x}^\top \boldsymbol{Q}_i \boldsymbol{x} + \sum_{j=1}^{N} \boldsymbol{u}_j^\top \boldsymbol{R}_{ij} \boldsymbol{u}_j \, \mathrm{d}t.$$

A two-player and a three-player inverse LQ differential game are considered exemplarily.

Figures C.1 and C.2 shows a 3D map for analyzing the dimensions of the matrix $\boldsymbol{M}_i$ for inverse LQ differential games with $N = 2$ and $N = 3$, respectively, with symmetric and diagonal cost function matrices and different numbers of states n and controls m_i. These are analogous to Figure 5.1 which showed the case $N = 1$. The number of equations (rows of $\boldsymbol{M}_i$) and the number of parameters M_i (columns of $\boldsymbol{M}_i$) are shown as a function of the number of states n and the number of controls m_i.

In Figure C.1a and C.2a, the number of parameters M_i is always greater than the number of equations $n\, m_i$ such that the solution set of player i is at least one-dimensional. In Figures C.1b and C.2b, we observe that there are combinations of n and m_i which lead to $n\, m_i \geq M_i$. The black line denotes the cases where $n\, m_i = M_i - 1 < M_i$ which shows that the kernel is guaranteed to exist and is one-dimensional. Therefore, from this line to the left, the solution set of player i can be expressed by $\ker(\boldsymbol{M}_i)$, while the area which is on the right side of the line represents the cases where solutions may be found by applying the results of Theorem 5.3.

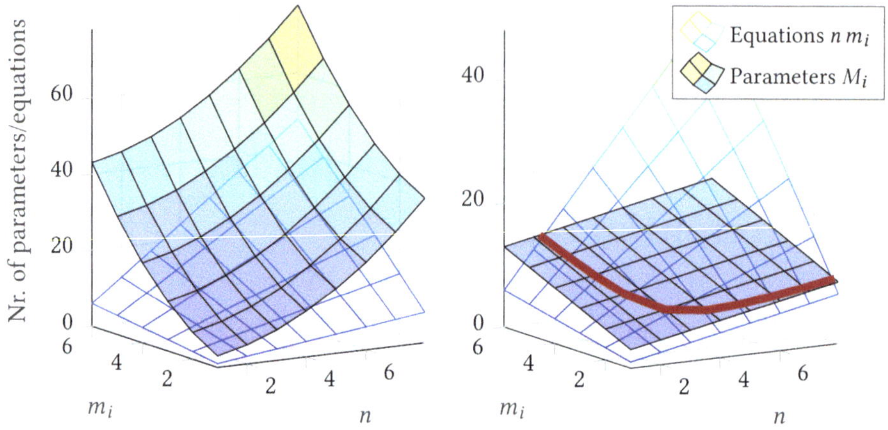

(a) Symmetrical cost function matrices **(b)** Diagonal cost function matrices

Figure C.1: Number of parameters and equations depending on the number of states and controls in a two-player inverse LQ differential game. The red thick line denotes the case where $n\,m_i = M_i - 1$.

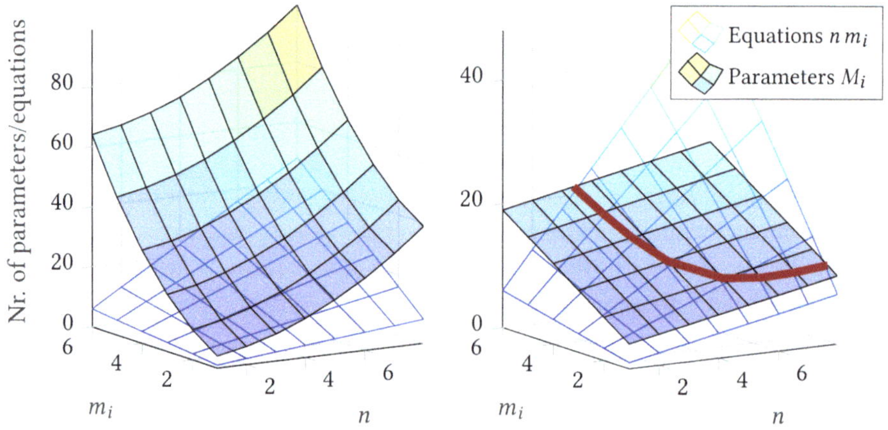

(a) Symmetrical cost function matrices **(b)** Diagonal cost function matrices

Figure C.2: Number of parameters and equations depending on the number of states and controls in a 3-player inverse LQ differential game. The red thick line denotes the case where $n\,m_i = M_i - 1$.

D Inverse Cooperative Dynamic Games Based on Maximum Entropy Inverse Reinforcement Learning

In this chapter, the probability density function given by (6.17) is leveraged such that a method to identify cost function parameters out of a solution of the dynamic game in the sense of Pareto is developed. Similar to the results of Chapter 6 the unbiasedness of the estimation is proved.

D.1 Preliminaries

In this appendix, Pareto efficient solutions are considered which can be described by a global cost function given by the sum of weighted player cost functions. Several global cost functions are possible depending on the selected weighting parameters to build the sum (cf. Section 3.6.3). One particular global cost function is given by the sum of uniformly weighted player cost functions defined as follows.

Definition D.1 (Global Cost Function as Uniformly Weighted Sum)
The uniformly weighted sum of all player cost functions is given by

$$J_\Sigma = \sum_{i=1}^{N} J_i = \sum_{i=1}^{N} -\boldsymbol{\theta}_i^\top \boldsymbol{\mu}_i =: -\boldsymbol{\theta}_\Sigma^\top \boldsymbol{\mu}_\Sigma \tag{D.1}$$

with

$$\boldsymbol{\theta}_\Sigma = \begin{bmatrix} \boldsymbol{\theta}_1^\top & \dots & \boldsymbol{\theta}_N^\top \end{bmatrix}^\top \tag{D.2a}$$

and

$$\boldsymbol{\mu}_\Sigma = \begin{bmatrix} \boldsymbol{\mu}_1^\top & \dots & \boldsymbol{\mu}_N^\top \end{bmatrix}^\top . \tag{D.2b}$$

The following assumption is introduced in order to be able to obtain Pareto efficient solutions.

Assumption D.1 (Convexity of the Global Cost Function)
The cost functions J_i are convex for all $i \in \mathcal{P}$.

Remark D.1:
It can be noted that

$$\arg\min_{\gamma} J_{\Sigma}(\boldsymbol{\gamma}) = \arg\min_{\gamma} \sum_{i=1}^{N} \frac{1}{N} J_i(\boldsymbol{\gamma}), \tag{D.3}$$

where $\boldsymbol{\gamma} := \{\boldsymbol{\gamma}_1, \dots, \boldsymbol{\gamma}_N\}$, holds since multiplying any cost function J_{Σ} with a constant factor $c \in \mathbb{R}^+$ (here $1/N$) does not alter the solution of the optimization problem. Therefore, under Assumption D.1 and with the results of Theorem 3.3, the minimizer of J_{Σ} describes a Pareto efficient solution of a cooperative game.

D.2 Identification Method and Unbiasedness of the Estimation

Sections 6.4 and 6.5 presented how to find cost function parameters which explain observed trajectories which arise from a noncooperative game with OL and FB Nash equilibrium strategies. This was done by means of a MLE based on a probability density function. This section presents a similar procedure such that parameters can be found which explain trajectories corresponding to a cooperative game with equally weighted cost functions as in Definition D.1.

The inverse dynamic game method is based on the density (6.17) with naturally arises with the maximum entropy principle as described in Section 6.3. The first step consists in rewriting (6.17) using (D.1) and (D.2), leading to

$$\begin{aligned} \mathrm{p}(\zeta \,|\, \boldsymbol{\theta}_{\Sigma}) &= \frac{\exp\left(\boldsymbol{\theta}_{\Sigma}^{\top} \boldsymbol{\mu}_{\Sigma}(\zeta)\right)}{\displaystyle\int_{\zeta} \exp\left(\boldsymbol{\theta}_{\Sigma}^{\top} \boldsymbol{\mu}_{\Sigma}(\zeta)\right) \mathrm{d}\zeta} \\ &= \frac{\exp\left(-J_{\Sigma}(\zeta)\right)}{\displaystyle\int_{\zeta} \exp\left(-J_{\Sigma}(\zeta)\right) \mathrm{d}\zeta}. \end{aligned} \tag{D.4}$$

This allows the definition of a likelihood function analogous to the one introduced in Definition 6.5. In this case, we denote the likelihood as

$$\mathcal{L}(\boldsymbol{\theta}_\Sigma \mid \mathcal{D}) = \prod_{l=1}^{n_t} \mathrm{p}\left(\tilde{\zeta}_l \mid \boldsymbol{\theta}_\Sigma\right), \tag{D.5}$$

where $\mathrm{p}\left(\tilde{\zeta}_l \mid \boldsymbol{\theta}_\Sigma\right)$ is obtained by evaluating (D.4) at $\tilde{\zeta}_l$, $l \in \{1, ..., n_t\}$.

The following theorem represents the main result concerning the identification of cost functions in an inverse cooperative dynamic game with Pareto efficient solutions.

Theorem D.1 (Unbiasedness of the Identification of Pareto Efficient Solutions)
Let n_t trajectories $\mathcal{D} = \{\tilde{\zeta}_1, \ldots, \tilde{\zeta}_{n_t}\}$ fulfilling Assumption 6.1 be available. Then, the MLE with respect to the observed trajectories, i.e.

$$\hat{\boldsymbol{\theta}}_\Sigma = \arg\max_{\boldsymbol{\theta}_\Sigma} \ln \mathcal{L}\left\{\boldsymbol{\theta}_\Sigma \mid \mathcal{D}\right\} \tag{D.6}$$

where $\mathcal{L}\left\{\boldsymbol{\theta}_\Sigma \mid \mathcal{D}\right\}$ is obtained by evaluating the likelihood function (D.5) *at $\tilde{\zeta}_l$, $l \in \{1, ..., n_t\}$, leads to parameters $\hat{\boldsymbol{\theta}}_\Sigma$ such that the resulting probability density function $\mathrm{p}\left(\zeta \mid \boldsymbol{\theta}_\Sigma^*\right)$ leads to an expectation of the cost function values $J_\Sigma\left(\zeta, \boldsymbol{\theta}_\Sigma^*\right)$ which is equal to the one corresponding to the probability density function with original parameters $\mathrm{p}\left(\zeta \mid \boldsymbol{\theta}_\Sigma^*\right)$, i.e.*

$$\mathbb{E}_{\mathrm{p}(\zeta \mid \boldsymbol{\theta}_\Sigma^*)}\left\{J_\Sigma\left(\zeta, \boldsymbol{\theta}_\Sigma^*\right)\right\} = \mathbb{E}_{\mathrm{p}\left(\zeta \mid \hat{\boldsymbol{\theta}}_\Sigma\right)}\left\{J_\Sigma\left(\zeta, \boldsymbol{\theta}_\Sigma^*\right)\right\}. \tag{D.7}$$

Proof:
The proof is analogous to the proof of Theorem 6.1. □

The results of Theorem D.1 imply that the expectation of the global costs (under the original parameters) produced by trajectories generated by the probability density functions with original and estimated parameters are equal. Note that this result is weaker than the one required in (6.7) as it considers only the overall costs. Nevertheless, for a cooperative game, it is enough to describe observed trajectories completely.

Remark D.2:
Similar to the results of Chapter 6, solving the optimization problem (D.6) *demands the possibility of evaluating the likelihood function $\mathcal{L}\left\{\boldsymbol{\theta}_\Sigma \mid \mathcal{D}\right\}$ and therefore the probability density function* (D.4) *at the trajectories $\tilde{\zeta}_l$, $l \in \{1, ..., n_t\}$. The denominator in* (D.4) *includes an integral over all trajectories $\tilde{\zeta}$ which are feasible with respect to the system dynamics and an initial state. An approach analogous to the one presented in Section 6.6 can be applied in this case.*

Remark D.3:

The result $\hat{\boldsymbol{\theta}}_\Sigma$ *of* (D.6) *contains the cost function parameters of all players in one single vector according to* (D.2). *Assuming that the number of features* M_i *is known for every player* $i \in \mathcal{P}$, *an individual parameter set* $\hat{\boldsymbol{\theta}}_i$ *can be determined by means of* (D.2a) *out of* $\hat{\boldsymbol{\theta}}_\Sigma$. *This is done by using the relation*

$$\hat{\boldsymbol{\theta}}_i = \hat{\boldsymbol{\theta}}_\Sigma(l_i^s : l_i^e). \tag{D.8}$$

with

$$l_i^s = 1 + \sum_{\alpha=1}^{i} M_{\alpha-1} \quad \text{and} \quad l_i^e = \sum_{\alpha=1}^{i} M_\alpha, \tag{D.9}$$

with $M_0 = 0$ *and where* $\hat{\boldsymbol{\theta}}_\Sigma(l_i^s : l_i^e) \in \mathbb{R}^{l_i^e - l_i^s + 1}$ *denotes a vector that contains the entries* l_i^s *to* l_i^e *of the vector* $\hat{\boldsymbol{\theta}}_\Sigma$.

The presented method is capable of identifying cost function parameters which explain trajectories corresponding to an optimal solution based on uniformly weighted player cost functions, which is one of the Pareto efficient solutions belonging to the Pareto frontier. Pareto efficient solutions can be obtained by minimizing the sum of cost functions of all players which are nevertheless not necessarily equally weighted (see Definition 3.9). The presented method would not be able to estimate the original parameters $\boldsymbol{\theta}_i^*$, but would be able to determine parameters $\hat{\boldsymbol{\theta}}_i$ which are also capable of describing the trajectories in this scenario. A simulation example where the effectiveness of the presented inverse dynamic game method is demonstrated can be found in [IBKH20].

E Supplementary Simulation Results

This chapter gives supplementary results of the simulative evaluation of the inverse dynamic game methods performed in Chapter 7.

E.1 Inverse Nonlinear Open-Loop Dynamic Game

E.1.1 Robustness to Measurement Noise

Figures E.1 – E.4 show the trajectory estimation results for different SNR values of the observed trajectory used for the inverse dynamic game methods. The estimated trajectories are determined by solving the dynamic game with the parameters $\hat{\boldsymbol{\theta}}_i$, $i \in \mathcal{P}$, identified by each of the considered methods, i.e. the parameters from Tables 7.3, 7.5 and 7.7 are used. The noisefree case is presented in Figure 7.4 and the 30 dB results are shown in Figure 7.8.

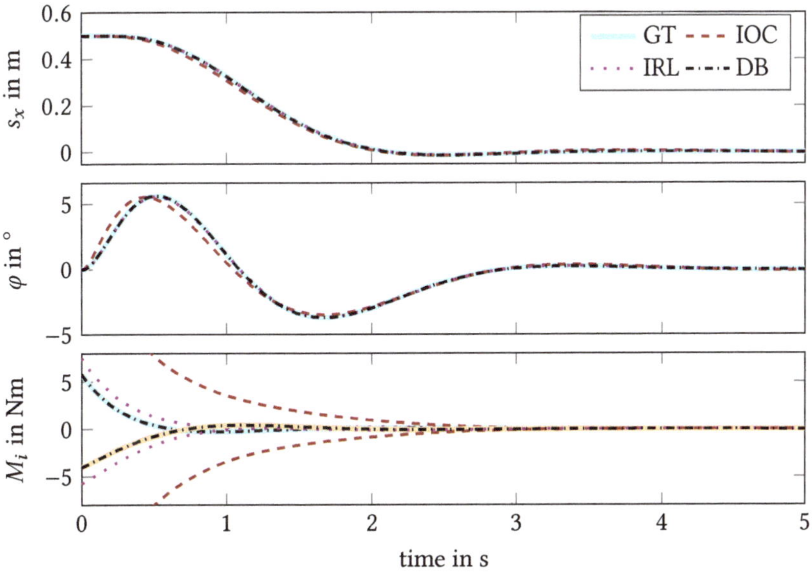

Figure E.1: Observed trajectories and estimations based on mean identification results of all methods, SNR = 20 dB

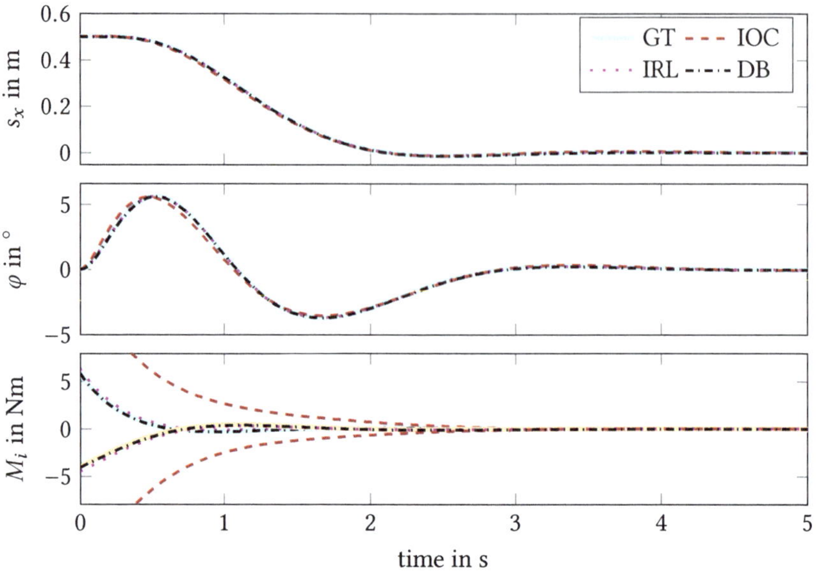

Figure E.2: Observed trajectories and estimations based on mean identification results of all methods, SNR = 25 dB

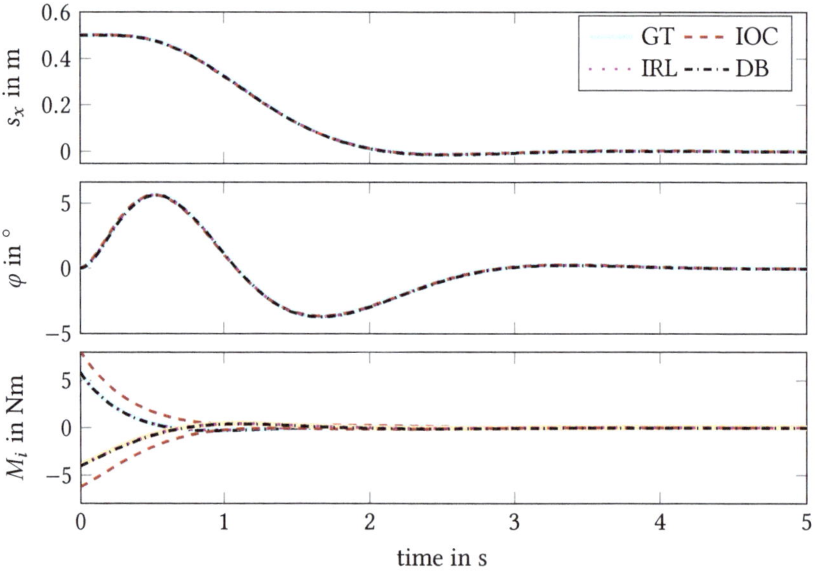

Figure E.3: Observed trajectories and estimations based on mean identification results of all methods, SNR = 35 dB

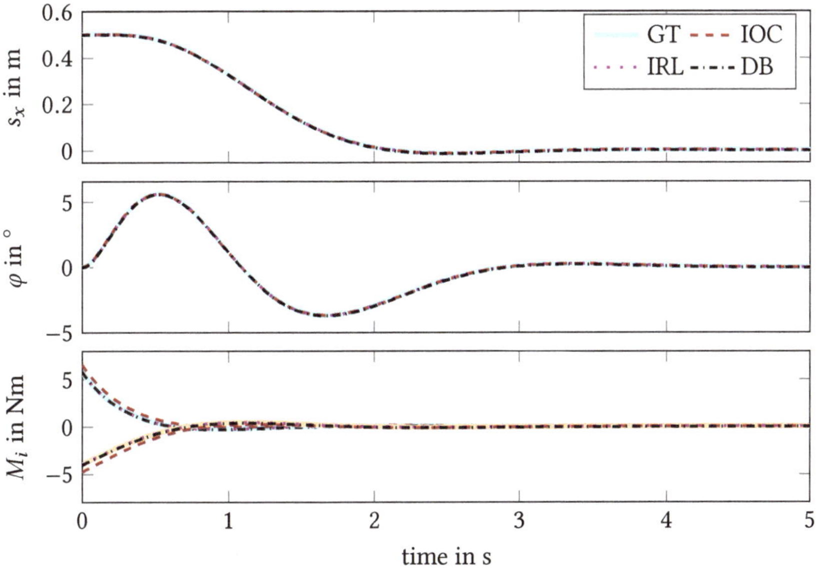

Figure E.4: Observed trajectories and estimations based on mean identification results of all methods, SNR = 40 dB

E.1.2 Robustness to Basis-Function Mismatch

Figures E.5 and E.6 show the comparison of the trajectories which result from the dynamic games solved with the parameters $\hat{\boldsymbol{\theta}}_i$, $i \in \mathcal{P}$ identified by each of the considered methods. The identification is based on observed trajectories generated in Section 7.4.1 and different basis functions (cases II and III) as given in Table 7.9.

E.2 Inverse LQ Feedback Differential Game

E.2.1 Robustness to Measurement Noise

The following Tables E.1 to E.6 list the mean values of the identified parameters corresponding to the matrices $\hat{\boldsymbol{Q}}_i$ and $\hat{\boldsymbol{R}}_{ii}$, $i \in \mathcal{P}$, over all 100 identification procedures conducted in Section 7.5.3, where the robustness of the inverse dynamic game methods to measurement noise is evaluated.

The following Figures E.7 – E.10 show the comparison of the trajectories which result from the dynamic games solved with the mean of the parameters $\hat{\boldsymbol{\theta}}_i$, $i \in \mathcal{P}$, identified by each of the considered methods and based on the observed trajectories with different SNR values. The noisefree case is presented in Figures 7.11 and 7.12 and the 20 dB results are shown in Figure

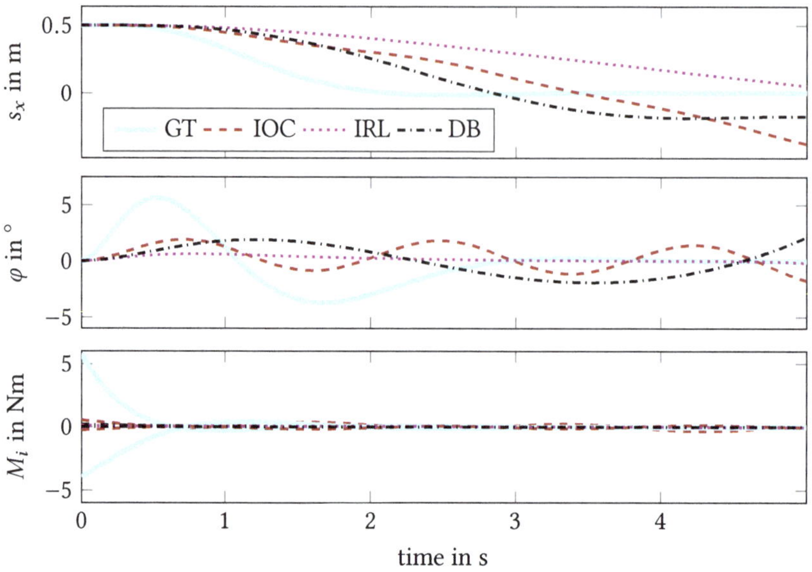

Figure E.5: Inverse open-loop dynamic game results for all methods in the basis function mismatch case II.

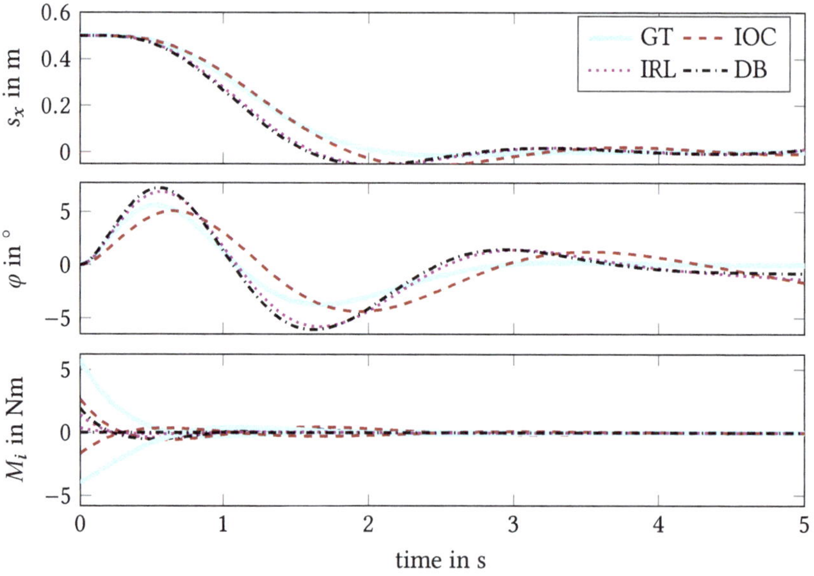

Figure E.6: Observed trajectories and estimations based on identification results of all methods in the basis function mismatch case III.

Table E.1: Mean values of the cost function matrices Q_i identified with IOC

SNR in dB	$\hat{Q}_{1,\text{mean}}$	$\hat{Q}_{2,\text{mean}}$	$\hat{Q}_{3,\text{mean}}$
20	(0.88, 0.62, 1.28, 0.65)	(0.80, 0.70, 1.02, 1.11)	(1.29, 0.74, 0.60, 0.55)
25	(0.95, 0.50, 1.69, 0.85)	(0.80, 0.68, 1.01, 1.65)	(1.29, 0.82, 0.52, 0.77)
30	(0.98, 0.43, 1.91, 0.95)	(0.82, 0.68, 1.00, 1.88)	(1.11, 0.93, 0.51, 0.91)
35	(1.00, 0.41, 1.98, 0.98)	(0.85, 0.67, 1.00, 1.96)	(1.04, 0.98, 0.50, 0.97)
40	(1.00, 0.40, 2.00, 0.99)	(0.93, 0.63, 1.00, 1.98)	(1.01, 1.00, 0.50, 0.99)
∞	(1.00, 0.40, 2.00, 1.00)	(1.00, 0.60, 1.00, 2.00)	(1.00, 1.00, 0.50, 1.00)

Table E.2: Mean values of the cost function matrices R_{ii} identified with IOC

SNR in dB	$\hat{R}_{1,(22),\text{mean}}$	$\hat{R}_{2,(22),\text{mean}}$	$\hat{R}_{3,(22),\text{mean}}$
20	0.99	0.85	1.97
25	1.00	0.95	1.97
30	1.01	0.99	1.99
35	1.01	0.99	1.99
40	1.00	1.00	2.00
∞	1.00	1.00	2.00

Table E.3: Mean values of the cost function matrices Q_i identified with IRL

SNR in dB	$\hat{Q}_{1,\text{mean}}$	$\hat{Q}_{2,\text{mean}}$	$\hat{Q}_{3,\text{mean}}$
20	(0.98, 0.20, 2.06, 0.94)	(0.93, 0.63, 1.00, 2.12)	(1.09, 0.95, 0.51, 0.90)
25	(0.99, 0.32, 2.02, 0.98)	(0.96, 0.62, 1.00, 2.07)	(1.02, 0.99, 0.51, 0.96)
30	(1.00, 0.38, 2.00, 0.99)	(0.96, 0.62, 1.00, 2.02)	(1.00, 1.00, 0.50, 0.99)
35	(1.00, 0.39, 2.00, 1.00)	(0.99, 0.61, 1.00, 2.00)	(1.00, 1.00, 0.50, 1.00)
40	(1.00, 0.40, 2.00, 1.00)	(1.00, 0.60, 1.00, 2.00)	(1.00, 1.00, 0.50, 1.00)
∞	(1.00, 0.40, 2.00, 1.00)	(1.00, 0.60, 1.00, 2.00)	(0.99, 1.00, 0.50, 1.00)

Table E.4: Mean values of the cost function matrices R_{ii} identified with IRL

SNR in dB	$\hat{R}_{1,(22),\text{mean}}$	$\hat{R}_{2,(22),\text{mean}}$	$\hat{R}_{3,(22),\text{mean}}$
20	0.98	1.11	2.01
25	0.99	1.05	2.01
30	1.00	1.02	2.00
35	1.00	1.00	2.00
40	1.00	1.00	2.00
∞	1.00	1.00	2.00

7.15 and 7.16. As it can be inferred from Figures E.9 and E.10, the trajectory comparison for the cases 35 dB and 40 dB yields no visually recognizable improvement. These are not

Table E.5: Mean values of the cost function matrices Q_i identified with the DB method

SNR in dB	$\hat{Q}_{1,\text{mean}}$	$\hat{Q}_{2,\text{mean}}$	$\hat{Q}_{3,\text{mean}}$
20	$(1.17, 0.37, 2.22, 1.01)$	$(1.32, 0.47, 0.99, 2.02)$	$(1.16, 0.92, 0.57, 1.46)$
25	$(1.08, 0.37, 2.15, 1.00)$	$(1.25, 0.49, 1.00, 2.02)$	$(1.10, 0.95, 0.53, 1.22)$
30	$(1.05, 0.39, 2.08, 1.00)$	$(1.20, 0.51, 1.00, 2.01)$	$(1.12, 0.94, 0.52, 1.13)$
35	$(1.03, 0.39, 2.04, 1.00)$	$(1.16, 0.53, 1.00, 2.02)$	$(1.08, 0.96, 0.51, 1.06)$
40	$(1.01, 0.40, 2.02, 1.00)$	$(1.09, 0.56, 1.00, 2.00)$	$(1.05, 0.97, 0.50, 1.04)$
∞	$(1.00, 0.40, 2.00, 1.00)$	$(1.04, 0.58, 1.00, 2.00)$	$(1.02, 0.99, 0.50, 1.00)$

Table E.6: Mean values of the cost function matrices R_{ii} identified with the DB method

SNR in dB	$\hat{R}_{1,(22),\text{mean}}$	$\hat{R}_{2,(22),\text{mean}}$	$\hat{R}_{3,(22),\text{mean}}$
20	1.55	0.98	2.16
25	1.18	0.99	2.01
30	1.10	1.00	2.04
35	1.15	1.00	2.02
40	1.03	1.00	2.01
∞	1.00	1.00	2.00

explicitely shown here as the result are practically identical to the noisefree case from Figures 7.11 and 7.12.

E.2.2 Robustness to Basis Function Mismatch

Figures E.11 – E.16 show the comparison of the observed trajectories with the ones which result from the dynamic games solved with the parameters $\hat{\theta}_i$, $i \in \mathcal{P}$ identified by each of the considered methods. The identification is based on observed trajectories generated in Section 7.4.1 and incomplete basis functions (cases II to IV) as given in Table 7.17.

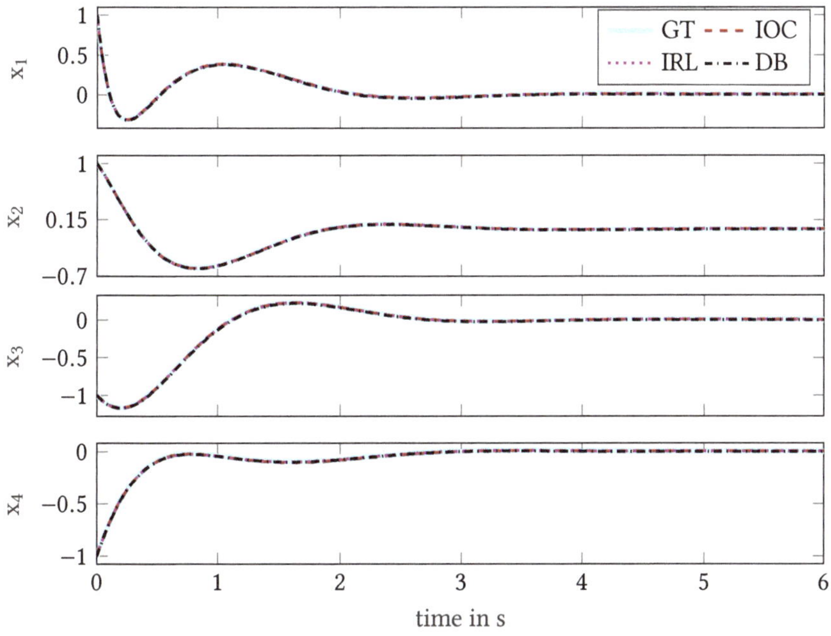

Figure E.7: Ground truth and estimated state trajectories of the inverse LQ feedback dynamic game with each method. The identification was conducted using noise-corrupted trajectories with SNR = 25 dB.

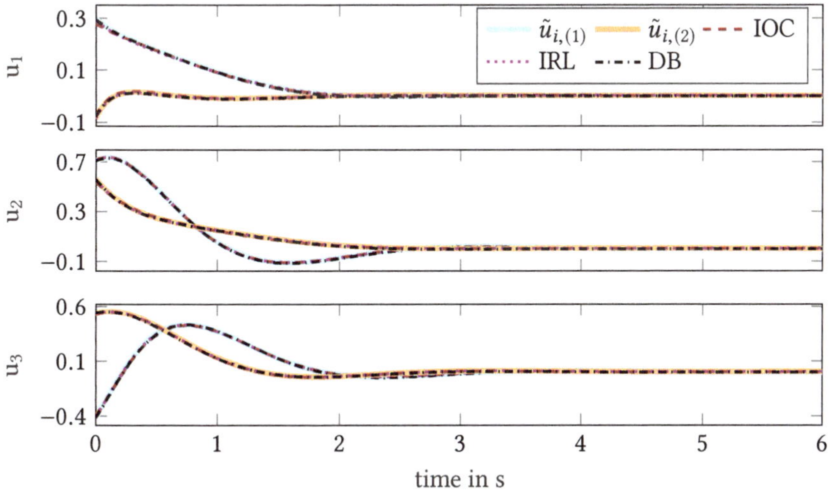

Figure E.8: Ground truth and estimated control trajectories of the inverse LQ feedback dynamic game with each method. The identification was conducted using noise-corrupted trajectories with SNR = 25 dB.

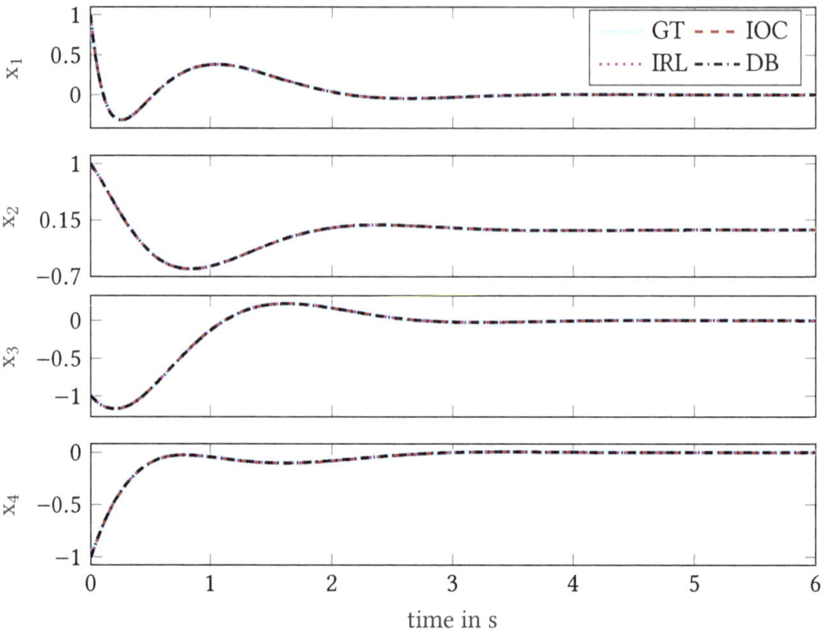

Figure E.9: Ground truth and estimated state trajectories of the inverse LQ feedback dynamic game with each method. The identification was conducted using noise-corrupted trajectories with SNR = 30 dB.

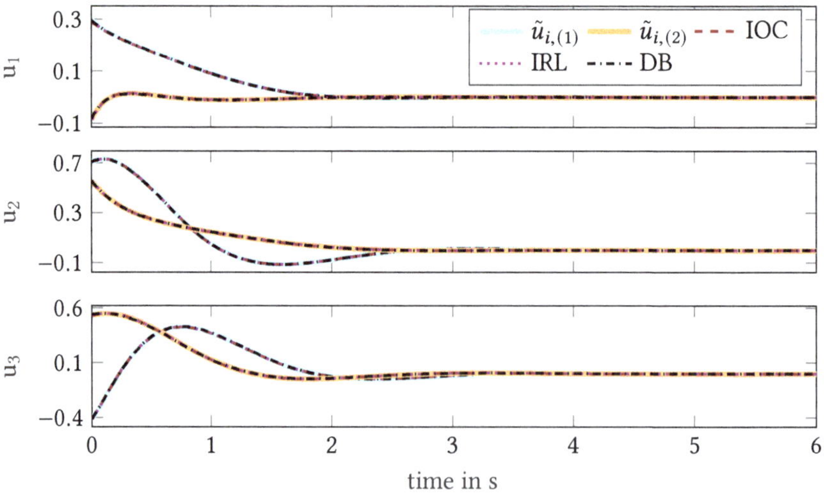

Figure E.10: Ground truth and estimated control trajectories of the inverse LQ feedback dynamic game with each method. The identification was conducted using noise-corrupted trajectories with SNR = 30 dB.

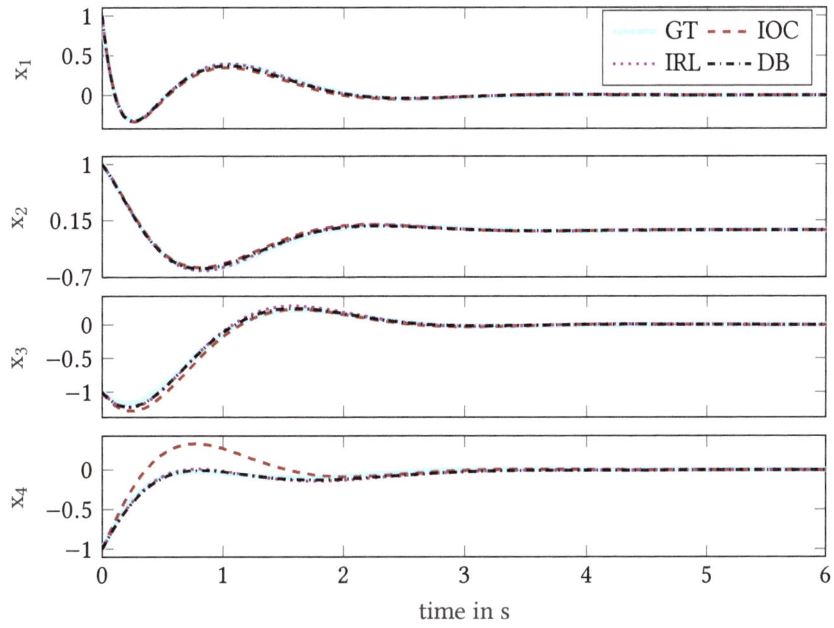

Figure E.11: Ground truth and estimated state trajectories of the inverse LQ feedback dynamic game with each method. The identification was conducted with the wrong assumption that $Q_{i,(j,j)} = 0$ for $i \in \{1, 2\}$ and $j \in \{3, 4\}$ (case II).

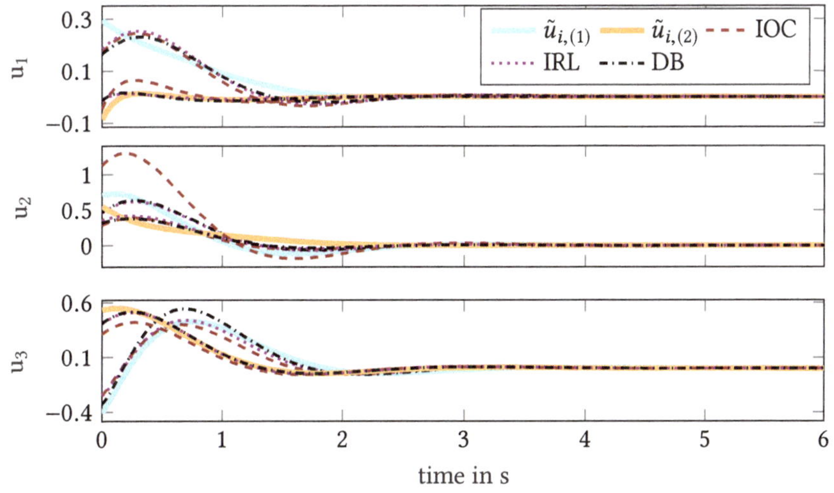

Figure E.12: Ground truth and estimated state trajectories of the inverse LQ feedback dynamic game with each method. The identification was conducted with the wrong assumption that $Q_{i,(j,j)} = 0$ for $i \in \{1, 2\}$ and $j \in \{3, 4\}$ (case II).

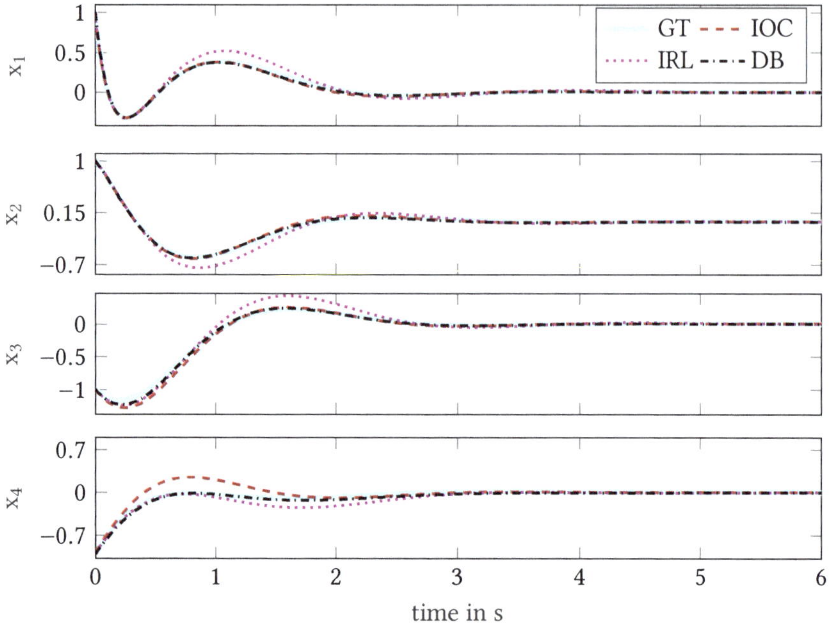

Figure E.13: Ground truth and estimated state trajectories of the inverse LQ feedback dynamic game with each method. The identification was conducted with the wrong assumption that $Q_{i,(j,j)} = 0$ for $i \in \{1, 2\}$ and $j \in \{2, 3, 4\}$ (case III).

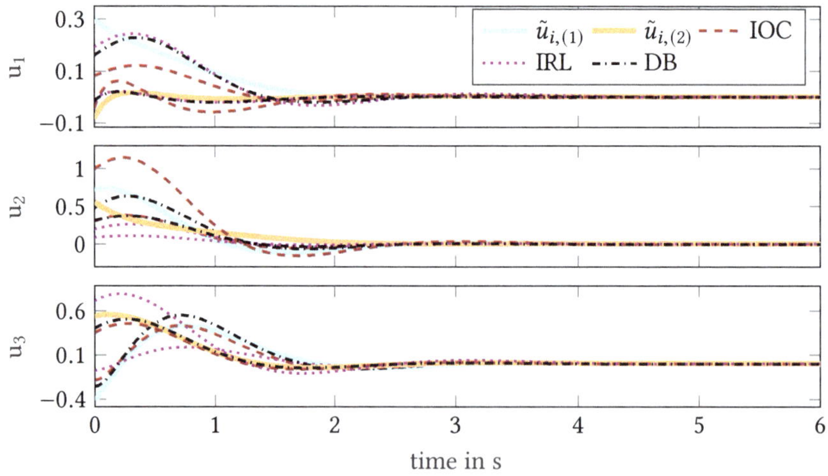

Figure E.14: Ground truth and estimated state trajectories of the inverse LQ feedback dynamic game with each method. The identification was conducted with the wrong assumption that $Q_{i,(j,j)} = 0$ for $i \in \{1, 2\}$ and $j \in \{2, 3, 4\}$ (case III).

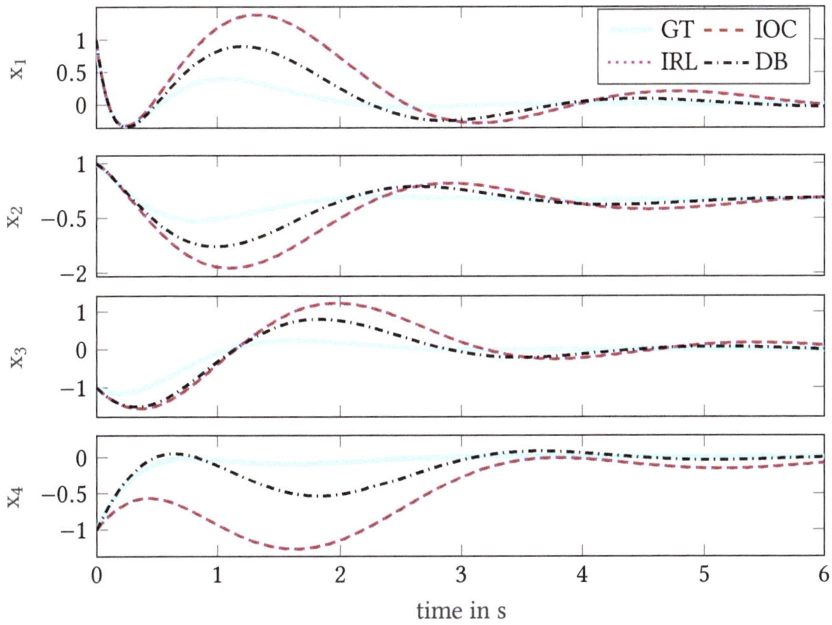

Figure E.15: Ground truth and estimated state trajectories of the inverse LQ feedback dynamic game with each method. The identification was conducted with the wrong assumption that $\boldsymbol{Q}_i = \mathbf{0}$ for $i \in \{1, 2\}$ (case IV).

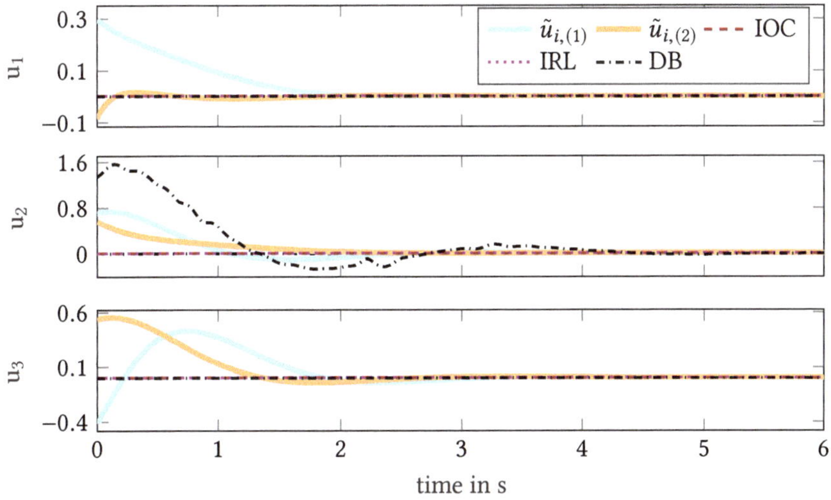

Figure E.16: Ground truth and estimated state trajectories of the inverse LQ feedback dynamic game with each method. The identification was conducted with the wrong assumption that $\boldsymbol{Q}_i = \mathbf{0}$ for $i \in \{1, 2\}$ (case IV).

F Supplementary Results of the Application in Shared Control

This section gives further information on the results of Chapter 8.

F.1 Further Details on the Experimental Setup

This section provides details on the parameters of the two steering wheels and on the developed control structure which realizes their virtual coupling are presented.

F.1.1 Steering Wheel Parameters

The following Table F.1 lists the parameters of the two steering wheels belonging to the cooperative steering system.

Table F.1: Steering wheel parameters

Parameter	SW1		SW2		Description
Θ_j	0.04	$\mathrm{kg\,m^2}$	0.054	$\mathrm{kg\,m^2}$	Rotational inertia
c_j	0.573	Nm/rad	0.573	Nm/rad	Spring constant
d_j	0.430	$\mathrm{Nm \cdot s/rad}$	0.430	Nm/rad	Damping constant

F.1.2 Steering Wheel Coupling Control

The steering wheels were coupled using a control algorithm which emulates a spring-damper element between them. This kind of coupling was first presented in [LDFH14], where it was also used in a study for analyzing haptic interaction between humans.

Figure F.1 shows the control loop of the cooperative steering system. The controller calculates a torque $M(t)$ which is equally distributed on each steering wheel. The aim is to regulate the difference $e_e(t) = e_{\mathrm{des}} - e_{\mathrm{meas}}(t)$ towards zero, where $e_{\mathrm{meas}}(t) = \varphi_1(t) - \varphi_2(t)$ is the difference of measured steering angles and $e_{\mathrm{des}} = 0$ is the desired angle difference.

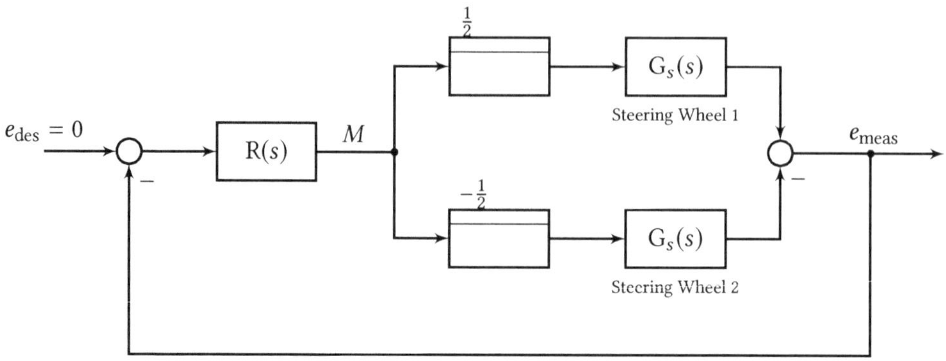

Figure F.1: Control structure of the cooperative steering system

The controller $R(s)$ is designed as a proportional-derivative (PD) controller with a low-pass filter positioned before the derivative part in order to supress measurement noise. The structure of the PD controller is illustrated in Figure F.2. The controller behavior is defined by the transfer function

$$R(s) = \frac{M(s)}{E(s)} = \frac{sK_D}{1 + T_p s} + K_P, \tag{F.1}$$

where $M(s)$ and $E(s)$ denote the Laplace transform of the torque $M(t)$ and the control error $e_e(t)$, respectively. The variables K_P and K_D denote the coefficients of the proportional and the derivative terms, respectively. The variable T_p denotes the time constant of the first-order lag filter. The values of these parameters are given in Table F.2.

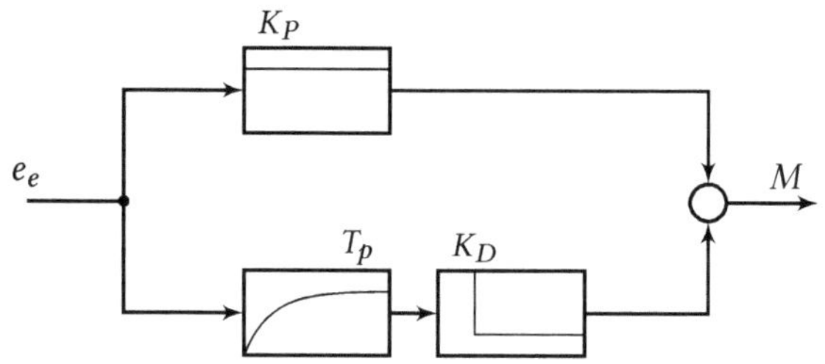

Figure F.2: PD controller used for the coupling of the steering wheels

Table F.2: PD controller parameters

Parameter	Value
K_P	1.96
K_D	0.175
T_p	1.825 ms

F.2 Supplementary Tables of the Shared Control Identification Results

The following Tables F.3 to F.5 give all trajectory estimation errors for all subject pairs which were obtained using the IOC, IRL and DB methods, respectively.

Table F.3: Cooperative steering experiment: Error between measured trajectories and trajectories obtained with the IOC method

Subject pair	δ^{x}	δ^{u_1}	δ^{u_2}
1	70.014	56.907	56.436
2	127.429	60.718	85.649
3	100.466	86.705	59.141
4	133.885	105.515	78.556
5	91.833	44.793	65.951
6	81.696	44.930	71.280
7	253.893	130.618	86.561
8	111.529	71.462	74.479
9	182.829	59.913	100.267
10	129.138	97.584	72.642
11	249.695	90.685	192.784
12	271.043	111.514	126.208
13	80.681	81.442	43.414
14	142.491	99.035	88.994
15	107.503	90.802	65.915
16	196.169	68.491	98.087
17	134.954	74.166	73.785
18	123.916	90.849	89.037
19	113.636	67.701	108.814
20	126.179	77.609	110.219
21	93.533	57.297	93.647
22	109.570	197.063	141.135
23	171.691	125.290	82.074
24	178.529	73.403	67.797
25	246.650	132.693	88.445
mean	145.158	87.887	88.853
median	127.429	81.442	85.649
SD	58.632	33.530	30.747

Table F.4: Cooperative steering experiment: Error between measured trajectories and trajectories obtained with the IRL method

Subject pair	δ^{x}	δ^{u_1}	δ^{u_2}
1	53.719	58.563	58.703
2	101.236	100.596	90.611
3	129.557	98.956	63.395
4	84.566	104.799	75.475
5	94.571	48.288	72.978
6	99.932	50.992	69.750
7	175.364	118.096	80.992
8	104.913	76.127	93.537
9	108.529	72.967	77.475
10	77.535	70.975	66.250
11	127.078	86.291	125.088
12	190.018	102.359	97.408
13	103.761	76.280	49.144
14	82.052	118.415	100.529
15	89.313	87.232	61.959
16	193.824	60.500	112.702
17	100.953	97.991	79.588
18	87.889	93.470	96.903
19	86.435	60.791	97.528
20	108.915	72.372	103.536
21	92.718	59.813	100.144
22	85.351	218.646	149.104
23	153.241	121.517	79.267
24	141.748	93.011	62.485
25	137.070	115.503	77.055
mean	112.411	90.582	85.664
median	101.236	87.232	79.588
SD	35.611	34.569	22.678

Table F.5: Cooperative steering experiment: Error between experimentally measured trajectories and trajectories obtained with the DB approach

Subject pair	δ^{x}	δ^{u_1}	δ^{u_2}
1	44.224	57.090	55.853
2	101.985	61.651	77.223
3	73.849	67.249	53.884
4	96.832	102.442	69.903
5	77.725	66.479	67.224
6	85.592	60.670	73.956
7	116.169	105.058	78.649
8	92.483	70.193	70.751
9	91.977	62.021	78.398
10	68.201	76.293	64.553
11	89.672	80.611	120.724
12	94.118	81.256	87.326
13	58.457	76.280	38.762
14	85.817	93.916	83.197
15	83.445	85.230	59.801
16	107.501	58.896	86.753
17	99.877	70.717	70.081
18	79.049	75.170	82.617
19	73.561	54.234	89.718
20	87.378	59.884	85.506
21	74.819	53.756	83.710
22	143.719	76.626	57.984
23	95.018	107.129	73.908
24	100.738	86.762	63.648
25	95.582	114.574	71.243
mean	88.711	76.168	73.815
median	89.672	75.170	38.762
SD	19.372	17.459	15.723

F.2.1 Statistical Test Results

The following Tables F.6 and F.7 give the p-values corresponding to the right-tailed Wilcoxon signed-rank test conducted to the data sets of NSAE of states and controls, respectively. In Table F.6, the hypothesis is always rejected with a significance level of $\alpha = 0.01$. The right-tailed property leads to the validity of the alternative hypothesis which states that $\delta^{\boldsymbol{x}}_{\text{median, row}} - \delta^{\boldsymbol{x}}_{\text{median, column}} > 0$. The same holds for Table F.7 with the exception of the NSAE of the controls obtained by the IOC and IRL methods. The hypothesis H_0 cannot be rejected and thus their difference is not statistically significant.

Table F.6: p-values of the Wilcoxon signed-rank test with H_0 : "$\delta^{\boldsymbol{x}}_{\text{median, row}} - \delta^{\boldsymbol{x}}_{\text{median, column}}$ comes from a distribution with zero median".

	IOC	IRL	DB
IOC	–	$1.249 \cdot 10^{-4}$	$1.639 \cdot 10^{-6}$
IRL	–	–	$1.085 \cdot 10^{-4}$

Table F.7: p-values of the Wilcoxon signed-rank test with H_0 : "$\delta^{\boldsymbol{u}}_{\text{median, row}} - \delta^{\boldsymbol{u}}_{\text{median, column}}$ comes from a distribution with zero median" (** denotes the failure of hypothesis rejection).

	IOC	IRL	DB
IOC	–	0.594**	$6.995 \cdot 10^{-5}$
IRL	–	–	$1.597 \cdot 10^{-5}$

References

Public References

[AB11] N. Aghasadeghi and T. Bretl. Maximum entropy inverse reinforcement learning in continuous state spaces with path integrals. In *2011 IEEE/RSJ International Conference on Intelligent Robots and Systems (IROS)*, pages 1561–1566. IEEE, 2011.

[AB14] N. Aghasadeghi and T. Bretl. Inverse optimal control for differentially flat systems with application to locomotion modeling. In *2014 IEEE International Conference on Robotics and Automation (ICRA)*, pages 6018–6025. IEEE, 2014.

[ABBE16] P. Arcidiacono, P. Bayer, J. R. Blevins, and P. Ellickson. Estimation of dynamic discrete choice models in continuous time with an application to retail competition. *The Review of Economic Studies*, 83(3):889–931, 2016.

[ACM+18] D. A. Abbink, T. Carlson, M. Mulder, J. C. F. de Winter, F. Aminravan, T. L. Gibo, and E. R. Boer. A Topology of Shared Control Systems—Finding Common Ground in Diversity. *IEEE Transactions on Human-Machine Systems*, 48(5):509–525, October 2018.

[AKFIJ12] H. Abou-Kandil, G. Freiling, V. Ionescu, and G. Jank. *Matrix Riccati Equations in Control and Systems Theory*. Birkhäuser, December 2012.

[AM89] B. D. O. Anderson and J. B. Moore. *Optimal control: linear quadratic methods*. Prentice Hall information and system sciences series. Prentice Hall, Englewood Cliffs, N.J, 1989.

[AMR95] U. M. Ascher, Robert M. M. Mattheij, and R. D. Russell. *Numerical solution of boundary value problems for ordinary differential equations*. Number 13 in Classics in applied mathematics. Society for Industrial and Applied Mathematics, Philadelphia, 1995.

[AN04] P. Abbeel and A. Y. Ng. Apprenticeship learning via inverse reinforcement learning. In *Proceedings of the twenty-first international conference on Machine learning*, page 1. ACM, 2004.

[ARARU+11] S. Albrecht, K. Ramírez-Amaro, F. Ruiz-Ugalde, D. Weikersdorfer, M. Leibold, M. Ulbrich, and M. Beetz. Imitating human reaching motions using physically inspired optimization principles. In *2011 11th IEEE-RAS International Conference on Humanoid Robots*, pages 602–607, October 2011. ISSN: 2164-0572.

[ÅW95] K. J. Åström and B. Wittenmark. *Adaptive control.* Addison-Wesley, Reading, Mass, 2nd edition, 1995.

[Bas74] T. Basar. A counterexample in linear-quadratic games: Existence of nonlinear nash solutions. *Journal of Optimization Theory and Applications*, 14(4):425–430, October 1974.

[BBL07] P. Bajari, C. L. Benkard, and J. Levin. Estimating dynamic models of imperfect competition. *Econometrica*, 75(5):1331–1370, 2007.

[Bel66] R. Bellman. Dynamic Programming. *Science*, 153(3731):34–37, July 1966.

[BGP15] D. Bertsimas, V. Gupta, and I. Paschalidis. Data-driven estimation in equilibrium using inverse optimization. *Mathematical Programming*, 153(2):595–633, November 2015.

[BH75] A. E. Bryson and Y. Ho. *Applied optimal control: optimization, estimation, and control.* CRC Press, 1975.

[BO99] T. Basar and G. J. Olsder. *Dynamic Noncooperative Game Theory: Second Edition.* SIAM, 1999.

[BOW09] D. A. Braun, P. A. Ortega, and D. M. Wolpert. Nash equilibria in multi-agent motor interactions. *PLoS Computational Biology*, 5(8):e1000468, 2009.

[Boy94] S. P. Boyd, editor. *Linear matrix inequalities in system and control theory.* Number 15 in SIAM studies in applied mathematics. Society for Industrial and Applied Mathematics, Philadelphia, 1994.

[BPC+06] C. L. Bottasso, B. I. Prilutsky, A. Croce, E. Imberti, and S. Sartirana. A numerical procedure for inferring from experimental data the optimization cost functions using a multibody model of the neuro-musculoskeletal system. *Multibody System Dynamics*, 16(2):123–154, August 2006.

[Bre78] J. Brewer. Kronecker products and matrix calculus in system theory. *IEEE Transactions on Circuits and Systems*, 25(9):772–781, September 1978.

[Bre11] A. Bressan. Noncooperative Differential Games. *Milan Journal of Mathematics*, 79(2):357–427, December 2011.

[BSLK97] C. Barbu, R. Sepulchre, W. Lin, and P.V. Kokotovic. Global asymptotic stabilization of the ball-and-beam system. In *Proceedings of the 36th IEEE Conference on Decision and Control*, volume 3, pages 2351–2355 vol.3, December 1997.

[BVBB14] C. G. Bolívar-Vicenty and G. Beauchamp-Báez. Modelling the Ball-and-Beam System From Newtonian Mechanics and from Lagrange Methods. In *12th LACCEI Latin American and Caribbean Conference for Engineering and Technology*, 2014.

[Cas80] J. Casti. On the general inverse problem of optimal control theory. *Journal of Optimization Theory and Applications*, 32(4):491–497, December 1980.

[CC72] C. Chen and J. Cruz. Stackelberg solution for two-person games with biased information patterns. *IEEE Transactions on Automatic Control*, 17(6):791–798, December 1972.

[CFG89] C. Carraro, J. Flemming, and A. Giovannini. The tastes of european central bankers. In *A European Central Bank?: Perspectives on Monetary Unification after Ten Years of the EMS*, pages 162–185. Cambridge University Press, 1989.

[CS17] V. T. Chackochan and V. Sanguineti. Modelling Collaborative Strategies in Physical Human-Human Interaction. In J. Ibáñez, J. González-Vargas, J. M. Azorín, M. Akay, and J. L. Pons, editors, *Converging Clinical and Engineering Research on Neurorehabilitation II*, volume 15, pages 253–258. Springer International Publishing, Cham, 2017.

[CT06] T. M. Cover and J. A. Thomas. *Elements of information theory.* Wiley-Interscience, Hoboken, N.J, 2nd edition, 2006.

[DFJ85] E. Dockner, G. Feichtinger, and S. Jørgensen. Tractable classes of nonzero-sum open-loop Nash differential games: Theory and examples. *Journal of Optimization Theory and Applications*, 45(2):179–197, February 1985.

[Doc00] E. J. Dockner. *Differential games in economics and management science.* Cambridge University Press, 2000.

[EBS00] J. C. Engwerda, W. A. van den Broek, and J. M. Schumacher. Feedback Nash equilibria in uncertain infinite time horizon differential games. *Proceedings of the 14th International Symposium of Mathematical Theory of Networks and Systems, MTNS 2000*, pages 1–6, 2000.

[EHAAM16] H. El-Hussieny, A. A. Abouelsoud, S. F. M. Assal, and S. M. Megahed. Adaptive learning of human motor behaviors: An evolving inverse optimal control approach. *Engineering Applications of Artificial Intelligence*, 50:115–124, April 2016.

[Eng01] S. E. Engelbrecht. Minimum Principles in Motor Control. *Journal of Mathematical Psychology*, 45(3):497–542, June 2001.

[Eng05] J. Engwerda. *LQ Dynamic Optimization and Differential Games.* John Wiley & Sons, 2005.

[Epp13] J. F. Epperson. *An Introduction to Numerical Methods and Analysis*. John Wiley & Sons, Incorporated, 2nd edition, 2013.

[ER11] J. C. Engwerda and P. V. Reddy. A Positioning of Cooperative Differential Games. In *Proceedings of the 5th International ICST Conference on Performance Evaluation Methodologies and Tools*, 2011.

[FFH17] M. Flad, L. Fröhlich, and S. Hohmann. Cooperative Shared Control Driver Assistance Systems Based on Motion Primitives and Differential Games. *IEEE Transactions on Human-Machine Systems*, 47(5):711–722, October 2017.

[FH91] T. Flash and E. Henis. Arm Trajectory Modifications During Reaching Towards Visual Targets. *Journal of Cognitive Neuroscience*, 3(3):220–230, July 1991. Publisher: MIT Press.

[FK88] T. Fujii and P.P. Khargonekar. Inverse problems in H_∞ control theory and linear-quadratic differential games. In *Proceedings of the 27th IEEE Conference on Decision and Control, 1988*, pages 26–31 vol.1, 1988.

[FK96] R. Freeman and P. Kokotovic. Inverse Optimality in Robust Stabilization. *SIAM Journal on Control and Optimization*, 34(4):1365–1391, July 1996.

[FMM+18] I. A. Faruque, F. T. Muijres, K. M. Macfarlane, A. Kehlenbeck, and J. S. Humbert. Identification of optimal feedback control rules from micro-quadrotor and insect flight trajectories. *Biological Cybernetics*, 112(3):165–179, June 2018.

[FN84] T. Fujii and M. Narazaki. A complete optimality condition in the inverse problem of optimal control. *SIAM journal on control and optimization*, 22(2):327–341, 1984.

[Fuj87] T. Fujii. A new approach to the LQ design from the viewpoint of the inverse regulator problem. *IEEE Transactions on Automatic Control*, 32(11):995–1004, November 1987.

[Gal11] J. Gallier. *Geometric Methods and Applications*, volume 38 of *Texts in Applied Mathematics*. Springer, New York, NY, 2011.

[Gam99] R. V. Gamkrelidze. Discovery of the Maximum Principle. *Journal of Dynamical and Control Systems*, 5(4):437–451, October 1999.

[Gu08] D. Gu. A Differential Game Approach to Formation Control. *IEEE Transactions on Control Systems Technology*, 16(1):85–93, January 2008.

[HdlCIR19] J. Herrera de la Cruz, B. Ivorra, and A. M. Ramos. An Algorithm for Solving a Class of Multiplayer Feedback-Nash Differential Games. *Mathematical Problems in Engineering*, 2019:1–14, May 2019.

[HFKB15] D. Huang, A. Farahmand, K. M. Kitani, and J. A. Bagnell. Approximate MaxEnt Inverse Optimal Control and Its Application for Mental Simulation of Human Interactions. In *Twenty-Ninth AAAI Conference on Artificial Intelligence*, volume 15, page 29th, 2015.

[HKZ12] A. Haurie, J. B. Krawczyk, and G. Zaccour. *Games and dynamic games.* World Scientific Publishing Company, Singapore, 2012. OCLC: ocn780435424.

[HMRAD16] D. Hadfield-Menell, S. J. Russell, P. Abbeel, and A. Dragan. Cooperative inverse reinforcement learning. In *Advances in Neural Information Processing Systems*, pages 3909–3917. Curran Associates, Inc., 2016.

[HS14] M. Hermann and M. Saravi. *A First Course in Ordinary Differential Equations: Analytical and Numerical Methods.* Springer India, New Delhi, 2014.

[HSB12] K. Hatz, J. Schlöder, and H. Bock. Estimating Parameters in Optimal Control Problems. *SIAM Journal on Scientific Computing*, 34(3):A1707–A1728, January 2012.

[HSK92] J. Hauser, S. Sastry, and P. Kokotovic. Nonlinear control via approximate input-output linearization: the ball and beam example. *IEEE Transactions on Automatic Control*, 37(3):392–398, March 1992.

[Isa99] R. Isaacs. *Differential games: a mathematical theory with applications to warfare and pursuit, control and optimization.* Courier Corporation, 1999.

[JAB13] M. Johnson, N. Aghasadeghi, and T. Bretl. Inverse optimal control for deterministic continuous-time nonlinear systems. In *52nd IEEE Conference on Decision and Control*, pages 2906–2913, December 2013.

[JAK89] L. Jódar and H. Abou-Kandil. Kronecker products and coupled matrix Riccati differential systems. *Linear Algebra and its Applications*, 121:39–51, August 1989.

[Jay57] E. T. Jaynes. Information Theory and Statistical Mechanics. *Physical Review*, 106(4):620–630, May 1957.

[JK73] A. Jameson and E. Kreindler. Inverse Problem of Linear Optimal Control. *SIAM Journal on Control*, 11(1):1–19, February 1973.

[JKL+19] W. Jin, D. Kulić, J. F. Lin, S. Mou, and S. Hirche. Inverse Optimal Control for Multiphase Cost Functions. *IEEE Transactions on Robotics*, 35(6):1387–1398, December 2019.

[Kal64] R. E. Kalman. When is a linear control system optimal? *Journal of Fluids Engineering*, 86(1):51–60, 1964.

[Kal09] J. F. Kalaska. From Intention to Action: Motor Cortex and the Control of Reaching Movements. In D. Sternad, editor, *Progress in Motor Control*, number 629 in Advances in Experimental Medicine and Biology, pages 139–178. Springer US, 2009.

[KB09] T. G. Kolda and B. W. Bader. Tensor Decompositions and Applications. *SIAM Review*, 51(3):455–500, August 2009.

[Kir04] D. E. Kirk. *Optimal control theory: an introduction.* Courier Corporation, 2004.

[KKD14] R. Kamalapurkar, J. R. Klotz, and W. E. Dixon. Concurrent learning-based approximate feedback-Nash equilibrium solution of N-player nonzero-sum differential games. *IEEE/CAA Journal of Automatica Sinica*, 1(3):239–247, July 2014.

[KM11] J. W. Krakauer and P. Mazzoni. Human sensorimotor learning: adaptation, skill, and beyond. *Current Opinion in Neurobiology*, 21(4):636–644, August 2011.

[KPRS13] M. Kalakrishnan, P. Pastor, L. Righetti, and S. Schaal. Learning objective functions for manipulation. In *2013 IEEE International Conference on Robotics and Automation (ICRA)*, pages 1331–1336. IEEE, 2013.

[KRJ+18] I. C. Konstantakopoulos, L. J. Ratliff, M. Jin, S. S. Sastry, and C. J. Spanos. A Robust Utility Learning Framework via Inverse Optimization. *IEEE Transactions on Control Systems Technology*, 26(3):954–970, May 2018.

[KS15] V. Kuleshov and O. Schrijvers. Inverse Game Theory: Learning Utilities in Succinct Games. In E. Markakis and G. Schäfer, editors, *Web and Internet Economics*, number 9470 in Lecture Notes in Computer Science, pages 413–427. Springer Berlin Heidelberg, December 2015.

[KT99] M. Krstic and P. Tsiotras. Inverse optimal stabilization of a rigid spacecraft. *IEEE Transactions on Automatic Control*, 44(5):1042–1049, 1999.

[Kuč73] V. Kučera. A review of the matrix Riccati equation. *Kybernetika*, 9(1):42–61, 1973.

[KVML18] B. Kiumarsi, K. G. Vamvoudakis, H. Modares, and F. L. Lewis. Optimal and Autonomous Control Using Reinforcement Learning: A Survey. *IEEE Transactions on Neural Networks and Learning Systems*, 29(6):2042–2062, June 2018.

[KWB11] A. Keshavarz, Y. Wang, and S. Boyd. Imputing a convex objective function. In *2011 IEEE International Symposium on Intelligent Control (ISIC)*, pages 613–619. IEEE, 2011.

[LBC18] X. Lin, P. A. Beling, and R. Cogill. Multiagent Inverse Reinforcement Learning for Two-Person Zero-Sum Games. *IEEE Transactions on Games*, 10(1):56–68, March 2018.

[LDFH14] J. Ludwig, G. Diehm, M. Flad, and S. Hohmann. Optimal interaction structure of human drivers cooperation: A pilot study. In *2014 IEEE International Conference on Systems, Man, and Cybernetics (SMC)*, pages 3593–3598, 2014.

[LHF14] D. C. Li, Y. Q. He, and F. Fu. Nonlinear inverse reinforcement learning with mutual information and Gaussian process. In *2014 IEEE International Conference on Robotics and Biomimetics (ROBIO 2014)*, pages 1445–1450, December 2014.

[LK12] S. Levine and V. Koltun. Continuous inverse optimal control with locally optimal examples. *Proceedings of the 29th International Conference on Machine Learning (ICML-12)*, pages 41–48, 2012.

[LPK11] S. Levine, Z. Popovic, and V. Koltun. Nonlinear inverse reinforcement learning with gaussian processes. In *Advances in Neural Information Processing Systems*, pages 19–27, 2011.

[Luk71] D. L. Lukes. A Global Theory for Linear-Quadratic Differential Games. *Journal of Mathematical Analysis and Applications*, 33(1):96–123, 1971.

[LZ18] Y. Lin and W. Zhang. Necessary/sufficient conditions for Pareto optimum in cooperative difference game. *Optimal Control Applications and Methods*, 39(2):1043–1060, 2018.

[MA73] P. Moylan and B. Anderson. Nonlinear regulator theory and an inverse optimal control problem. *IEEE Transactions on Automatic Control*, 18(5):460–465, October 1973.

[Mar91] F. H. C. Marriott. *A dictionary of statistical terms.* Longman Scientific & Technical, Harlow, 5th edition, 1991.

[Med78] J. Medanic. Closed-loop Stackelberg strategies in linear-quadratic problems. *IEEE Transactions on Automatic Control*, 23(4):632–637, August 1978.

[MFH17] S. Mosbach, M. Flad, and S. Hohmann. Cooperative longitudinal driver assistance system based on shared control. In *2017 IEEE International Conference on Systems, Man, and Cybernetics (SMC)*, pages 1776–1781, October 2017.

[MFP17a] T. L. Molloy, J. J. Ford, and T. Perez. Inverse Noncooperative Differential Games. *2017 IEEE 56th Annual Conference on Decision and Control*, pages 5602–5608, 2017.

[MFP17b] T. L. Molloy, J. J. Ford, and T. Perez. Inverse Noncooperative Dynamic Games. *IFAC-PapersOnLine*, 50(1):11788–11793, July 2017.

[MGP+18] T. L. Molloy, G. S. Garden, T. Perez, I. Schiffner, D. Karmaker, and M. V. Srinivasan. An Inverse Differential Game Approach to Modelling Bird Mid-Air Collision Avoidance Behaviours. *IFAC-PapersOnLine*, 51(15):754–759, January 2018.

[MHB16] J. Mainprice, R. Hayne, and D. Berenson. Goal Set Inverse Optimal Control and Iterative Re-planning for Predicting Human Reaching Motions in Shared Workspaces. *IEEE Transactions on Robotics*, 32(4):897–908, 2016.

[MHLK17] W. Ma, D. Huang, N. Lee, and K. M. Kitani. Forecasting Interactive Dynamics of Pedestrians with Fictitious Play. In *2017 IEEE Conference on Computer Vision and Pattern Recognition (CVPR)*, pages 4636–4644, Honolulu, July 2017.

[ML14] H. Modares and F. L. Lewis. Linear Quadratic Tracking Control of Partially-Unknown Continuous-Time Systems Using Reinforcement Learning. *IEEE Transactions on Automatic Control*, 59(11):3051–3056, November 2014.

[MSA17] T. Mylvaganam, M. Sassano, and A. Astolfi. A Differential Game Approach to Multi-agent Collision Avoidance. *IEEE Transactions on Automatic Control*, 62(8):4229–4235, August 2017.

[MTFP16] T. L. Molloy, D. Tsai, J. J. Ford, and T. Perez. Discrete-time inverse optimal control with partial-state information: A soft-optimality approach with constrained state estimation. In *2016 IEEE 55th Conference on Decision and Control (CDC)*, pages 1926–1932, 2016.

[MTL10] K. Mombaur, A. Truong, and J. Laumond. From human to humanoid locomotion—an inverse optimal control approach. *Autonomous Robots*, 28(3):369–383, April 2010.

[MZ18] M. Menner and M. N. Zeilinger. Convex Formulations and Algebraic Solutions for Linear Quadratic Inverse Optimal Control Problems. In *2018 European Control Conference (ECC)*, pages 2107–2112, 2018.

[Nai03] D. S. Naidu. *Optimal control systems*. Electrical engineering textbook series. CRC Press, Boca Raton, FL, 2003.

[Nas51] J. Nash. Non-Cooperative Games. *Annals of Mathematics*, 54(2):286–295, 1951.

[NC61] Y. Nubar and R. Contini. A minimal principle in biomechanics. *The bulletin of mathematical biophysics*, 23(4):377–391, December 1961.

[NF04] D. Nori and R. Frezza. Linear optimal control problems and quadratic cost functions estimation. In *Proceedings of the Mediterranean Conference on Control and Automation*, volume 4, 2004.

[NKJ+10] S. Natarajan, G. Kunapuli, K. Judah, P. Tadepalli, K. Kersting, and J. Shavlik. Multi-Agent Inverse Reinforcement Learning. In *2010 Ninth International Conference on Machine Learning and Applications*, pages 395–400, December 2010.

[NR00] A. Y. Ng and S. Russell. Algorithms for Inverse Reinforcement Learning. *Inproceedings of the 17th International Conference on Machine Learning*, pages 663–670, 2000.

[NS07] G. Neu and C. Szepesvári. Apprenticeship learning using inverse reinforcement learning and gradient methods. In *Proceedings of the Twenty-Third Conference on Uncertainty in Artificial Intelligence*, pages 295–302, 2007.

[NW06] J. Nocedal and S. J. Wright. *Numerical optimization.* Springer series in operations research. Springer, New York, 2nd edition, 2006.

[OR94] M. J. Osborne and A. Rubinstein. *A course in game theory.* MIT Press, Cambridge, Mass, 1994.

[Par14] V. Pareto. *Manual of Political Economy: A Critical and Variorum Edition.* OUP Oxford, May 2014. First published in Italian in 1906.

[PCC+15] M. C. Priess, R. Conway, J. Choi, J. M. Popovich, and C. Radcliffe. Solutions to the Inverse LQR Problem With Application to Biological Systems Analysis. *IEEE Transactions on Control Systems Technology*, 23(2):770–777, March 2015.

[PGLD13] S. Pressé, K. Ghosh, J. Lee, and K. A. Dill. Principles of maximum entropy and maximum caliber in statistical physics. *Reviews of Modern Physics*, 85(3):1115–1141, 2013.

[PHL14] E. Pauwels, D. Henrion, and J. B. Lasserre. Inverse optimal control with polynomial optimization. In *53rd IEEE Conference on Decision and Control*, pages 5581–5586, 2014.

[PJJB12] A.-S. Puydupin-Jamin, M. Johnson, and T. Bretl. A convex approach to inverse optimal control and its application to modeling human locomotion. In *2012 IEEE International Conference on Robotics and Automation (ICRA)*, pages 531–536. IEEE, 2012.

[PR15] A. M. Panchea and N. Ramdani. Towards solving inverse optimal control in a bounded-error framework. In *2015 American Control Conference (ACC)*, pages 4910–4915. IEEE, 2015.

[PR17] A. M. Panchea and N. Ramdani. Inverse Parametric Optimization in a Set-Membership Error-in-Variables Framework. *IEEE Transactions on Automatic Control*, 62(12):6536–6543, December 2017.

[PRBF18] A. M. Panchea, N. Ramdani, V. Bonnet, and P. Fraisse. Human Arm Motion Analysis Based on the Inverse Optimization Approach. In *2018 7th IEEE International Conference on Biomedical Robotics and Biomechatronics (Biorob)*, pages 1005–1010, August 2018.

[PSS+16] M. Pfeiffer, U. Schwesinger, H. Sommer, E. Galceran, and R. Siegwart. Predicting actions to act predictably: Cooperative partial motion planning with maximum entropy models. In *2016 IEEE/RSJ International Conference on Intelligent Robots and Systems (IROS)*, pages 2096–2101, Daejeon, South Korea, October 2016. IEEE.

[RA07] D. Ramachandran and E. Amir. Bayesian inverse reinforcement learning. In *International Joint Conferences on Artificial Intelligence*, volume 7, pages 2586–2591, 2007.

[RBS16] L. J. Ratliff, S. A. Burden, and S. S. Sastry. On the Characterization of Local Nash Equilibria in Continuous Games. *IEEE Transactions on Automatic Control*, 61(8):2301–2307, August 2016.

[RBZ06] N. D. Ratliff, J. A. Bagnell, and M. A. Zinkevich. Maximum margin planning. In *Proceedings of the 23rd international conference on Machine learning*, pages 729–736. ACM, 2006.

[RGZH12] T. S. Reddy, V. Gopikrishna, G. Zaruba, and M. Huber. Inverse reinforcement learning for decentralized non-cooperative multiagent systems. In *2012 IEEE International Conference on Systems, Man, and Cybernetics (SMC)*, pages 1930–1935. IEEE, 2012.

[Rub06] S. J. Rubio. On Coincidence of Feedback Nash Equilibria and Stackelberg Equilibria in Economic Applications of Differential Games. *Journal of Optimization Theory and Applications*, 128(1):203–220, January 2006.

[Rus98] S. Russell. Learning agents for uncertain environments. In *Proceedings of the eleventh annual conference on Computational learning theory*, pages 101–103. ACM, 1998.

[SC88] S. Siegel and N. J. Castellan. *Nonparametric Statistics for the Behavioral Sciences*. MacGraw-Hill, 2nd edition, 1988.

[SC12] G. F. Stocco and G. Cybenko. Inverse game theory: learning the nature of a game through play. In E. M. Carapezza, editor, *Sensors, and Command, Control, Communications, and Intelligence (C3I) Technologies for Homeland Security and Homeland Defense XI*, page 835905. International Society for Optics and Photonics, May 2012.

[SH69a] A. W. Starr and Y. C. Ho. Further properties of nonzero-sum differential games. *Journal of Optimization Theory and Applications*, 3(4):207–219, April 1969.

[SH69b] A. W. Starr and Y. C. Ho. Nonzero-Sum Differential Games. *Journal of optimization theory and applications*, 3:184–206, 1969.

[Sha48] C. E. Shannon. A mathematical theory of communication. *The Bell System Technical Journal*, 27(3):379–423, July 1948.

[SKR00] L. F. Shampine, J. Kierzenka, and M. W. Reichelt. Solving Boundary Value Problems for Ordinary Differential Equations in Matlab with bvp4c. *Tutorial Notes, MATLAB File Exchange*, page 27, 2000.

[ŠKZK17] A. Šošić, W. R. KhudaBukhsh, A. M. Zoubir, and H. Koeppl. Inverse Reinforcement Learning in Swarm Systems. In *Proceedings of the 16th Conference on Autonomous Agents and MultiAgent Systems*, AAMAS '17, pages 1413–1421, 2017.

[Son89] E. Sontag. A 'universal' construction of Artstein's theorem on nonlinear stabilization. *Systems & Control Letters*, 13(2), 1989.

[Sta52] H. von Stackelberg. *The Theory of the Market Economy.* Oxford University Press, 1952.

[Tad13] S. Tadelis. *Game theory: an introduction.* Princeton University Press, Princeton ; Oxford, 2013.

[Tha67] F. Thau. On the inverse optimum control problem for a class of nonlinear autonomous systems. *IEEE Transactions on Automatic Control*, 12(6):674–681, 1967.

[TJ02] E. Todorov and M. I. Jordan. Optimal feedback control as a theory of motor coordination. *Nature Neuroscience*, 5(11):1226–1235, November 2002.

[TM90] S. Tsutsui and K. Mino. Nonlinear strategies in dynamic duopolistic competition with sticky prices. *Journal of Economic Theory*, 52(1):136–161, October 1990.

[TMP16] D. Tsai, T. L. Molloy, and T. Perez. Inverse two-player zero-sum dynamic games. In *2016 Australian Control Conference (AuCC)*, pages 192–196, November 2016.

[Tod04] E. Todorov. Optimality principles in sensorimotor control. *Nature Neuroscience*, 7(9):907–915, September 2004.

[TW19] A. Turnwald and D. Wollherr. Human-Like Motion Planning Based on Game Theoretic Decision Making. *International Journal of Social Robotics*, 11(1):151–170, January 2019.

[TZ11] A. V. Terekhov and V. M. Zatsiorsky. Analytical and numerical analysis of inverse optimization problems: conditions of uniqueness and computational methods. *Biological Cybernetics*, 104(1-2):75–93, February 2011.

[Var70] P. Varaiya. N-person nonzero sum differential games with linear dynamics. *SIAM Journal on Control*, 8(4):441–449, 1970.

[VNM47] J. Von Neumann and O. Morgenstern. *Theory of games and economic behavior.* Princeton Univ. Press, 1947.

[Wan07] X. Wang. *On various equilibrium solutions for linear quadratic noncooperative games.* PhD thesis, The Ohio State University, 2007.

[Wee01] A. J. T. M. Weeren. A Geometric Approach to Infinite Horizon Linear Quadratic Differential Games. *IFAC Proceedings Volumes*, 34(20):35–40, September 2001.

[WOP16] M. Wulfmeier, P. Ondruska, and I. Posner. Maximum Entropy Deep Inverse Reinforcement Learning. *arXiv preprint arXiv:1507.0488*, March 2016.

[Zha11] F. Zhang. *Matrix theory: basic results and techniques.* Universitext. Springer, New York, 2nd ed edition, 2011.

[ZLH19] H. Zhang, Y. Li, and X. Hu. Inverse Optimal Control for Finite-Horizon Discrete-time Linear Quadratic Regulator Under Noisy Output. In *2019 IEEE 58th Conference on Decision and Control (CDC)*, pages 6663–6668, December 2019.

[ZMBD08] B. D. Ziebart, A. L. Maas, J. A. Bagnell, and A. K. Dey. Maximum Entropy Inverse Reinforcement Learning. In *Proceedings of the 23rd National Conference on Artificial intelligence*, volume 3, pages 1433–1438, 2008.

[ZMSFZ16] S. Zazo, S. V. Macua, M. Sánchez-Fernández, and J. Zazo. Dynamic Potential Games With Constraints: Fundamentals and Applications in Communications. *IEEE Transactions on Signal Processing*, 64(14):3806–3821, July 2016.

[ZPCP17] J. Zhang, S. Pourazarm, C. G. Cassandras, and I. Ch. Paschalidis. Data-driven Estimation of Origin-Destination Demand and User Cost Functions for the Optimization of Transportation Networks. *IFAC-PapersOnLine*, 50(1):9680–9685, July 2017.

[ZZWZ16] D. Zhao, Q. Zhang, D. Wang, and Y. Zhu. Experience Replay for Optimal Control of Nonzero-Sum Game Systems With Unknown Dynamics. *IEEE Transactions on Cybernetics*, 46(3):854–865, March 2016.

Own Publications and Conference Contributions

[IBKH20] J. Inga, E. Bischoff, F. Köpf, and S. Hohmann. Inverse Dynamic Games Based on Maximum Entropy Inverse Reinforcement Learning. *arXiv preprint arXiv:1911.07503v2*, 2020.

[IBM+19] J. Inga, E. Bischoff, T. L. Molloy, M. Flad, and S. Hohmann. Solution Sets for Inverse Non-Cooperative Linear-Quadratic Differential Games. *IEEE Control Systems Letters*, 3(4):871–876, 2019.

[IEFH18] J. Inga, M. Eitel, M. Flad, and S. Hohmann. Evaluating Human Behavior in Manual and Shared Control via Inverse Optimization. In *2018 IEEE International Conference on Systems, Man, and Cybernetics (SMC)*, pages 2699–2704, 2018.

[IFDH15] J. Inga, M. Flad, G. Diehm, and S. Hohmann. Gray-Box Driver Modeling and Prediction: Benefits of Steering Primitives. In *2015 IEEE International Conference on Systems, Man, and Cybernetics (SMC)*, pages 3054–3059, 2015.

[IFH19] J. Inga, M. Flad, and S. Hohmann. Validation of a Human Cooperative Steering Behavior Model Based on Differential Games. In *2019 IEEE International Conference on Systems, Man, and Cybernetics (SMC)*, 2019.

[IKFH17] J. Inga, F. Köpf, M. Flad, and S. Hohmann. Individual human behavior identification using an inverse reinforcement learning method. In *2017 IEEE International Conference on Systems, Man, and Cybernetics (SMC)*, pages 99–104, 2017.

[KIB+19] A. Kastner, J. Inga, T. Blauth, F. Köpf, M. Flad, and S. Hohmann. Model-Based Control of a Large-Scale Ball-on-Plate System With Experimental Validation. In *2019 IEEE International Conference on Mechatronics (ICM)*, volume 1, pages 257–262, 2019.

[KIR+17] F. Köpf, J. Inga, S. Rothfuß, M. Flad, and S. Hohmann. Inverse Reinforcement Learning for Identification in Linear-Quadratic Dynamic Games. *IFAC-PapersOnLine*, 50(1):14902–14908, 2017.

[MIF+20] T. L. Molloy, J. Inga, M. Flad, J. J. Ford, T. Perez, and S. Hohmann. Inverse Open-Loop Noncooperative Differential Games and Inverse Optimal Control. *IEEE Transactions on Automatic Control*, 65(2):897–904, 2020.

[PIK+20] O. Petrosian, J. Inga, I. Kuchkarov, M. Flad, and S. Hohmann. Optimal Control and Inverse Optimal Control with Continuous Updating for Human Behavior Modeling. *IFAC-PapersOnLine*, 53(2):6670–6677, 2020.

[RIK+17] S. Rothfuß, J. Inga, F. Köpf, M. Flad, and S. Hohmann. Inverse Optimal Control for Identification in Non-Cooperative Differential Games. *IFAC-PapersOnLine*, 50(1):14909–14915, 2017.

Workshop Contributions

[IH16] J. Inga and S. Hohmann. Modellierung und Identifikation mithilfe eines erweiterten schaltenden autoregressiven Systems. In *GMA-Fachausschuss 1.30 Workshop "Modellbildung, Identifikation und Simulation in der Automatisierungstechnik"*, Anif, Austria, September 2016.

[IH18] J. Inga and S. Hohmann. Identifikation in Differentialspielen basierend auf inverser Optimalsteuerung. In *52. Regelungstechnisches Kolloquium*, Boppard, Germany, February 2018.

[Ing17] J. Inga. Human Behavior Identification Using Inverse Reinforcement Learning. In *Robotics and Autonomous Systems (RAS) Seminar*, Queensland University of Technology, Brisbane, November 2017.

[IRKH17] J. Inga, S. Rothfuß, F. Köpf, and S. Hohmann. Inverse Optimierung in linear-quadratischen dynamischen Spielen. In *GMA-Fachausschuss 1.50 Workshop "Grundlagen vernetzter Systeme"*, Günzburg, Germany, May 2017.

Supervised Theses

[Bei17] Cassandra Beik. Modellierung von Bewegungsprimitiven auf Basis von parametrisierten Funktionalen. Master Thesis, Karlsruhe Institute of Technology (KIT), 2017.

[Bis18] Esther Sophie Bischoff. Entwicklung von Lösungskonzepten für inverse dynamische Spiele. Master Thesis, Karlsruhe Institute of Technology (KIT), 2018.

[Bom17] Matthias Bomke. Entwicklung von dynamischen Optimierungsalgorithmen für kooperative Regelungssysteme basierend auf Pfadintegralen. Master Thesis, Karlsruhe Institute of Technology (KIT), 2017.

[Che19] Muqian Chen. Implementierung und Vergleich von Identifikationsalgorithmen für dynamische Spiele. Bachelor Thesis, Karlsruhe Institute of Technology (KIT), 2019.

[Cre19] Andreas Creutz. Online-Identifikation menschlichen Verhaltens auf der Grundlage inverser dynamischer Spiele. Master Thesis, Karlsruhe Institute of Technology (KIT), 2019.

[Eit18] Michael Bernhard Martin Eitel. Untersuchung menschlichen Verhaltens in kooperativen Szenarien mittels inverser Optimierung. Bachelor Thesis, Karlsruhe Institute of Technology (KIT), 2018.

[Hei18] David Heiming. Entwicklung einer Regelung für Force-Feedback-Bedienelemente eines Ball-auf-Platte-Systems. Bachelor Thesis, Karlsruhe Institute of Technology (KIT), 2018.

[Kas18] Adam Kastner. Entwicklung eines Regelungskonzepts für die automatische Balancierung einer Kugel auf einer Platte. Bachelor Thesis, Karlsruhe Institute of Technology (KIT), 2018.

[Köp16] Florian Köpf. Entwicklung eines Inverse Reinforcement Learning Verfahrens für die Modellierung menschlicher Bewegungen. Master Thesis, Karlsruhe Institute of Technology (KIT), 2016.

[Már18] Alejandro Márquez. Entwicklung einer Regelung für die Kinematik eines Ball-auf-Platte-Systems. Bachelor Thesis, Karlsruhe Institute of Technology (KIT), 2018.

[Mey16] Fabian Meyer. Evaluation menschlicher Bewegungen mittels inverser Optimierungsverfahren. Bachelor Thesis, Karlsruhe Institute of Technology (KIT), 2016.

[Ort16] Tobias Ortelt. Entwurf und Konstruktion eines Ball-auf-Platte-Systems. Bachelor Thesis, Karlsruhe Institute of Technology (KIT), 2016.

[Rie18] Bernhard Riester. Inverse Optimierung für die Identifikation des menschlichen Verhaltens bei einer kooperativen Folgeregelung. Master Thesis, Karlsruhe Institute of Technology (KIT), 2018.

[Rot16] Simon Rothfuß. Inverse Optimization for Identification of Human Behavior in a Cooperative Scenario. Master Thesis, Karlsruhe Institute of Technology (KIT), 2016.

[Sah19] Nasri Sahloul. Untersuchung der haptischen Interaktion zweier Partner anhand eines Ball-auf-Platte-Systems. Bachelor Thesis, Karlsruhe Institute of Technology (KIT), 2019.

[Sun19] Peng Sun. System Theoretic Analysis of Inverse Dynamic Games. Master Thesis, Karlsruhe Institute of Technology (KIT), 2019.

[Xu17] Ziwei Xu. Modellierung von menschlichen Armbewegungen im 3D-Raum. Bachelor Thesis, Karlsruhe Institute of Technology (KIT), 2017.

Karlsruher Beiträge zur Regelungs- und Steuerungstechnik (ISSN 2511-6312)
Institut für Regelungs- und Steuerungssysteme

Band 01 Diehm, Gunter
Identifikation des menschlichen Bewegungsverhaltens auf der Basis von Primitiven.
ISBN 978-3-7315-0608-9

Band 02 Flad, Michael
Kooperative Regelungskonzepte auf Basis der Spieltheorie und deren Anwendung auf Fahrerassistenzsysteme.
ISBN 978-3-7315-0610-2

Band 03 Eckert, Marius
Modellbasierte Identifikation fraktionaler Systeme und ihre Anwendung auf die Lithium-Ionen-Zelle.
ISBN 978-3-7315-0690-4

Band 04 Krebs, Stefan
Intervallbeobachter für lineare parametervariante Systeme und deren Anwendung auf die Asynchronmaschine.
ISBN 978-3-7315-0857-1

Band 05 Kaspar, Stephan
Fahrdynamikuntersuchungen eines Elektrofahrzeugs mit Einzelrad-Hinterradantrieb.
ISBN 978-3-7315-0916-5

Band 06 Sauter, Patrick S.
Modellierung und zentrale prädiktive Regelung von multimodalen Energieverteilnetzen.
ISBN 978-3-7315-0963-9

Band 07 Kupper, Martin
Verteilte Zustandsschätzung fraktionaler Systeme und ihre Anwendung auf Lithium-Ionen-Batteriesysteme.
ISBN 978-3-7315-0971-4

Band 08 Merkert, Lennart
Optimal Scheduling of Combined Heat and Power Generation Considering Heating Grid Dynamics.
ISBN 978-3-7315-1056-7

Die Bände sind unter www.ksp.kit.edu als PDF frei verfügbar oder als Druckausgabe bestellbar.

Band 09 Ludwig, Julian
Automatisierte kooperative Transition einer Regelungsaufgabe zwischen Mensch und Maschine am Beispiel des hochautomatisierten Fahrens.
ISBN 978-3-7315-1069-7

Band 10 Inga Charaja, Juan Jairo
Inverse Dynamic Game Methods for Identification of Cooperative System Behavior.
ISBN 978-3-7315-1080-2